Anonymous

Dental headlight

Anonymous

Dental headlight

ISBN/EAN: 9783337269630

Printed in Europe, USA, Canada, Australia, Japan

Cover: Foto ©berggeist007 / pixelio.de

More available books at **www.hansebooks.com**

THE DENTAL HEADLIGHT.

| VOL. 13. | NASHVILLE, TENN., JANUARY, 1892. | No. 1. |

Original Communications. . . . .

ODONTALGIA.

BY L. G. NOEL, M.D., D.D.S., NASHVILLE, TENN.

TOOTHACHE patients are generally the most unwelcome visitors to the office of the modern dentist. This statement will sound strange and unreasonable to that class of the laity who indulge this luxury, and who think of the dentist as a gentleman of elegant leisure, lounging idly about his handsome apartments, waiting for some one to call upon him to extract a tooth. True, it is the business of the dentist to relieve suffering; but those who neglect their teeth until pain drives them to the dentist are not his best patrons. On the other hand, the successful treatment of these cases often wins for us not only the confidence and esteem of the sufferer, but the patronage of his family and friends. A man may have lived in blissful ignorance of the profession of dentistry, and all that pertains to the care of the teeth; but let him once suffer a genuine attack of this "hell o' a' diseases," as Burns has so aptly styled it, and let him fly to the dentist for relief, finding him (as he always should) "a very present help in time of trouble," and that man will always think of the dentist as a very useful member of society.

Odontalgia may arise from a variety of causes, and in citing these we cannot do better than to give Mr. Tomes's classification:

"1. *Morbid Conditions of the Tooth Pulp.*—Under this head would be included irritation, acute and chronic inflammation of the pulp, pressure from confined matter in the pulp cavity, and deposit of secondary dentine in its substance. Probably, also, the exposure of sensitive dentine gives pain, by setting up irritation of the pulp; as does, also, caries in its early stages.

"2. *Morbid Conditions of the Periosteum and Exostosis.*—Under the second head would come inflammation of the periosteum; acute and chronic alveolar abscess, in its various forms, and those lesions which are mainly manifested by alterations in the fangs of the teeth—such as roughening by absorption, or increase by exostosis.

inflammation shall have extended from the pulp to the pericementum and intra-alveolar periosteum, we find that cold fluids give relief, and the application of heat induces pain; the tooth is elongated from a thickening of its pericementum and the alveolar lining, by inflammation; and there will be acute and manifest soreness upon percussion. When in the first stage, as above described —sensitive to cold, with no considerable pericementitis, and no tenderness upon percussion—we may hope to save the vitality of the pulp, and bring it to tolerate the presence of a filling; but after the case has progressed to the stage last described, with pain from the application of heat, and soreness upon the occlusion of the opposing teeth, we may unhesitatingly proceed to devitalize the nerve.

Many cases come into the hands of the dentist, where the pulp is merely in a state of irritability from the fracture of a portion of the crown of the tooth or the presence of a small carious cavity. In either case the irritation is conveyed to the pulp through the dentinal fibrillæ, whose peripheral extremities are constantly wrought upon by chemical and thermal agents, until acute inflammation of the pulp may be established. In the case of carious cavities, the presence of septic matter in which microbes flourish, we have the most fruitful and pernicious source of irritation. Miller and Andrews have demonstrated the presence of microbes in the dentinal tubuli, and how these, by rapid proliferation, break down the structure of the dentine. The treatment clearly indicated will be the excision of all carious dentine, and the application of antiseptics, germicides, and mild escharotics. Having completed the excavation, an application of wood creosote should be made to the cavity, and a gutta-percha or oxyphosphate filling introduced. If oxyphosphate is deemed best, it will generally be best to coat the surface of the cavity with chloro-percha before filling. This temporary plug should remain until all pulp irritability has passed, and the tooth is no longer painful to the contact of warm and cold foods and fluids. The hasty introduction of a metal filling may set up acute pulpitis, and result in death of the pulp and alveolar abscess. The excavation of such cavities will usually be attended with great pain, but in the molar and lower bicuspid teeth this may be greatly alleviated by applying for a few moments a little powdered nitrate of silver. In other cases Herbst's obtundent may be applied. In cases of pulp irritation from fracture thorough cauterization with powdered nitrate of silver will be found our best therapeutic measure, where the fracture is out of sight. The discoloration it causes would of course contraindicate its use upon the front teeth. Many fractures so treated will resist decaying agents for years; indeed, fractured surfaces often undergo a spontaneous cure by the hardening of the dentine by the calcification of the dentinal fibers, but the timely application of an escharotic may save the vitality of the pulp and prevent alveolar abscess—possibly the loss of the tooth. The aching tooth may be frequently difficult to locate, the patient often attributing the pain to sound teeth, remote from the real offender. Where the cavities are hidden and located upon the necks of the teeth a small jet of cold water from a pointed syringe, thrown in between the teeth, will be found a valuable diagnostic measure. Many patients have wandered from dentist to dentist, vainly seeking relief because of their inability to locate such cavities. It is not strange that such patients give their confidence and patronage to the person giving relief.

For the relief of inflamed and irritable dentine pure wood creosote is our sheet anchor, and in cases of acute inflammation of the pulp from exposure by caries it is our most valuable remedial agent. Where the opening through the pulp wall is large, and the carious tissues have become washed out, leaving the pulp directly exposed to all extraneous irritants, the course of the inflammation is usually violent but brief, soon running into death and sloughing of the nerve, followed by alveolar abscess if the carious cavity becomes closed with food. It is seldom such a case comes into the hands of the dentist in time for treating and capping the pulp; but if the exposure is made by lifting the leathery mass of organic matter from the nerve with an excavator, in an effort to prepare the cavity, we may hope to save the nerve by proper treatment. Whether we shall attempt this must be determined by the conditions above referred to. If the tooth is elongated, sore upon percussion, and sensitive to warm or hot fluids and foods, the attempt will be useless. If the contrary conditions are found, we should content ourselves at first sitting with excavating the carious dentine, applying creosote, and closing the cavity with cotton and gum sandarach. Oil of cloves is a useful remedy, and soothing to inflamed and exposed nerves; and cocaine also gives a measure of relief; but none of these are so reliable as wood creosote. Carbolic acid would perhaps take second rank as a remedy for toothache, but in some instances it seems to act as an irritant.

ALVEOLAR ABSCESSES.*

BY W. J. MORRISON, D.D.S., NASHVILLE, TENN.

ALVEOLAR abscesses result from inflammation, having its seat in the apical space, proceeding to the formation of pus; therefore the location of alveolar abscess is always in its inception the apical space, no matter where it may extend afterward. If this be not the case, it is not an alveolar abscess, though it be an abscess within the alveolus of a tooth.

If an abscess should occur on the side of a root of a tooth as the result of injury, and if it be not the effect of the death of the pulp of the tooth, it is properly a traumatic alveolar abscess.

If such an abscess occurs from any of the diseases that attack the sides of the root of the tooth, and if it be thought well to designate it as an alveolar, the word should, in all cases, be accompanied by an adjective expressing the fact. This is necessary to accuracy.

Like many other affections within the oral cavity, the serious effect upon the patient has not received the careful consideration of the medical fraternity which it demands. But this is not to be wondered at when, from experience, we soon find that the advice given to patients by many of the general practitioners in diseases of the teeth and gums displays an ignorance that is equaled only by the voodoo medicine man of darkest Africa, and their treatment of a patient affected with an abscess of this nature has a parallel only in the practice by faith and mind cure advocates, who, when a patient applies to them and is suffering the torments of the damned with a fever at 100° to 103°, tell him to say to himself, "My tooth don't hurt, my tooth don't hurt, my tooth don't hurt," and it will get well; and if he can say these charmed words long enough—that is, until the pus breaks down the bony walls—the patient will in all probability get relief. We find many of the general practitioners administering to the patient for neuralgia; giving opiates, recommending hot applications, and in some cases lancing the gums; and finally a cure is effected after the patient has suffered for days and nights in the same manner as the patient who was cured by the mind cure advocate. Yet these kettles will call the pots black.

For me to enter into the minute details of the formation of pus, the causes of the absorption of the bony walls, the burrowing of pus,

---

* Read before the Nashville Academy of Medicine.

the great destruction of bone which occurs before the surface is reached, the effect on the peridental membrane, the intense pain to the patient, the effect on the gums, the modification of the symptoms by abatement of intense pain, followed by the swelling of the face, the eye being often closed and the jaws stiffened, would be needlessly consuming your time with a review of subjects that are already familiar to you.

The danger of blood poison and the fistular discharge upon the face are subjects worthy of your consideration. While the breaking down of the floor of the antrum and the discharge of pus into this cavity is a complication of this disease that often baffles the skill of the specialist. The extraction of a tooth is seldom necessary and seldom demanded for the cure of abscesses. yet, in case of an advanced form with an inferior incision, and where trouble and antrum complications are caused. extraction is, in nearly all cases. indicated, as is also the case where the alveolar process has necrosed. With these exceptions the treatment of alveolar abscesses in the vast majority of cases presents but little difficulty. About the same line of treatment will be indicated in severe acute cases as in the treatment of chronic alveolar abscesses.

Constitutional treatment should not be neglected; poultices applied to the face should be strictly forbidden. No matter whether the tooth be little decayed or whether it be decayed at all, a free access should be obtained to the canals and these should be well cleaned with a broach.

The enlargement of the canals in the teeth with any kind of drill is not to be recommended. The chances are that such a method will be productive of more harm than good. Those canals that need enlarging are the ones that will be most dangerous to enlarge. In the treatment of many diseases of this nature many dentists have their hobbies, and others use remedies they know not why, and some others will be strong believers in the germ theory and will use medicines that are merely disinfectants or escharotics.

I will not undertake to review the many antiquated treatments that are still practiced in this advanced age of scientific research, nor will I consume your valuable time in discussing the *over* treatment of this disease. using arguments to sustain it that are not worthy of consideration in any learned society. but I will only say that for the treatment for alveolar abscesses—that is, the medicinal treatment—I use the peroxide of hydrogen. apply the rubber dam. turn on a current of hot air (not *warm*. but *hot*). pump into the root canals a two per cent alcoholic solution of bichloride of mercury, dry with hot air again, pack the cavity with pellets of cotton saturated with the above solution, seal with wax, and dismiss the patient for three or four days, according to the severity of the case. Where the case is not of too long standing, a second treatment will as a rule effect a cure.

In chronic cases with fistula opening the medicinal treatment is the same, except an aqueous solution of 1 to $\frac{1}{1000}$ bichloride of mercury is used as a wash, and dressing the root canals as above.

It would make this paper too tedious to enter into the minute detail of extirpating the root of the tooth and the removal of diseased bone in cases where this was necessary, nor will I encroach upon the reserved rights of those who can take the exact measure of a tooth, and, by mechanical appliances and calculations, determine the exact length of a wooden peg or a lead point that will be required to fill the canal exactly to its apex. Even the most tortuous root canals have no horrors for them, as it only makes their calculations more interesting because of the complications. But I *do* reserve the right to say that in all teeth back of the second bicuspids to use copper wire in connection with chloro-percha.

The stiffness, yet adaptability of the wire in tortuous cases, the therapeutical action of the sulphate of copper upon the tooth structure, the hermetic sealing of the chloro-percha combined with the preserving qualities of the chloroform, make these agents most desirable in nerve canal filling.

THE PERIDENTAL MEMBRANE.*

BY MARVIN M'FERRIN, M.D., D.D.S., NASHVILLE, TENN.

THE peridental membrane comprises that tissue intervening between the root of the tooth and the bony walls of its alveolus. It serves to fix the tooth in its position, and protect from shock in sudden or forcible occlusion—passive functions which are performed by the fibrous elements. These fibers, designated as the principal fibers, form the bulk of the tissue of the membrane.

They are fixed in the cementum of the tooth's root on the one side, on the other in the bone which forms the wall of the alveolus, or in the gum; stretching across the intervening space in various

---

*Read before the Nashville Academy of Medicine.

directions, and so suspending the tooth in its socket. These fibers are of the white or inelastic connective tissue variety, and we never find them crossing the intervening space horizontally, in the shortest direction, but always obliquely, so as to accommodate movement of the root without the fibers being lacerated or stretched.

The pericementum is the matrix of development of the osteoblasts, which build up portions of the alveolar walls, and of the cementoblasts, which construct the cementum.

These cells are supposed to be received into the fibrous meshes of this membrane, from the blood supply, as leucocytes or amœboid cells, and here undergo their development or that differentation which fits them for the building up of bone on the one side and cementum on the other.

During this development they become allied to their respective places, or harmoniously adjusted to their individual tissues—i. e., the surface of the bone and the surface of the cementum. Besides the osteoblasts and cementoblasts, the membrane presents various cellular elements—such as fibrinoblasts, for the augmentation or renewal of its fibrous tissues; osteoclasts for removal of the walls or portions of the walls of the alveolus, for the accommodation of changes in the position of the teeth, or of the cementum for change of the form of the tooth's root. Beside the so-called principal fibers is a tissue of very fine fibers of ordinary fibrous connective tissue, which seems to pervade the entire membrane—the interfibrous or indifferent tissue, as Dr. Black terms it. The fibers of this tissue run in a course diagonal to the principal fibers; and, while pursuing no very definite direction, have a tendency to lie horizontal to the cementum. This tissue offers investment to the blood vessels and nerves, in addition to the tissue which properly belongs to their walls, but seems not to attach itself to the cementum or bone. The vascular supply of the root membrane is derived from three sources: from the vessel destined for the pulp of the tooth, from the vessels of the bone, and from the vessels of the gum. From the main source at the apical space, from the artery which enters the root canal, we have from four to six or eight branches which pass down along the sides of the root supplying the membrane, and anastomosing with the arteries of the Haversian canals and also with the arteries of the gum. By this arrangement of the circulation, even though an abscess or an accident may cut off the blood supply from the apical space, it is well maintained from the gum tissues and alveolar walls.

The sensory function of the membrane is supplied by nerves entering it much in the same manner as the blood vessels do. The membrane is abundantly supplied with nerves which enter through the walls of alveolus, at the apical space, gingival border, and over the rim of the alveolus. The tooth pulp and the tissue which becomes the peridental membrane come from the same source, and were once continuous over the whole base of the pulp; hence the nearly identical manner in which their vascular and nervous supply is received. The principal bundles of nerves enter by the apical space, and then divide; a portion of which pass through the apical foramen, and a number pass down the sides of the root, supplying the membrane and cementum and inosculating with those from the Haversian canals and from the gum tissues. It is through this supply of nerves that the peridental membrane becomes the organ of touch for the tooth. giving it such exquisite sensibility when it is the seat of inflammation; yet this rich supply of nerves and blood vessels give the membrane a great recuperative power, which is manifested in its marked tendency to recover from severe injuries. The lymphatics occur chiefly in the form of rows of cells insinuated between the fibers. They anastomose freely with each other, and form a network over the whole of the roots of the tooth, close to but not in contact with the cementum. The individual cells are like those of the lymphatic glands, and in the larger groups they are seen to be enveloped in a very delicate limiting membrane. These glands are largely affected in inflammation of this membrane, and seem to be the seat of this very destructive affection. Also, they seem to be the first affected in salivation with mercury, and when the teeth become sore from constitutional condition, some agent in the blood seems to affect them.

A very common form of pericementitis we find beginning in the apical space, consequent on the death of the pulp, or beginning when the pulp is irreparably inflamed. The tissue involved in the inflammation, being encased between unyielding walls in such a way as to hinder its expansion when engorged by the influx of blood. soon becomes exquisitely painful. As a rule, we have the death of the pulp preceding the beginning of the inflammation in the apical space, usually by some days, or until putrescence of the pulp has proceeded so far as to give rise to poisonous material which escapes into the tissue by way of the apical foramen. In many cases this is delayed for weeks or months, or, as we sometimes see, may not occur at all. No tooth, however, with an empty pulp chamber, is safe from apical pericementitis. Pus is usually formed in the apical space very quickly—within twenty-four hours—but sometimes is delayed for several days. With the formation of pus the case becomes one of acute alveolar abscess. . . . As the pericementum is endowed with nerves of tactile sensation, we always have pain on pressure on the tooth when this membrane is inflamed. As the pulp has no nerves of touch when it becomes diseased, the tooth is not tender unless the inflammation has passed through the apical foramen, thus ushering in apical inflammation. It is characteristic of the diseases of organs having no nerves of touch to refer pain to associate or distant parts, as in affections of the pulp to the ear or different parts of the face. This peculiarity occurs so frequently in diseases of the organs having no sense of touch as to make it a general law of symptomatology.

The dental pulp is especially sensitive to thermal changes, and this sensibility is markedly increased in its diseases. There seems to be but one condition in which thermal change causes marked pain in pericementitis, and this is when the pulp chamber is filled with gas in such a way as to cause pressure on the tissues of the apical space. In this case heat will give rise to an expansion of gas, increasing the pressure and pain, while the application of cold will relieve it. This, then, is diagnostic, for in affections of the pulp both heat and cold cause pain. The cause of apical pericementitis is some irritating agent which finds its way into the tissues of the apical space, by way of the apical foramen. This agent is usually furnished by products of the decomposition of the pulp of the tooth.

Selections. . . . .

ORIGIN OF PUS.*

Of all the topics that could possibly engage our attention, the history of pus, in the light of modern science, is the one which comes most directly to the front in the general survey of the broadest underlying progressive principles of our pathological science.

If we define the evolution of pus or suppuration to be one of the results of inflammation, to be one of its most deplorable consequences, most to be dreaded, most particularly to be prevented, then shall we look to the consummation and fulfillment of every means at command to abort the development of so sad a result.

What indeed is now the present order of the day in every department of the healing art? Visit the wards of the hospitals of to-day and witness the preparations made in undisguised apprehension of the advent of this formidable complication, the production of pus. See the equipments in readiness for the observance of the entire technique of antiseptic treatment. If a tumor is to be removed or an amputation performed; if a cataract is to be extracted, or the parturient act has just been accomplished normally and safely, behold the array of solutions, of the bichloride, carbolic acid, eucalyptus, iodoform, and aristol at hand to meet the possible exigencies of the occasion. With such, constituting as they do the most approved germicides at command, cavities are to be flushed, wounded surfaces irrigated, instruments, and even the surgeon's hands, disinfected through repetitions of most tedious ablutions, a surgical toilette for both patient and surgeon, the purport of which would have proved wholly inexplicable to the most veteran of surgeons of a very late period.

Now the significance of all this parade of modern methods of treatment is a resolute and determined warfare against this purulent termination of inflammation, a veritable antiseptic combat, in which the enlightened surgeon contends against the outburst of *pus*, under the avowed supposition that its presence implies the presence of bac-

---

* Read before the Southern Dental Association, at Moorehead City, N. C., August, 1891.

teria, while these contribute, in the opinion of many, to increase irritation of the tissues, inducing the birth and multiplication of pus
cells. Whether this theory of pyogenesis be correct, and whatever may
be the relation that exists between the development of bacteria and
that of pus, we shall not just now discuss, further than to direct attention to what, in the present state of our science, is to be regarded
as the real biological meaning of pus.

Modern research has revolutionized our views respecting the true
nature of this morbid fluid. As the word "pus" implies, the ancients
regarded it as a putrefactive product from the tissues, the resultant
of a general disintegration and subsequent putrescent change of the
organic detritus of the part—erosive and destructive in character,
breaking down healthy structures, resolving them into decay, and
when "cradled, cribbed and confined" within restricted limitations,
carrying destruction always in its wake. Another theory ascribed
its evolution to a chemical change brought about by the development of sulphuretted hydrogen combining with certain ammoniacal
emanations forming a hydrosulphate of ammonia, which explained
the offensive odor of the product under certain circumstances. But
we shall not stop to refute these antiquated views, by showing that
pus is sometimes reparative, not erosive and destructive in its operations; that it is often a simple exaggerated nutritive or secretory
process, contributing a healthful rather than a morbid step toward
cicatrization; nor shall we remind you that pus is often sweet to
the taste and void of anything approximating putrescence. Neither
shall I detain you with the familiar phases of its history, repeating
that pus is an alkaline product of a certain specific gravity, containing fatty substances and salts; a serous fluid charged with albumen; a liquid, in a word, strikingly like, if not identical with, the
serum of the blood, in which float numbers of spheroidal corpuscles,
called pus cells; that its physical appearance varies with the special
variety of the product, being creamy white, sweetish in taste and
smell when said to be *laudable pus;* yet often in its admixture with
blood, mucus or serum, variable in its color or consistency; tena-
cious, viscid, or watery, when it is then designated as sanious, ichor-
ous, cheesy, or gummy. These facts concerning a product so
studied and known require no rehearsal on this occasion. Permit
me rather to call attention to the striking feature of these pus cells
or corpuscles, above mentioned as floating in this fluid, as they are
now known to give character to the entire product spoken of as pus
—a character so distinctive as to enable one to recognize pus in the

smallest conceivable quantity wherever existing in the tissues or fluids of the body. No more pertinent proof could be given of this fact, than to recount the remarkable diagnosis made many years ago by Dr. Donne of an incipient abscess of the breast, at a period in its development when no skill on the part of any surgeon could possibly have discovered it. Mr. Donne, while examining a specimen of human milk, at the Maternity Hospital in Paris, chanced to observe the presence of a few scattered pus cells under a microscope, mixed with the innumerable milk globules in this otherwise healthy milk, and at once declared that there must be an abscess in the breast. The woman was perfectly well at the time, with no trace of trouble in the gland; yet at a short subsequent date a large abscess was actually developed, confirming Mr. Donne's diagnosis.

The cardinal point, then, in the study and recognition of the veritable nature of pus will be the interpretation of the pus cell. The corpuscular element of a fluid, when any such exists, affords the clearest evidence of the nature, origin, and purpose of the product, whether this be secretory or excretory. The greatest confusion once obtained respecting the true individuality of this pus corpuscle. It was affirmed that no distinction was possible between a mucous or pus globule. They were supposed to be one and the same. It was not until a forced resemblance to the white cell of the blood became generally observed that any definite idea existed respecting this question. The problem came gradually to be solved through the conjoined labors of several experimenters. Thus Addison, nearly half a century ago, pointed out the identity of pus cells and leucocytes of the blood, and even foreshadowed their migration from the blood vessels. This identity between these cells was later on again brought out by Waller in the fullest enumeration of their histological features and behavior under various reagents.

Yet these views were reluctantly accepted, since it was impossible to understand how a leucocyte, circulating with blood disks in the vessels of a part, could escape through their walls. Then came the declaration of Cohnheim, who minutely described the actual migration of leucocytes by their amœboid endowments through the attenuated walls of the capillaries within the area of the inflamed territory; and announcing that he had clearly seen the accomplishment of this extraordinary phenomenon while engaged in observing the changes occurring in the webfoot of a frog, where artificial inflammation had been induced. This positive statement, however, in nowise appeared to account for the rapid multiplication of such cells

in any ordinary case of profuse suppuration, but rather seemed to
militate against the fact, since such a drain upon the blood as must
necessarily ensue in the elimination of all of its white corpuscles
would at once destroy its vitality. It was then ascertained that
upon the emigration of these leucocytes, these began forthwith to
proliferate within the interstices of the tissues, and more than this,
that they induced a similar multiplication of the connective tissue
corpuscles of the part; so that in due time the entire region of inflammatory action became invaded by an infinite number of rapidly
multiplying nucleated cells, some of which had escaped from the
vessels as emigrants within the diseased territory; while others derived from the connective tissue might be said to be natives to the
manor born. This now explains fully the copious discharges of pus
oftentimes witnessed.

But this is not the whole of the process of true suppuration. It
must be remembered that a transudation of plastic exudates from
the vessels of the inflamed part was known to be one of the earliest
phenomena of inflammation. Now into this effusion, known always
to take place, leucocytes find their way. This intervascular infiltration of the plasma of the blood becomes then a fluid matrix in which
nutritive material is furnished for the subsequent growth and generation of a multitudinous progeny of cells derived, not alone from
the blood cells, but from the tissue cells also.

Pus, thus considered in its incipient or nascent development, is a
product, be it observed, from the blood itself, and, indeed, resembles
in its earliest stages the vital organic processes of nutritive secretion to a marked degree. It is only when untrammeled in its progress and exaggerated in its vital (though perverted) activity, that
the phenomenon takes the character of a truly morbid process. Thus
in following out the destiny of the separate elements of this product
during their several stages, we come to a clear and perfect apprehension of the *three* distinct terminations of inflammation: by *resolution, organization,* and *suppuration.*

First, let us suppose the initiative step expressed in the effusion
of exudative lymph be at once arrested, then absorption takes place
before any danger reaches the tissues, and inflammation is said to
terminate in *resolution;* but should the emigration of leucocytes
have occurred to a limited extent only conjointly with diffusible
lymph and the fibrillation of its fibrin, then a slow but sure step
toward tissue formation occurs, and now inflammation results in *organization.* Should, however, this cell proliferation prove in the as-

cendency, and the degenerated and metamorphosed corpuscles reach their destructive influence upon surrounding parts, then does the inflammatory invasion end finally in *suppuration*.

Such is the value of modern microscopical research in unraveling the nature and origin of pus.

It seems strange that so few observers were able at first to verify this exodus of the white corpuscles of the blood. Indeed, Duval discussed at length the accuracy of the assertion and ridiculed the idea of cells measuring the one-half-thousandth of an inch passing through coats of the blood vessels. He believed that the preparatory step in Cohnheim's experiments upon frogs opened certain lymphatics, and the constant flow of lymph cells grouped these cells together along the outsides of the vessels, simulating the appearances which had so completely deceived him.

That no one may doubt, in the present state of our knowledge on this point, that the white blood cells do escape into the tissues through the vascular walls, we shall detail the recent experiment of Schafer as presenting irrefragable evidence of the fact.

This curious experiment consists in "feeding the white corpuscles," as Schafer terms it, and an account of the same will conclude our remarks for the present on this interesting topic.

The white corpuscles enjoy amœboid movements, and therefore exhibit the tendencies of all amœboid organisms—that is, they take in their interior any small particles of insoluble materials floating within their range.

To obtain white cells in sufficient numbers to be able to institute the curious experiment of feeding them, Schafer puts frog's blood into a small glass tube, hermetically closes both ends, and lays the capillary tube aside for one hour. Examined after this period, a clot has been formed, has contracted, and the lymph around the clot is found full of leucocytes. Placing a drop of this colorless fluid from the clot, surcharged with white cells, under the microscope, he now proceeds to feed them by introducing a drop or two of Indian ink, which has been well rubbed up in a normal solution of table salt. The black particles of the Indian ink scattered through the fluid are soon ingulfed into the protoplasm of the amœboid white cell, after the manner of other amœboid organisms, the intussusception occurring slowly, until almost every white cell becomes filled with little black particles which can never after be discharged, as these cells do not discharge their cargo.

This property of the white corpuscles he thus utilizes. The insoluble particles of an ink solution he injects into the blood vessels of an animal, and then inflammation is induced at a special part. He afterward finds in the intervascular spaces of the inflamed part leucocytes containing similar particles of Indian ink. The corpuscles consequently must have come from the blood, for such particles have not the power of passing through the coats of the blood vessels unless carried by the white corpuscles.

In this succinct history of pus as we now understand it I have endeavored to show and prove that as the product of inflammation it especially consists in the wandering from their vascular abodes of certain saline and albumininous elements; first, that pave the way for the subsequent cell extrusion, cell distention and proliferation of the white and connective tissue corpuscles, and that the entire process may be an exaggeration of the normal physiological forces at play within the inflamed area of a part. (Dr. J. B. Patrick, in *Southern Dental Journal.*)

THE CARE OF CHILDREN'S TEETH.*

To write something that would be entitled to consideration by this body is an attainment much 'to be desired; but to write something upon the subject indicated by the caption of my paper, calculated to excite discussion and action on the part of the dental profession and the people of the United States, would be an honor more exalted.

The care of the teeth of children is to-day the most important question that addresses itself to the American people. The almost universal presence of disease in the mouths of the children of our common country is a startling fact that appeals to all who give health and hygiene any attention whatever. I feel certain that some of the statements I may make in reference to this question will subject me to severe criticism at the hands of some, but this amounts to nothing when I feel certain that these statements are correct.

That the teeth of this country are, to a greater or less extent, becoming more susceptible to decay and disease is to my mind apparent, which fact is more attributable to want of function than any other one thing. Teeth do not only begin to decay earlier in life than was formerly the case, but they continue to succumb to vari-

---

* Read before the Southern Dental Association, at Morehead, N. C., August, 1891.

ous diseases to a much later period in human existence than they did originally. And that many affections of the teeth are more virulent and rapid in the processes of destruction than they used to be is a fact that should be given due weight in this discussion.

This position is sustained by the observation that many of our therapeutic agents are losing their value, comparatively speaking. From want of functional activity we find the roots of the teeth tardy in their development, so that in many cases extensive decay has occurred before the apical foramen is closed and the tooth pulp ceased to be a persistent one.

That the presence of large cavities in the teeth, whether filled or not, *does* retard the proper development of the roots in many cases, I feel perfectly certain, just as the presence of large cavities, and especially large cavities in children's teeth with amalgam fillings in them, retards the normal absorption of the roots of the first twenty prior to their removal. Hence metals should not be used in the treatment of children's teeth, *especially amalgam.* A very small amalgam filling in a child's tooth will, to a greater or less extent, retard if not prevent the physiological process of absorption of the root.

After the above statements by way of explanation of what we have to say in regard to the treatment or management of all cases coming under our care, we will give a few rules that should be enforced by the dental profession:

1. All children should have a competent dentist to superintend the eruption of the first set of teeth.

2. In all cases of infantile trouble, where the diagnosis is doubtful, a dental expert should be called in consultation to determine whether dental irritation is a serious factor or not.

3. All children, after the eruption of the first set of teeth, which is usually completed about the twenty-fourth month, should be subjected to from one to four examinations each year, the frequency of such examinations to be determined by the physiological make-up of each individual case, and given such attention as to prevent and control dental caries, which is about the only affection we have to combat in the management of the first set of teeth, unless gross neglect after the crisis of the eruption has passed has allowed the little sufferers to pass into a state pitiable indeed to behold.

4. No child should be allowed to enter the time of erupting the permanent teeth with caries preying upon all or any of the temporary set, as the fluids of the mouth would be liable to transmit to

2

the second set the germs of disease, or at least all the predisposing conditions to dental caries would be augmented.

5. I would most earnestly insist that all children should have a skillful dentist to conduct the removal of deciduous teeth, *and this should be accomplished in pairs.*

By thus conducting the shedding of the milk teeth, most of the irregularities of the permanent ones would be prevented.

In the treatment of children's teeth, we feel certain that for incipient caries on the proximate surfaces, separating with the Authur corundum disks, or file to that extent that all small cavities are removed so thoroughly that no trace of disease is left. In cases in which the cavities are too deep to admit of removal by the use of the disk, the decay should be reamed out and the cavity filled with some of the various cements, this to be removed as often as necessary to keep the teeth in a good sanitary condition.

In more advanced cases, where we not only have small cavities on the approximate surfaces, but deep-seated decay in the crowns or grinding surfaces, they should be well prepared and filled with some of the preparations as indicated above.

One other class of cases to which I wish to invite attention is a more advanced and complicated type: children patients who present themselves with one or more dead teeth to treat and look after, with or without open sinuses or abscesses. Such cases should be treated until a cure is effected, filled and ground down with corundum stones, completely nonantagonizing them so that the nonabsorption of their roots may be supplemented by the process of exfoliation. A clinical fact we would do well to note is that when an abscess has developed as a result of the death of the pulp in a deciduous tooth, the eruption of the permanent one is accomplished much earlier as a rule than nature intends they should appear.

The care of children's teeth, or the treatment of the first set, which is composed of twenty, ten in each jaw, has not received the attention at the hands of authors and practitioners that the magnitude and importance of the subject demands. As a humble yet enthusiastic advocate of the claim of our profession to public recognition, I am compelled to say that if dentistry ever takes that high position that it is entitled to, and accomplishes the amount of good that it is capable of accomplishing for humanity, this department of practice must be more closely attended to in the future than it has been in the past.

The fact that of the scholastic population of our country ninety-five per cent. or a little more, are affected with dental caries is an appalling statement, and from a healthful standpoint, bodes more evil to the American people than any other fact.

If I were asked to make a last statement as to what I believed would be the most beneficial to future generations, or to make a last request, and that request be granted, I would state that the application of the art of denttstry to the wants of the *whole* people would do *more* good than any other hygienic procedure that could possibly be instituted, and request that all people, and especially the people of America, be supplied with a sufficient number of good, competent dentists to keep the mouths of at least the children of our country in proper condition.

One of the first things to be looked to in the management of the child is the undivided and absolute support of the parents or those having charge of the little patients. To attempt to serve a little child whose father and mother is all the time pulling back and questioning the propriety or correctness of the dentist in attempting to keep disease out the mouth, affords a difficulty that will prevent the most skillful dentist from accomplishing the desired result. Let me say just here that the fathers and mothers of America are more derelict in their duty in regard to the care of the teeth of their children than in any other particular, and I here state that it is my firm conviction that this dereliction is largely responsible for the large death rate in the children in this country. The startling fact that fully one-third or a little more of the deaths of this country occur with children under five years of age is appalling, and makes one shudder to contemplate. Hence the emphasis with which we have attempted to call attention to the few points in this short and hurriedly written paper.

In conclusion allow me to say that, while there are some good medicines to be used in the treatment of cases of the young, most of the relief afforded is by good surgery based upon proper diagnosis, followed by good sanitation.

In all cases, not to cleanse the mouth and teeth after taking food is a sin against good health, and should not be tolerated by good society.

I cannot conclude this short paper without entering my protest against the habit of eating between meals, especially such things as candies and sweetbreads furnished by the little fruit-stands at the street corners and mouths of alleys of our cities; such places being, in my judgment, a greater menace to public health and welfare than the so much abused whisky traffic in this country. (Dr. J. Y. Crawford, in *Southern Dental Journal.*)

PHYSIOLOGICAL ACTION OF OBTUNDENTS.*

THIS paper has been written with a view of making clear and harmonizing a few conflicting ideas regarding the action of obtundents of sensitiveness of the dentinal fibril.

Owing to conflicting theories, it will be fair to assume that this tissue, whose sensibility we desire to obtund, is of the nature of simple protoplasm, with (in this instance) an exalted sensitive function. It contains albumen, and is therefore coagulable; it is made up largely of water, and may be desiccated; it is incased in a tube, and when it does not entirely fill this tube, a fluid fills this space. This fluid may be removed, and the temperature of the fibril may be reduced. These are the conditions upon which we are to work.

The dentinal fibril does not possess a blood circulation, and on that account the nutritive movements are slow—so slow that any agent which produces local anæsthesia in other parts in a few minutes will be very slow in acting to any depth in the dentinal fibril; so slow, indeed, that we need hardly look in this direction for a practical agent. Cocaine, as a typical representative of this class of agents, has not been effective for this reason.

Any agent which is to act upon the sensitive function of the dentinal fibril must be more powerful than cocaine; it must not coagulate albumen; it must have an affinity for water; it must possess a penetrating property and insinuate itself into the tubule, for it will not be carried in by circulation, and but very slowly by any nutritive movement. Until we find an agent possessing these qualities, we need not expect to obtund sensitive dentine by suspending the fibril's irritability.

The most satisfactory results have been obtained in other directions. The dentinal fibril has a definite composition. All the proteids are in a definite proportion, and we have reason to believe that the proportion is very delicately balanced. If the structure of the fibril be changed, its irritability ceases, and it can no longer communicate.

The change of structure most easily accomplished is the coagulation of its albumen and the withdrawal of its water. If the albumen is coagulated, it is as effectual in checking neural movement as though the albuminous ingredient had been withdrawn; indeed, the coagulation of albumen itself renders it harder, and its presence

---

* Read before the American Dental Association, at Saratoga.

would prevent any exhibition of life more readily than if it were removed. The coagulation of albumen, unless produced by heat, is somewhat self-limiting; so that, since heat of 160° F. is not allowable for sensitive dentine, the action of present known coagulants is not entirely satisfactory. Before coagulation can be a practical success, we must have an agent which penetrates to some depth.

Penetrating escharotics, such as arsenic and the like, since they burn the tissue beyond recovery and endanger the pulp, should not be considered.

The other method of changing the structure of the dentinal fibril is by the removal of the water.

The water is in a twofold relation with the fibril—that which is a constituent of the fibril itself, and that which fills any inequalities between the fibril and Neuman's sheath. One is a constituent, and the other a condition.

The water, which is a constituent, is always present; it is always found where there is life and motion; it is always a necessary condition for neural activity; it makes up three-quarters of the entire body, and about ninety per cent. of the dentinal fibril. Constituting such a large proportion of the fibril, it is evident that there would be a proportionate change in the structure and physical character if the water were removed.

The size would be decreased; the albumen, if coagulated, would become harder; and in this desiccated condition it would be practically dead, incapable of performing nutrition and function.

The water, which fills any inequalities between the fibril and its investments, may be accessory in the transmission of sensation. In this relation it may transmit vibrations to the fibril, or even to the fibriloblast.

One of the most delicate nerves of special sense—the auditory— receives its impressions through the endolymph in which its terminals float; so, if the water surrounding the fibrils should be removed, the fibril's transmitting power may be affected.

The watery contents of the tubule are withdrawn by evaporation, and by bringing an agent in contact with it for which it has an affinity. Evaporation is produced by raising the temperature and subjecting the fibril to a current of air. This is practically accomplished by the repeated blasts from a hot-air syringe.

The water surrounding the fibril is comparatively easily abstracted, but to loosen the molecular grip in the fibril itself requires more force.

There are many agents which have an affinity for water, but not all are suitable for use in the cavity of a tooth.

As compared with coagulation as a means of changing the structure, dehydration is more thoroughly accomplished with the present known methods, and hence the results are more satisfactory; so much so that I suppose the majority of operators use dehydration almost entirely as a means of obtunding sensitive dentine.

We have an agent which, when used on sensitive dentine, both withdraws the water and coagulates the albumen, and a more effective single medicament for sensitive dentine we have not. I refer to chloride of zinc—not a fluid when its affinity for water has been satisfied, but a crystal which will deliquesce when placed in the cavity.

The last method by which the fibril-transmitting power may be lessened is by changing the temperature. Of all the obtundents of sensitive dentine, extreme cold is the most effective; not that it is any better than perfect coagulation or dehydration, but because the temperature of dentine may be reduced more thoroughly than the albumen may be coagulated or the water withdrawn in ordinary practice.

The reduction of the temperature is best accomplished by the use of volatile agents, such as sulphuric ether, chloride of methyl, or nitrous oxide.

It will be observed that the fibril's function may be completely suspended by coagulating all the albumen, by withdrawing all its water, or by lowering its temperature to a certain point. Entire coagulation of its albumen is practically impossible at present, the entire removal of the moisture almost impossible, and the reduction of temperature is the only one which acts to any desired depth.

Recently there have been introduced three new methods having the same principle of action, which are misleading—viz.: the thermal obtunder of Small, the Milton obtunder, and the Richmond obtunder. These are instruments and methods which use an elevation of temperature, and the vapor of an alcohol or any essential oil, or a combination of these, upon the dentine. The effect of throwing a blast of hot air upon a tooth is to heat it. By heating the tooth evaporation of its dentine takes place. The fibril is dried up, and there is an outward movement as the water vaporizes and escapes from the tubule.

The application of a vaporizing agent in a warm or hot place is essentially one of desiccation. The heat is the only virtue in this, unless the agent has an affinity for water, when it may aid in carrying off the water vapor. The vapor of alcohol is effective on this account, and I think those who have recommended the use of oils or any agent which has not an affinity for water are laboring under 'a delusion.

Unfortunately, the agents which are most effective are most dangerous to the pulp, and in our selection of an agent for the case at hand this is to be borne in mind. (L. B. Custer, in *International Dental Journal.*)

A WORD TO STUDENTS ON THE IMPORTANCE OF MECHANICAL DENTISTRY.

LOOKING back on the examinations of past years, one cannot help noticing the greater range of subjects that are nowadays required to satisfy the examiners for the L.D.S. (Eng.) diploma. Indeed, it is only of late years that the question of mechanical dentistry has found its way into the papers in any serious form; and this is the more to be wondered at, seeing what splendid mechanics some of the older dentists were. But no, for some reason or other this matter was almost totally ignored—at any rate in the practical part of the examination. Lately, however, it has been considered of so much importance that the hospitals have been obliged to provide laboratories where the students may acquire the degree of excellence necessary to pass their examination. Now this brings us to the all-important question of pupildom. One may say that there are three classes of pupils: One, and let us hope the largest, that *will* work; another, who think that the only way to acquire a competent knowledge of mechanical dentistry is by manufacturing articles of *vertu* for his lady friends; and, lastly, the pupil who takes unto himself the manner of a chrysalis, only to singe his wings when he has developed into the full bloom of a medical student. Three years is none too long to master all the minutiæ of our work, even though constant attention be paid throughout the whole of that period. For the time being let us consider dentistry as a trade. In other trades pupils do not start on their own account directly they are out of their articles, whereas a dentist as often as not does; for after all the hospital is only another period of apprenticeship; sometimes we even find the two periods run into one—the hospital and the workroom attended at one and the same time. Surely we, who are supposed to be masters of practically a trade and a profession,

have more knowledge and a greater delicacy of touch to acquire
than a smith or a carpenter. Ask one of these. "Of what use is
an apprentice at the end of four years?" and when you hear his reply think that some dentists are considered competent to practice
after four, and only four, years of study. It is an established fact
that many men are apt, too apt, to consider the mechanical side of
our work as beneath their dignity, and that operative work possesses charms that the workroom does not; still, for all that, dentists
and embryo dentists should bear in mind that the term dentistry includes both branches of the work, and that both the operator and
mechanic have equal rights to the name. Certainly much of your
future success will depend on the careful study of laboratory matters; it trains the hand as nothing else can. A good mechanic will
be found, generally speaking, to be a good operator; and the British public are better judges of mechanical than operative work.
During the years of their apprenticeship pupils should remember
that it does not fall to every one's lot to practice in London, and
that many are forced by circumstances over which they have no
control to practice in the country, where no work is done for the
profession. Then vain regrets for lost time and opportunities take
the place of the much-vaunted gold stoppings that once were considered the sum total of a dental education.

The technical part of an industrial pursuit can be *learned;* principles
alone can be *taught.* To learn the trade of husbandry, the agriculturist must serve an apprenticeship to it; to inform his mind in the
principles of science he must frequent a school specially devoted to
this subject. It is impossible to combine the two; they must be
taken successively, and careful study devoted to each.—(Guy Harper,
in *Dental Record.*)

THE STANDARD OF PRELIMINARY EDUCATION THAT SHOULD BE INSISTED UPON.*

HAD the title of this article been "The Standard of Preliminary
Education That Would Be Desirable," much more might be thought of
and mentioned as comprising the intellectual equipment of the pupil
about to enter upon his professional study; but since the title reads
as it does, it forces us to consider only such accomplishments as seem
essential to his success at the very start.

* Read before the Odontographic Society of Chicago, October 1, 1891.

Considering the brief course of instructions in our dental colleges, even on the present three-year plan, much valuable time appears to be wasted in many cases where matriculants are not prepared, by rigid school discipline in habits of study and close concentration of thought for a complete and thorough grasping of the scientific subjects into which they are compelled to plunge during the first year.

Here, at least, a trained mind counts for more than a trained hand. The statement has been made by one in charge of a special department in one of our dental schools in Chicago that the greatest difficulty in bringing a class to a uniform degree of proficiency lies in the fact that their preliminary training has been so totally dissimilar. It seems to be a question of former environment. One student, we will say, has just emerged from some high school or other literary institution; and though coming with a mind charged with a number of abstract facts, knows really nothing, as his theory is as yet unseasoned by experience. Others, after quitting their tasks in school, have spent a number of years on farms, or in the mercantile pursuits of various kinds, and having allowed their minds to go to weeds, find it difficult to learn anything from the lectures. This might indicate the necessity of establishing a preparatory school in which such men could receive a uniform training which would show good results in the present method of class teaching.

The standard that should be insisted upon, in our opinion, would be of such a character that a written examination would be considered only a partial test of eligibility, to be credited for what it is worth and no more.

A private interview with the Dean of the Faculty, in which some knowledge of the candidate's general intelligence, good sense, former occupation, age, moral attributes, personal motives and reasons for his determining on the study of dentistry, would assist in deciding as to the natural fitness of the applicant and could be supplementary to the written examination.

As to the latter test the present requirements as published in the college announcements include simply a good English education, which is rather indefinite, as this might mean a thorough elementary schooling or it might extend to higher mathematics and natural sciences. One college is more explicit and indicates one of the following subjects at the option of the candidate, viz.: Latin, German, or Physics.

We think a knowledge of the general principles of physics should be obligatory, as none of the natural sciences has so much to do with the various phenomena which are presented in the study and treatment of diseased teeth.

The other branches as laid down in the announcements seem comprehensive enough if only a rigid examination on them be held. A thorough mastery of the essentials, taken with a well-trained memory, means a better mind for future good than a more extensive but uncertain knowledge. (Charles J. Merriman, in *Dental Review*.)

A METHOD OF MAKING A CROWN.

If you will kindly give space to a brief description of how to make a porcelain crown for bicuspids (especially), or molars. I will be glad to contribute the following:

After having fitted a cap, with one or more posts as the case demands, to the root, a saddleback tooth, Figure 1. is selected, ground and adjusted to place and soldered to cap, Figure 2. Then it is re-

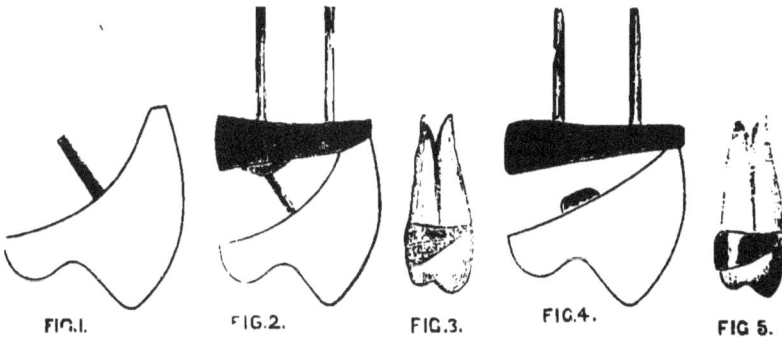

FIG.I. FIG.2. FIG.3. FIG.4. FIG 5.

moved from investment, thoroughly cleaned of articles of marble dust or plaster, and an easily fusing body is packed under crown and around the pin, completely filling vacant spaces and contouring to the desired shape and fused. You then have a crown when completed as represented by Figure 3.

The same crown can be made by backing the saddleback tooth with gold, adjusting and filling the vacant place with gold solder. See Figures 4 and 5. But it is not quite so neat as when the body is used instead.

If one fits himself up with a Parker furnace (I do not know of a better one; there may be, however) for soldering. he will have little use for a blowpipe, save a small blowpipe for light soldering, and not only the above-described crown, but any other crown can be easily made, or any piece of bridge work.

I am living in a small town where illuminating gas is not used, yet with a gasoline generator, bought of the Buffalo Dental Manufacturing Company, and a Parker furnace I have no trouble in soldering any piece of work. With any care at all, there is no danger of cracking the teeth. I have repeatedly heated a crown, soldered, and cooled it off in about ten or fifteen minutes without cracking it. With this little furnace there is no use of any more investment than merely to hold the parts together till soldered.

The drawings explain themselves. (A. P. Johnstone, D.D.S., in *Dental Review*.)

DRIVEN CRAZY BY NITROUS OXIDE GAS.

Mrs. Ella Targett, of 784 Amsterdam Avenue, a woman of large frame, is in the insane pavilion at Bellevue. Her outbreaks give her keepers the greatest trouble, and she is almost constantly in a strait-jacket. She is the wife of George Targett, an agent for several apartment houses, and was taken by him to the hospital on July 16.

When he brought her there, he said that he believed his wife's insanity had been brought on by a sitting at the dentist's, at which she had nineteen teeth drawn under laughing gas. Mr. Targett said: "She left the house early in the afternoon, but did not return until about 9 o'clock at night. She seemed utterly broken down when she reached the house, and was in a fever. For an hour I could not get a word from her as to the experience she had undergone. The aunt declared that she had first asked the dentist if it would be perfectly safe to administer the anæsthetic, and he answered her that it would. A physician who had attended my wife before was called in. He said she ought not to have taken gas. She grew steadily worse, and the insanity under which she now suffers has rapidly developed. She has never been out of her mind before. We have been married eleven years."

Dr. Stuart Douglas, under whose care Mrs. Targett is at Bellevue, said yesterday that her condition is extremely serious. "She screeches in horror at the blood and bloody visions pictured by her disordered mind, and writhes and twists with furious strength in the hands of those who seek to restrain her. The suicidal mania is strongly developed." (*Philadelphia Times*, July 19.)

WHAT ARE THE ESSENTIALS IN THE PROPER PREPARATION OF OUR DENTAL STUDENTS?

FIRST, a good common English education. This much, and perhaps less, has filled the presidential chair of the United States. Second, a proper attention to, and knowledge of, what is known as prosthetic or mechanical dentistry. To this should, of course, be added a partial insight into collateral branches, such as chemistry and metallurgy, anatomy. physiology, pathology, therapeutics, materia medica.

None of our best physicians could pass a critical examination in *all* of these branches. They might become specialists in some one or other of these branches, but never in all.

Anatomy is a lifelong study. We have scarcely a half dozen all around anatomists in the country. Now the question, really and honestly considered, is: How much anatomy is essential to the proper and scientific practice of dentistry? Should a dental graduate practitioner be required to know more of it than is necessary to his work specialty, and so on with the rest of the collateral branches?

A pharmaceutical graduate should be taught all of materia medica that is known, for this is his specialty. A pathologist and microscopist make these subjects their specialty, and of course should be thoroughly educated in that direction. An accurate knowledge of anatomy in its surgical sense is indispensable to the surgeon— that is his specialty. I am not condemning in any way the pursuit of knowledge. Every man, to the best of his ability, should improve himself in all directions; but should he do this at the expense of the more important studies? The time which can be spared to dental education forbids this to the ordinary student. A man cannot and should not practice two professions., (Dr. W. C. Klatte, in *Items of Interest.)*

INTERNATIONAL MEDICAL CONGRESS.

THE meeting will be held in September, 1893, so says the *Medical Record*. The following sections have been provided: Anatomy, Physiology, Pathology, Therapeutics, Clinical Medicine, Surgery, Obstetrics, Psychological Medicine, Ophthalmology, Dermatology and Syphilography, Forensic Medicine, Hygiene; and Dental Surgery is not in it, as usual. without a petition or some sort of begging for a place. 1881, by petition; 1884, no section—a subsection in surgery was established: 1887, left out. but afterward by request made a section; 1890, provided for in the beginning; 1893. ——, unless by request. This is because there is not in all Italy an organized dental profession, and there is not at this time a dental department in any teaching body in the whole kingdom. Dr. John E. Grevers. of Amsterdam, once said in a letter to the *Dental Review* that there were countries in which it would not be possible, from the native talent alone, to organize a dental section. This prediction is about to be realized, unless the foreigners in Italy take hold of the section; and few, if any of them, hold the medical degree. It is unfortunate that the meeting in Chicago and the one in Rome should come so close to each other, but we will have to do the best we can to bear up against the inevitable. (Editorial, *Dental Review*.)

ODDS AND ENDS.

I FILL the apex of roots to be crowned, if straight and convenient of access, with lead. Where it can be used and the discoloration will not appear, there is nothing like it. Just beat it out thin and cut into strips about one-sixteenth of an inch in width, and twist these. but not tightly. Then, if too large, gently roll between fine sand or emery paper. and this kind of a lead plug will easily be pressed to fit every space and foramen. Where I cannot get such access, I must use chloro-percha or gutta-percha. Where a fine broach cannot be used, I work as much iodol or iodoform into the root as possible, leaving the pulp chamber coated with iodoform, made to the consistency of paste with pure wood creosote; then I fill the best I can with chloro-percha, and have no trouble with them afterward. The extremely small canals are not going to give trouble treated in this way. and I have reason to believe they will ossify before the medicine is exhausted.

I have recently crowned (using the gold crown) two living bicuspids. They were broken down badly, but had good nerves. I am having thus far, and I verily expect, complete success in both.

In spite of all the slurs spoken against copper amalgam, on account of its color, there are some places where nothing else will do so much toward saving teeth; and, though I use it sparingly and carefully, I am going to continue its use where indicated.

In May *Items*, page 299, Dr. William E. Blakeney says: " It is a much better and cheaper plan to use cement instead of gold for packing joints of artificial teeth." I want to say that cement, however good after being vulcanized, will dissolve and wash out and leave a receptacle for food, bugs, and the like.

A gentleman finds fault that I have to use gold to fill my poor joints. Now I've not a word to say to those who can make *perfect* joints. I cannot, and my article was intended for those like myself who could not make a perfect joint by grinding. But by packing with gold foil we can make our poor joints appear to be continuous gum, where nothing can corrupt, nor bugs live in to smell.

Dr. France, of Milford, Del., has not been true to his profession if he has known and used this plan for thirty years and has never told it (see June *Items*, page 354). But I thank Dr. France for the idea of using tin foil. I'll try it; but I fear its color will condemn it, though I know it to be one of our most faithful and honest servants in the whole line of filling materials. (Dr. Lindley H. Henley, in *Items of Interest*.)

F. R. M. S.

MOST men do love to attach a long string of initials after their names, sometimes significant of real culture and erudition—and sometimes not. Among those which belong to the latter class are the letters F. R. M. S.—Fellow of the Royal Microscopical Society. A man who parades these habitually must be badly in need of something to distinguish him. The distinction is an English one, and costs just ten dollars annually—rather a stiff price. A Delavan D.D.S. was once sold for only twelve dollars, and that paid but once. While our English brethren confer the "distinction" of an F. R. M. S. as promiscuously as they do, the conditions not being as onerous as they were in the Delavan fraud, they will do well to be chary of their criticisms of American distinctions. (Exchange.)

Extracts.

RELATIVE CONDUCTIVITY OF FILLING MATERIALS.

In an article on this subject in the December number of the *Dental Review*, Thomas L. Gilmer, M.D., D.D.S., of Chicago, placed the conductivity of the different materials in the following order, 1,000 representing the best conductor:

Gold	1,000.
Lawrence amalgam	852.5
Copper amalgam	702.7
Oxyphosphate of zinc	584.27
Oxychloride of zinc	525.25
Gutta-percha	520.

He adds: "A test of the conducting qualities of alloys presents curious results. For instance, if one per cent. of silver, which is represented as a conductor by 1,000, be added to gold, the conducting quality of the alloy is changed from 980, which gold alone gives, to 840. If two metals be combined, one being the best known conductor and the other the poorest, the latter predominating, the conducting quality of the alloy formed is no better than if it did not contain a particle of the better conductor."

Works on physics give the conducting quality of the various metals used in dentistry, previous to their being transformed into shape suitable for filling materials, as follows, 1,000 representing the best conductor:

Silver	1,000.
Gold	981.
Copper, rolled	845.
Copper, cast	811.
Tin	422.
Platinum	380.

DISEASES OF SECOND DENTITION.

I think it can be easily shown that many of the morbid conditions and diseases of adolescence are due to the second dentition. In addition to the distinguished American physicians whose names have been mentioned, I can give a number of others who attribute serious and even fatal maladies to the second dentition. J. Russel

Reynolds says that there is no doubt that epilepsy may be caused by the second dentition. Erb and Hein mention teething as a cause of poliomyelitis anterior and spinal meningitis, and many other writers speak in a similar strain. Dr. Louis Starr has recently published a paper in the *Therapeutic Gazette* on the diseases of the second dentition and their treatment. Several writers, among others Dr. Mulveany, of Brooklyn, have called attention to the association of hip joint disease and the eruption of the first permanent molar teeth, the majority of cases of this complaint showing themselves between the fifth and seventh year. Dr. Mulveany asserted that he had never seen a case that did not begin during the period mentioned. He also gives some interesting cases of pseudo hip disease in small children, all of them evidently too young to simulate the complaint. In one of these so good a surgeon as Dr. Sayre is said to have been so completely deceived by the symptoms and history of the case as to have advised an exsection of the head of the femur and actually to have cut down with the intention of removing this part of the bone, only to find that the joint was in a healthy state. (Richard Cole Newton, M.D., in *International Dental Journal.*)

TIN.

I do not care to enter into the question of the use of tin and gold further than to mention the use of tin on cervical marginal edges. This practice seems to have become popular of late years, and represents one of the general acknowledgments of the difficulty experienced in preventing recurrence of decay in these localities. I must say, however, that I have been unable to see any really good reason for its use. Good margins can be made with it, and possible it may be a little easier to learn to manipulate it than noncohesive gold. This difference seems to me inconsiderable, and insufficient to counterbalance the objections to tin. Margins made of tin become dark and unsightly, and they often give the impression that decay has recurred, leading every one who may examine the mouth to question them. Tin is so soft that the point of an explorer may be easily thrust into it. Not unfrequently good margins have been seriously injured in this way. Certainly just as good margins may be made with noncohesive gold foil, to which these objections do not apply. (G. V. Black, in the *Cosmos.*)

DEVITALIZING PASTE.

Of all the formulas tried, none have proved so uniformly satisfactory as a combination of arsenious acid with the so-called Robinson's remedy, mixing sufficient arsenic with Robinson's remedy to make stiff paste. This paste is then worked into cotton fiber until it will hold no more, after which the cotton, so treated, is loosely twisted into a cord and cut up into pieces about the size of a small mustard seed. These little pieces form a most convenient means for making the devitalizing application. They are slightly adhesive, and one can be accurately applied over the point of exposure and sealed in position in the usual manner. Out of somewhat more than one hundred applications made as stated, there has not been a single case of distressing pain resulting. When applied to an aching pulp, the effect is sedative, the pain gradually subsiding until it disappears in about thirty minutes. Two explanations for its sedative property may be reasonably assigned. First, the Robinson's remedy is strongly alkaline, and the excess of alkali acts as a sedative to the pulp, a property which is shared by all the soluble alkalis and their carbonates; and second, when the arsenic is mixed with the Robinson's remedy it enters into combination with the free potassium hydrate to form potassium arsenite, which is soluble. The arsenic is therefore presented to the pulp in a soluble form requiring no effort upon the part of that organ for its absorption, and consequently its irritating effect is reduced; but whatever may be the exact explanation, the satisfactory action of the combination is beyond question, and has led me to use it exclusively and without hesitation in all cases of pulp exposure when devitalization is the correct procedure. (Dr. E. C. Kirk, in *Cosmos.*) ____

THE TOOTH OF THE FOOL.

Has the fool better teeth than the wise man? He ought to have, if the theory put forward by the President of the British Dental Association to account for the deterioration of the British tooth be correct. Smith Turner says: "Another competitor with the teeth for sustenance is the brain. Phosphorus is one of the great essentials to the center of the nervous system. In starvation certain organs are the latest to suffer, and are nourished and sustained at the expense of the rest of the body. One of these organs is notably the brain, and so we are entitled to infer that in any deficiency of the phosphates the ever-active brain, with its copious circulation and energetic metabolism, is more likely to appropriate its full, or approximately full, share of the phosphates than are the teeth, where hitherto we had hardly looked for the katabolic process."

It is, at any rate, some satisfaction to reflect that if the world is

3

becoming more toothless it is at the same time becoming wiser.
But will not the phrase " wisdom tooth " want amending if progress
in wisdom entails not the gain of teeth, but the loss of them? (*Pall
Mall Gazette.*)

A LITTLE LEARNING.

THE great trouble is that many men cannot imbibe a single
draught of knowledge without becoming craniologically intoxicated.
A single large idea overstimulates their intelligence, and they lose
mental balance. There is something more than mere metrical jin-
gle in the famous stanza from Pope:

> A little learning is a dangerous thing;
> Drink deep, or taste not the Pierian spring;
> There shallow draughts intoxicate the brain,
> And drinking largely sobers us again.

Intelligent and enthusiastic dentists who have the ear of their
confrères have discoursed learnedly upon microbes and ferments,
upon germicides and detergents, upon coagulants and noncoagu-
lants; while the younger men, who were perhaps taking their first
sip of the fascinating draught, or at least with less of experience and
knowledge, listened with open ears and wondering eyes. Nothing
was said of other conditions than that of septicism, because the
speaker took it for granted that all were as well informed as himself, and that they would add the modifying provisions. A single
factor seemed to be raised to a dominating position, and remedies
new even to the chemist were specifically recommended, without a
fair comprehension of their properties. Is it any wonder that in
following such teaching mortifying failures resulted? (From Dr. W.
C. Barrett's article on " Conservatism.")

SULPHO-CARBOLIC ACID A DISINFECTANT.

ACCORDING to E. Leplace, who experimented in the Hygienic Institute at Berlin, a mixture of crude sulphuric and crude carbolic
acids is an excellent disinfectant. Equal parts of sulphuric and 25
per cent. carbolic acid are heated for a short time, when the mixture
can be united with any proportion of water. In efficiency this disinfectant compares with a 0.1 per cent. corrosive sublimate solution,
and possesses the advantage of its relatively nonpoisonous property.
(*National Medical and Pharmaceutical Journal.*)

POWER OF MENTAL IMPRESSIONS.

In 1862 Mr. Woodhouse Braine was called upon to give chloroform to a nervous, hysterical girl for the purpose of having two tumors removed from the scalp. In order to accustom her to breathing through the inhaler before giving her chloroform, he placed it over her face and she at once began to breathe rapidly through it. In half a minute she said: "O, I feel it, I feel I am going off." Immediately after she was found to be insensible to pinching and her muscles were flaccid. Both tumors were removed without her having taken a drop of chloroform, and after the operation she declared she had not felt a particle of pain. The doctor very facetiously adds: "To the time she left the hospital she firmly believed in the potency of the anæsthetic which had been administered." ("Influence of the Mind upon the Body," Tuke). (From the *Scientific American*.)

Was it the mental impression or rapid breathing as demonstrated by Dr. Bonwill, of Philadelphia?—Ed.

DECAY OF A REPLANTED TOOTH.

Attention having been called by Prof. W. D. Miller to the question of decay in replanted teeth (*Dental Cosmos*, April, 1891, p. 253), the following report of a case in my own practice may be of interest:

More than fourteen years ago (February, 1877). I replanted for a lady the right upper lateral incisor on account of chronic pericementitis, after having filled a cavity in the mesial surface with gold. About a year subsequent to the replantation, caries took place upon the distal surface, and as I feared that the force incident to the insertion of a gold filling would be a source of irritation to the root, I filled it with cement. Since that time I have renewed the filling twice, the last time about four years ago.

The tooth, as in the case reported by Miller, has a chronic fistula opposite the apex of the root; nevertheless it remains comfortable and renders satisfactory service to the patient. (Dr. L. Weil, Munich, Germany, in *Cosmos*.)

PULLING A TOOTH RESTORES SIGHT.

A case which will attract attention in medical circles is reported to-day from Lamar, Ark. Mrs. Eliza Ryan, a widow eighty years old who has been totally blind for thirty years, had a tooth pulled from the upper jaw. The root of the tooth was nearly an inch in length. When it was extracted, Mrs. Ryan complained of intense pain in her eyes, and later cried out that she could see plainly, her sight having been restored. (*Chicago Evening News*.)

Editorial. . . . .

Happy New Year!

We make our New Year's salutations to our contributors and subscribers attired in a new and we think a much-improved dress, and wish them health and prosperity.

Oklahoma has a law regulating the practice of dentistry.

The *Dental Review* says: "Make a saturated solution of zinc sulphate in water and use it with the powder in oxychloride packages and see how hard it will become. Use it as a foundation for filling, or in pulpless tooth crowns."

The December number of the *Dental Record* (London) announces that "Mr. H. O. Merriam, D.D.S., Salem, Mass.," has been elected a corresponding member of the Odontological Society of Great Britain. "Mr." Merriam will no doubt feel honored.

Brazil has a law for the medical examination of persons about to marry to determine their fitness. It is a sanitary measure that is found to be necessary to stop the transmission of scrofula, which at one time threatened to destroy the strength of the people.

Aluminium wire is said to be useful for many purposes in the dental office, such as strengthening rubber plates, making canal points, or as a carrier for conveying iodine, aromatic acid, or any corrosive agent, excepting muriatic acid, it being soft, pliable, and clean.

The Royal College of Surgeons of England, in June, at an "Ordinary Meeting of the Council," admitted seventeen as licentiates in dentistry. Seventeen candidates were referred back to their professional studies. It seems that there is a greater need for limiting the number of accessions to the practice in England than in this country, or the "Council" is more exacting than our boards.

Dr. A. P. Johnstone, Chairman of the Committee on Appliances and Improvement for the meeting of the Southern Dental Association next July, spent a day recently in Nashville for conference with Dr. White. This display is in the best of hands, and will certainly be superior to any ever offered at an Association. Dr. Johnstone's varied experience peculiarly adapts him to this work, and we bespeak for him the hearty coöperation of all dealers and inventors.

The case of poor, young Baab, of New York, who was bitten by a patient while having a tooth pulled, that we mentioned in July *Items*, proved fatal. First it was lockjaw, as we then mentioned, and then typhus fever caused by the general diffusion of the poison. What a warning this should be to the careless insertion of the fingers between the jaws to make such results possible. While I am writing this note, a friend calls to say that his son, a young dentist of Camden, N. J., is suffering from a similar injury. Let us hope the result will not prove as serious. (Editorial in *Items*.)

Dr. Gordon White, President, announces the following as Chairmen for the various committees of the next meeting of the Southern Association. Let these gentlemen follow the example set by Mr. President, and enter early upon their labors.

Executive Committee.—Sid G. Holland, Atlanta, Ga.

Committee on Arrangements.—W. H. Richards, Knoxville, Tenn.

Dental Chautauqua.—W. H. Richards, Knoxville, Tenn.

Committee on Publication.—H. C. Herring, Concord, N. C.

Clinic Committee.—W. R. Clifton, Waco, Tex.

Operative Dentistry.—J. Rollo Knapp, New Orleans, La.

Prosthetic Dentistry.—George B. Clement, Macon, Miss.

Dental Education.—B. Holly Smith, Baltimore, Md.

Dental Hygiene.—D. R. Stubblefield, Nashville, Tenn.

Pathology and Therapeutics.—W. J. Barton, Paris, Tex.

Histology and Microscopy.—E. P. Beadles, Danville, Va.

Chemistry.—R. C. Young, Anniston, Ala.

Literature and Voluntary Essays.—G. F. S. Wright, Georgetown. S. C.

Appliances and Improvements.—A. P. Johnstone, Anderson, S. C.

DENTAL LEGISLATION.

AT the last meeting of the National Board of Dental Examiners much interest was manifested in the debate which took place upon the question of the desirability of more uniformity in laws regulating the practice of dentistry, and particularly in States which require graduates to submit to examinations. Schoolmen as a rule are indifferent more or less, as the terror of an Examining Board makes the student more industrious, but when the law is worded as that of Michigan, which specifies that the "requirements of the college granting the degree shall be equal to those of the University of Michigan to exempt from an examination." the intent is too manifest to require any explanation. Such legislation defeats the legitimate object of dental legislation, and sooner or later recoils, and with it such decision as was rendered in the courts in New Hampshire last spring. Conservative men have been advising against such radical measures. and predicting such results, but it seems that experience is the only schoolmaster many ever heed.

Surely the royal road to dentistry has become a difficult one. Some of those who are most anxious to throw obstacles in the way of those desiring to enter the profession found an easy side entrance themselves and slipped in. Once in. they want to surround the thing with a wall. ⸻

GUTTA-PERCHA AS A ROOT FILLING.

THE following extract is from an article entitled "Gutta-percha as a Filling for Root Canals," by Dr. Louis Ottofy, read before the First International Dental Congress, Paris, 1889. and recently printed in the *Dental Review:*

In using the solution gutta-percha and chloroform for root filling, it must be born in mind that during manipulation and before final sealing of the root canal the chloroform should be permitted to evaporate ; that the canal is not intended to be filled with liquefied gutta-percha, but that the greater portion—at least one-half or two-thirds—should be filled with some other material.

Can the canal be said to be filled with gutta-percha? The writer and many others have for years been following the practice. using lead points, gold foil, gold and copper wire. and have always believed they were filling the pulp canal with these materials. No, the philosopher-dentist was shooting at chloro-percha *per se* as a root canal filling, and not chloro-percha combined with other materials. Dr. Ottofy's attempts to prove that gutta-percha alone dissolved in chloroform could be relied upon as a root filling, yet shows conclusively that it cannot be relied upon.

NASHVILLE A PORT OF ENTRY.

MR. HERMAN HASSLOCK. of the firm of Hasslock & Ambrose, printers, for years the publishers of the DENTAL HEADLIGHT, Surveyor of Customs at this port, has issued the following instructions to all importers:

"To have goods appraised here it is required of importers that they be explicit in ordering the consignor to have the goods shipped, 'Under I. T. Bond. without appraisement, to Nashville, Tenn.' If the goods are shipped via New York, they should be consigned in care of either the Merchants' Dispatch. the Star Union Line. Cumberland Gap Dispatch, Great Southern Dispatch, or the American Express Company, or the Wells Fargo Express Company; if via New Orleans, they should be shipped in care of Illinois Central or New Orleans & Texas Railway. If the above suggestions are complied with by the shipper, the goods will come here for appraisement, thereby saving considerable to our merchants. The consignor should mail to the transportation company a consular invoice and bill of lading for presentation to the collector of port of first arrival, so as to make bond for transportation."

THE WORLD'S COLUMBIAN DENTAL MEETING.

ELSEWHERE will be found a report of the last meeting of the Executive Committee for this great meeting. It deserves a careful reading.

Prof. J. Taft, Chairman of State Committee, has appointed a Committee for Tennessee to stir up an interest and secure the co-operation of all good men of the profession in the State.

At an early date a circular letter will be mailed to many in the hope of getting such historical facts concerning the literature, discoveries, improvements, associations, and pioneers of the profession of the State as will be of value, to be put in such a form as to be preserved and published in the report of this committee. All who are in possession of such matter are requested to forward it to any of the committee as early as possible.

DENTAL PATENTS GRANTED DURING NOVEMBER.

WE acknowledge indebtedness to Messrs. Collamer & Co., attorneys and solicitors. Washington, D. C., for the following list of dental patents granted in November:

Dental Chair—G. W. Archer, Rochester, N. Y.

Dental Engine—J. T. Calvert, Spartanburg, S. C.

Dental Vulcanizer—T. J. Carrick, Baltimore, Md.

Dental Mallet—M. L. Bosworth, Warren, R. I.

Forming Dental Plates—M. R. Griswold, Hartford, Conn.

Rubber Dam Clamp—F. E. Hansen, Minneapolis, Minn.

DOGOLOGY.

A LETTER from a dental student to a friend in Nashville is authority for the statement that in one of the dental schools in the United States the freshmen are required to dissect dogs this winter. Comparative anatomy is valuable, but it would seem better to teach something of the human anatomy first. This is a wanton waste of time, and students of the institution would be justified in indignantly rebelling at such a pretense at scientific instruction.

Marriages and Obituaries. · · · · ·

MARRIED.

J. C. Pearson, D.D.S., and Miss Annie C., daughter of Mr. and Mrs. O. T. Grimes, of Jemison, Ala., December 16, 1891.

Samuel J. Lawrence, D.D.S., and Miss Maude V. Johnson, at the First Presbyterian Church, Fort Worth, Tex., December 22. 1890.

Granville L. Brown and Miss Clara Hyde Dickey, at the Baptist Church, Glasgow, Ky., Tuesday evening, December 15, 1891.

J. E. Andrews, D.D.S., and Miss Oya Allen, of Harrison, Ark., Tuesday, October 20, 1891. ——

DIED.

W. W. Kemper, M.D., D.D.S., at his home (Frederickstown, Mo.). December 8, 1891, from the effects of chloroform inhalation for the relief of neuralgia. Dr. Kemper was born in Frederickstown, Mo.. in 1859; graduated in medicine at St. Louis; took his degree in dentistry from Vanderbilt University in 1882, at which time he also received the Founder's Department Medal. For several years he followed the practice of dentistry exclusively. Three years ago he began to devote a part of his time to general medicine; and at the time of his sudden death had a practice that required all his time. He was unmarried.

In Yazoo City. Miss.. Thursday morning. December 3, 1891, Mrs. Lulie Hart Martin, wife of Dr. W. T. Martin, the well-known dentist of that place. From an obituary notice in the *Sentinel* we extract the following: "Mrs. Martin was a woman of more than ordinary character and scholarly attainments. Possessed of a bright and well-cultivated mind, she was a woman who would have shone in any circle. Charming and fluent in conversation; graceful, elegant, and gentle in manners; pure and lovely in heart, she was sought, admired, and loved by a host of friends, who are sorrowful indeed at her sad demise.

> 'Tis the infinite wonder of all
> That she should have let life's flowers fall.

Back to the scene of her childhood they carried her remains to the family graveyard at Harttown—back where the wild flowers bloom in luxuriance, and the noise and the din of the world will not disturb her peaceful sleep." Dr. Martin's professional friends throughout the South will join us in tendering him warmest sympathy.

Associations. · · · · · · · · · ·

AMERICAN DENTAL ASSOCIATION.

W. W. Walker, New York, N. Y., President; J. D. Patterson, Kansas City, Mo., First Vice President; S. C. G. Watkins, St. Clair, N. J., Second Vice President; Fred J. Levy, Orange, N. J., Corresponding Secretary; George H. Cushing, Chicago, Ill., Recording Secretary; A. H. Fuller, St. Louis, Mo., Treasurer.

The next meeting will be held at Niagara Falls, N. Y., on August 2. 1892.

SOUTHERN DENTAL ASSOCIATION.

Gordon White, Nashville, Tenn., President; E. L. Hunter, Enfield, N. C., First Vice President; J. T. Calvert, Spartanburg, S. C., Second Vice President; W. H. Marshall, Oxford, Miss., Third Vice President, D. R. Stubblefield, Nashville, Tenn., Corresponding Secretary; H. C. Herring, Concord, N. C., Recording Secretary; Henry E. Beach, Clarksville, Tenn., Treasurer.

The last Tuesday in July, 1892, and Harrogate, Tenn., were selected as the time and place of next meeting.

WORLD'S COLUMBIAN DENTAL MEETING.

W. W. Walker, Chairman, 67 West Ninth Street, New York City; J. S. Marshall, Treasurer, Argyle Building, Chicago. Ill.; A. O. Hunt, Secretary, Iowa City, Ia.

Executive Committee.—L. D. Carpenter, Atlanta, Ga.; J. Y. Crawford, Nashville, Tenn.; W. J. Barton, Paris, Tex.; J. Taft, Cincinnati, O.; C. S. Stockton, Newark, N. J.; L. D. Shepard, Boston. Mass.; W. W. Walker, New York City; A. O. Hunt, Iowa City, Ia.; H. B. Noble, Washington, D. C.; George W. McElhaney, Columbus. Ga.; J. C. Storey, Dallas, Tex.; M. W. Foster, Baltimore, Md.; A. W. Harlan, Chicago, Ill.; J. S. Marshall, Chicago, Ill.; H. J. McKellops, St. Louis, Mo.

The meeting will be held in Chicago in 1893.

ALABAMA DENTAL ASSOCIATION.

George Eubanks, President; C. L. Boyd, First Vice President; J. H. Hall, Second Vice President; J. H. Allen, Recording Secretary; G. M. Rousseau, Treasurer.

Place of meeting, 1892, Montgomery, second Tuesday in April.

FLORIDA DENTAL ASSOCIATION.

L. M. Frink. President; W. A. McQuarg, First Vice President; W. A. Snead, Second Vice President; L. F. Frink. Corresponding Secretary; C. P. Barrs, Recording Secretary; C. P. Carver. Treasurer. Executive Committee: B. B. Smith, W. McL. Dancy. J. H. Crossland, James Chase, and T. B. Hannah.

The ninth meeting will be held in Jacksonville, Fla., in May. 1892.

GEORGIA DENTAL ASSOCIATION.

W. G. Browne, President; S. M. Roach. First Vice President: W. W. Hill, Second Vice President; L. D. Carpenter, Corresponding Secretary; S. H. McKee, Recording Secretary; H. A. Lawrence. Treasurer.

The Board of Examiners consists of J. H. Coyle, A. G. Bouton. G. W. McElhaney. William C. Wardlaw, and D. D. Atkinson.

The Association is to meet in Rome, Ga., at such time as the Executive Committee may elect.

KENTUCKY DENTAL ASSOCIATION.

H. B. Tileston, Louisville. President; M. W. Steen. Augusta. Vice-President; J. F. Canine. Louisville. Treasurer: J. H. Baldwin. Louisville. Secretary.

MISSISSIPPI DENTAL ASSOCIATION.

D. B. McHenry. Grenada, President: A. A. Dillehay, Meridian. First Vice President; J. B. Rembert, Jackson, Second Vice President; A. A. Wofford, Columbus, Third Vice President; W. E. Walker, Bay St. Louis, Recording Secretary; P. H. Wright. Senatobia, Corresponding Secretary; L. A. Smith, Port Gibson. Treasurer.

TENNESSEE DENTAL ASSOCIATION.

D. R. Stubblefield. Nashville, President; S. B. Cook. Chattanooga. First Vice President; W. W. Jones, Murfreesboro. Second Vice President; P. D. Houston, Lewisburg, Recording Secretary: J. L. Mewborn, Memphis. Corresponding Secretary; H. E. Beach. Clarksville, Treasurer.

Place of meeting. Chattanooga, first Tuesday in July. 1892.

SYNOPSIS OF THE SARATOGA MEETING
OF THE EXECUTIVE COMMITTEE OF THE W. C. D. M.

THE roll call showed the following present: Drs. Walker, Hunt. Marshall, Crawford, Taft, Stockton, Shepard, Noble, McElhaney Harlan, McKellops.

The following Local Committee of Arrangements was named by the Nominating Committee, and confirmed: E. D. Swain, Chairman, 65 Randolph Street, Chicago; C. N. Johnson, Opera House Building, Chicago; D. B. Freeman, Fortieth Street, Drexel Boulevard, Chicago; E. S. Talbot, 125 State Street, Chicago: C. R. E. Koch, 3011 Indiana Avenue, Chicago.

W. W. WALKER,
A. W. HARLAN,
JOHN S. MARSHALL,
Nominating Committee.

On motion of Dr. Shepard, the following rules for the action of the Finance Committee were adopted:

Resolved, That the General Finance Committee be authorized to guarantee to contributors who would be eligible to membership as follows:

1. That every contributor of twenty (20) dollars should receive the transactions of the meeting if he does not attend; and if he attends and joins the meeting, his receipt for contribution shall be accepted by the Treasurer as full payment of the membership fee of ten (10) dollars.

2. That every contributor of thirty (30) dollars or upward shall have all the advantages of the contributor of twenty (20) dollars, and in addition shall receive free the Commemorative Medal which will be struck.

3. That every contributor of ten (10) dollars shall receive the transactions if he does not attend, but will be expected to pay the membership fee of ten (10) dollars in addition should he attend the meeting.

4. That any funds which may remain in the treasury after the payment of all obligations will be donated to the Dental Protective Association of the United States.

On motion of Dr. Harlan, the General Finance Committee was authorized to appoint a sub-Finance Committee for each State and Territory.

The Nominating Committee reported as follows on Committees Nos. 3, 12, and 13:

Committee on Exhibits.—Charles P. Pruyn, Chairman, 70 Dearborn Street, Chicago; Arthur E. Matteson, 3,700 Cottage Grove Avenue, Chicago; W. J. Martin, 181 West Madison Street, Chicago.

Committee on Membership.—Edmund Noyes, Chairman, 65 Randolph Street, Chicago; B. F. Luckey, Patterson, N. J.; E. S. Chisholm, Tuscaloosa, Ala.; C. M. Bailey, 28 Syndicate Block, Minneapolis, Minn.; Daniel N. McQuillen, Sixteenth and Chestnut Streets, Philadelphia, Pa.

Committee on Education and Literary Exhibits.—J. Y. Crawford, Chairman, Nashville, Tenn.; C. N. Pierce, 1415 Walnut Street, Philadelphia, Pa.; A. H. Fuller, 2602 Locust Street, St. Louis, Mo.; C.

A. Brackett, 102 Zeno Street. Newport, R. I.; B. H. Catching, Atlanta, Ga.

Dr. Stockton moved to confirm the nominations as made on Committees Nos. 3, 12, and 13.

Dr. Harlan moved that a new Committee No. 18, to consist of three members, be created, to be called a Committee on History of Dentistry in the United States. Carried.

Dr Shepard moved that it is the sense of this committee that it is desirable to change the time of meeting from August and September to the month of July. Carried.

Dr. Taft moved that a committee of three be appointed to fix the time of meeting, and the following were named: A. W. Harlan, J. S. Marshall, A. O. Hunt.

Dr. Marshall moved that the next meeting of this committee be held in Chicago, January 17, 1892. Carried.

After some discussion the following motion by Dr. McKellops was left for action at the next meeting:

Resolved, That any member of this committee having failed to attend three successive meetings will be considered to have vacated his position on the committee.

Adjourned.

Following is the list of committees appointed and confirmed to date:

General Finance Committee.—L. D. Shepard, Chairman, Boston, Mass.: T. W. Brophy, 96 State Street, Chicago, Ill.; A. L. Northrop,* New York City.

Committee on Exhibits.—Charles Pruyn, Chairman, 70 Dearborn St., Chicago, Ill.; Arthur E. Matteson, 3,700 Cottage Grove Avenue, Chicago, Ill.; W. J. Martin.* Chicago, Ill.

Committee on Conference with State and Local Societies.—J. Taft, Chairman, Cincinnati, O.

For Tennessee.—H. W. Morgan, Chairman, Nashville; B. S. Byrnes, Memphis; W. H. Richards, Knoxville; H. E. Beach, Clarksville.

Auditing Committee.—L. D. Shepard, Chairman, Boston, Mass.: R. R. Andrews, Cambridge, Mass.; Charles A. Meeker,† Newark, N. J.

Local Committee of Arrangements.—E. D. Swain,* Chairman, 65 Randolph St., Chicago, Ill.; C. N. Johnson, Opera House Building, Chicago, Ill.; D. B. Freeman, 4,000 Drexel Boulevard, Chicago, Ill.; E. S. Talbot, 125 State Street, Chicago, Ill.: C. R. E. Koch,* 3011 Indiana Avenue, Chicago, Ill.

---

* Declined.   † Not heard from.

Committee on Essays.—E. C. Kirk, Chairman, Philadelphia, Pa.; T. W. Wassall, Chicago, Ill.; A. H. Thompson, Topeka, Kan.; H. H. Johnson, 26 Second Street, Macon, Ga.; L. G. Noel, Nashville, Tenn. *Nominating Committee.*—W. W. Walker, New York City; A. W. Harlan, Chicago, Ill.; John S. Marshall, Chicago. Ill.

A. O. HUNT, *Secretary.*

PATHOLOGY AND THERAPEUTICS.

DISCUSSION IN THE SOUTHERN DENTAL ASSOCIATION.

DR. J. Y. CRAWFORD, of Nashville, Tenn.: The question I wish to speak of is the subject of dental caries. There was a good deal said last night in reference to the paper by the gentleman from Virginia. I was somewhat surprised to see the drift of the argument, leading us to the conclusion that the old ghost of heredity still lifts its objectionable head. While hereditary taint is a question to consider, I do believe, with all due regard to every man on the other side of the question, that it is made a scapegoat and loophole through which many mistakes of the medical world have slipped. Take, for instance, the practical question of life insurance; and there is many a man to-day whose family is exposed to want as a result of the misconception of this question of hereditary taint. One of the most successful insurance companies anywhere ignores the idea of the influence of hereditary taint as the direct production of disease, and I would to God that every other insurance company in the world would take the same high and broad view of the question. We don't inherit dental caries. It is very doubtful whether any individual ever inherited pulmonary trouble or not. I will prophesy that the remarkable hereditary tendency of the objectionable disease referred to will one of these days be questioned. My little bark of reputation may be wrecked on that, but it will go down honestly. Not that some peculiar tendency may be inherited, but that the parent will transmit to his child the specific and peculiar germ known as the tuberculous germ is universally doubted by the best critical minds in medicine. We don't inherit dental caries, but the susceptibility to it. And I want to say this much: that dental caries, as a specific disease localized, has not yet been settled from an ætiological standpoint.

Dr. H. W. Morgan: If I understood Dr. Crawford correctly, he made the statement that it is impossible for a devitalized tooth to decay. If this is true, how is it possible for Dr. Miller to produce, artificially, dental caries? He certainly does it, and so perfectly that when his specimens have been placed under a microscope it is impossible for some of the learned microscopists in this country to distinguish the difference between the caries that takes place in the mouth and that which he has been able to produce artificially.

Dr. Crawford: I don't say that it was identical. I don't admit the observations of Dr. Miller as being absolutely correct. I shall plead guilty to the charge that the destruction of the teeth after devitalization is not analagous to what it was before. I don't believe in Dr. Miller's experiment. He said a great many things I don't believe. The greater the man is, the greater blunder he is capable of making.

Dr. W. C. Wardlaw: It strikes me that we are not occupying such different platforms as we might think, from Dr. Crawford. At first I thought he was standing alone, and was questioning some very radical positions when he spoke about heredity; but when he goes on and explains his meaning, I understand and accept his position and agree with him. We inherit the susceptibility, the liability, and disposition to certain diseases, which under peculiar circumstances will develop them; and I believe that there are many who have such susceptibilities that go through life without trouble. So, when Dr. Crawford goes on with his explanation of the chemico-vital theory, I don't think he is speaking of the chemico-vital theory which was in acceptance several years ago. That, I think, has been sufficiently disproved; and when he goes to claiming for his chemico-vital theory two conditions—that of vitality and chemical action—we all agree with him.

Dr. H. E. Beach: I will not detain the Association perhaps more than two minutes. Dr. Crawford makes the assertion that teeth continue to grow during life. He said they grew and occupied more space as time moved on. I do not believe it. Teeth do grow after they have been developed in the mouth, but they don't grow externally. When the crown of a tooth has become fully developed, so far as that crown is concerned it has more substance than it ever has at any other time; it is larger than ever afterward, externally; that peculiar notched condition in the incisor teeth is an evidence of that fact. While they do grow, instead of growing larger they grow less; they wear out. This is the point I want to call attention to. I differ with the Doctor, and if you will give me five years to take the dimensions of teeth and demonstrate the fact, I think

I can prove it. The teeth of rodents continue to grow from the direction of the root. but I am speaking of human teeth. We are not sharks. If you break out a shark's tooth, another one immediately springs up under it. Did you ever see that in a human being's teeth?

Dr. Poole: It seems to me that physicians should know something about the injury done to teeth by the giving of certain medicines. We dentists should talk in our neighborhood on this subject and recommend our patients to remedy this evil by washing their mouths and taking medicine through quills to protect the teeth from acids. It is not a new thing that bad teeth are an indication of having bad health. There is no doubt that when one part of the body suffers from any disease the whole suffers. When a person has a very bad spell of fever, you would not see any effect on the teeth immediately; but six or twelve months later you will find the teeth injured. I think the vitality of the teeth is weakened like any other part of the body is weakened.

Dr. Smith: I want to say, in justice to the intelligent physician of the day. that I think we very often rub up against him, and do it in rather an apparently unkind way. I think we ought to keep hand in hand with the medical man, and do him no injustice. I think a large majority in the practice of medicine have a kindly regard for the human system, and very few would give medicine which would have a deleterious effect upon the teeth without trying to protect them. If we take such scientific men as Dr. Miller, whose reputation is established and whose position is insured as a scientific investigator, and call into question his theories, as my good brother Dr. Crawford did, when we call into question those theories that have been illustrated and proven beyond doubt, I think we are going in the wrong direction. I have not the temerity to tackle this man in my own strength. I love harmony and pleasure, and I am not going to get on him. I want to make it as smooth and easy for myself as possible. As to Dr. Crawford's theory of the development of teeth long after the spark of life begins to decline, I don't believe it. Dr. Morgan took a good point when he spoke of Dr. Miller's experiments, which Dr. Crawford did not sufficiently meet: the point in regard to the changes that had taken place in vitalized and devitalized teeth.

Dr. Wardlaw: Mr. President, I think Dr. Smith has struck it right when he says it does not make a very great difference to us whether the teeth continue to grow and develop or not. I see these imperceptible influences going on all the time, and we cannot properly appreciate the changes for months and years. I asked about the continual growth of the teeth of rodents, and he puts that aside as not having much influence. We all know there is a continual growth in the teeth of rodents, and their peculiar condition of life requires that. Why may not the same thing exist in a similar degree in the human teeth? I believe that may be the case.

Dr. Crawford: The statement that when a tooth passes through the gum and can be felt and handled, it is just as great in its various diameters as it will ever be, I do not believe. If all the books and all the dentists in the world say that, I know it is not so. How do the little grooves and notches on the lower jaw disappear? Does the tooth wear out? If it continued to wear out, it would wear out before a man was forty years of age. These notches grow out. When they disappear, what can you conclude has occurred? Just as squirrels' teeth grow and continue to grow and respond to the demands of nature, so does a child's tooth continue to grow and respond to the demands of its nature. We, as a profession, have made this mistake: we forget there are many scientific conclusions to arrive at; just like a man who was impressed, perhaps a century ago, that somewhere in the vastness of space there was another floating planet that had not been observed. He sat in his room alone and scientifically calculated that there was another floating planet; he asked astronomers to look, and what did they see? They saw a planet! No man had seen it, but by scientific reasoning they found it was there. If there were no process for the reproduction of the hoof of a horse, an ordinary horse would wear out hundreds of hoofs. If nature is not constantly reproducing and responding to the laws of metabolism, what will supply the demands for new growth? The only difference in a case of fracture and the process of reproduction and growth is that the conditions of the latter are more favorable; but when we have all the conditions favoring the union of teeth, they do grow together. If a person would come to me of twenty-one years of age with an incisor tooth broken by a blow, if it was a simple fracture of the tooth, I would hope to accomplish the feat of union by keeping it perfectly still. If it grows in one instance, it is a proof that it could, under proper circumstances, grow again. (*Southern Dental Journal.*)

GEORGE EUBANK, D.M.D..

THE DENTAL HEADLIGHT.

| VOL. 13. | NASHVILLE, TENN., APRIL, 1892. | NO. 2. |

Original Communications. • • • •

ADDRESS TO THE GRADUATING CLASS
OF THE DEPARTMENT OF DENTISTRY OF VANDERBILT UNIVERSITY.

BY C. S. STOCKTON, M.D., D.D.S., NEWARK, N. J.

Mr. Chancellor, Ladies, and Gentlemen: When I first received your kind invitation to make this address, I thought that Tennessee was too far away from New Jersey; but as I let the matter rest for a few days it did not seem so far, and there were some reasons why I would like to accept. One was that I would like to see your city and this far-famed institution. Another good reason was that I desired once again to meet and greet my good friend and your good friend, Dr. W. H. Morgan, than whom there is no grander specimen of a man within my knowledge of men. Then, too, I felt that a rest from the active cares of a busy practice would do me good; and so I am here, young gentlemen, to congratulate you upon the successful prosecution of your studies, and to welcome you to the ranks of a profession. The thought dominating your minds to-night is, undoubtedly: "Shall I succeed?" Therefore some portion of my address will bear upon this, and the other, an attempt to show that there is in the profession of dentistry more than the pulling, filling, or inserting of artificial teeth.

The new steamer "Normania," on her maiden voyage from Southampton to New York, encountered a field of floating ice-floes dotted with occasional bergs of great dimensions. As usual, when the warm air of the Southern ocean current strikes the frigid atmosphere of the floating icebergs, fogs shut down on the surface of the water. But the ship was kept at her top speed, racing through fog and ice for New York. The "Normania" could have taken a less hazardous course. She could have proceeded more cautiously, but it was imperative that she should break the record if possible. Finally she struck an iceberg, and nothing but a wonderful chance—humanly speaking—saved her from shipwreck. All

4   (49)

this is forgotten in the acclaim with which the gallant commander is greeted and the almost phenomenal swiftness of the ship is recorded. The cautious people who, with due regard to life and safety, take passage on ships that are not "greyhounds of the sea," are derided by the hurrying tourists who must "do" Europe in sixty days, and by the well-seasoned travelers to whom five hours or twenty hours saved in an ocean voyage is quite as valuable as an entire lifetime of ordinary folks on shore. "This is the age of hustling," they say, "and the swiftest is bound to get there."

It certainly is an age in which everybody appears to be in a hurry. Speed, at the expense of safety, comfort, and even long life, is supposed to be the chief end of man. The American habit of brag runs almost altogether into boastfulness of things accomplished "in no time." The Chicago man brags of the hog that goes in squealing at one end of a chute and comes out at the other end barreled and merchantable pork, all in the twinkling of an eye. And the Chicago demon of haste takes possession of a majority of our population. Hurry is infectious, and thousands of men move on the double-quick merely because their neighbors do. Business men snatch their news from the headings in the newspapers, their intellectual improvement from placards in railway stations, their religious incitements from fragments of chance conversations, and their bodily nourishment from flying lunch counters. "What a hurry we are all in!" they say, breathlessly, proud of living in the age of hustling; and then they plunge away again, as if immortality depended upon their being ahead of all competitors.

As a matter of cold fact, there is a great deal of humbug about all this hurrying and scurry through life. Our grandfathers, sneered at as slow old fellows by the present generation of "hustlers," accomplished more in their leisurely, deliberate, and careful way than their degenerate successors do, if we take into consideration the wonderful additions made in recent years to the machinery and appliances of trade and commerce. Those very respectable old fogies of past ages planned and executed great works, founded immense fortunes, and achieved magnificent and lasting results in every department of human society. Yet we know that they worked without hurry and without the multitudinous labor-saving appliances with which the modern man of business surrounds himself. As a matter of fact, too, it will be found that a majority of the crowds that daily rush and worry through the streets accomplish little or nothing in the world, and when their brief hour has been fretted away have nothing whatever to show for their tumultuous struggle. After awhile, perhaps, our people will learn that the age of hustling and all its feverish nonsense is worse than unproductive of grand results. They will presently discover that there is no occasion for steady maintenance of high-pressure speed, and that no solid results can come out of a life which is a perpetual scramble.

Yes, the scramble for success has seized upon us all; and if I give some of the salient points to-night, it may not be out of place in this assembly. The world is full of contrasts: Night and day, life and death, joy and pain, the resplendent sun, the song of birds, the sight of fair fields, are inspiringly beautiful; yet were it not for the night, we should know nothing of the mysterious grandeur and glory of the stars.

Men may gloat over their successes, but failures will come.

One man's ideas may be very high, another exceedingly low; one succeeds without effort; and such attainment would be disappointment to another; but the world applauds lustily the easy, brilliant performance. Success is the thought of life, rather than the patient working out of its details. We accept the standards which a false condition of society raises and glamour with the pomposity of wealth all those successful, who, like Gould and others, win through cunning and unscrupulousness.

There is no difficulty unsurmountable, no sorrow unbearable, no failure irreversible.

There must be adaptability between man and his work. Square men were never designed for round holes. To every man there is a place; but there are hundreds of men trying to make life's voyage, and though they have clear notions of how they ought to act, they are dragging about with them the anchor of ignorance and superficial knowledge.

Hard work is the great factor of success, and Gœthe truly said that there is no genius but hard work.

The busiest are the happiest. Work is the salt of life. A pair of shirt sleeves are a good coat of arms.

The napkin in which the slothful servant wrapped up his talent was the sweat cloth with which he ought, in his toil, to have been wiping the beads of perspiration from his brow.

Some men have too aristocratic conceptions of their dignity, and the snobbish notion of the meniality of work. That merchant never succeeded who was not content at some time, or in an emergency, to do the drudgery. A dainty, lily-white fingered dilettante, looking with holy abhorrence upon soiled hands or clothing, may indeed live a sort of effeminate, emasculated, wax-doll existence; but as for noble doing or succeeding in the strife of life, when Greek meets Greek, they are miserable nonentities. Mere leisured ease is contemptible beside the horny hands of the hodcarrier.

It is the man who, in the beginning of his career can intelligently lay out his life work, set himself to a task which shall require twenty or thirty of his best years, and devotedly persevere in it, who wins, who can endure all discouragements and adverse criticisms, and keep the end steadily in view. The man with moderate ability who plans and economizes his time and gets the whole worth of his existence is the victor; he lays out a campaign, not only for all summer, but for his whole life. For any man, especially a professional one, the closest, severest study is requisite to success. It was the motto of a celebrated chemist to examine all the contents of vials which others threw away. Only by the most continuous reading can the professional man keep abreast with the thoughts and discoveries of his chosen calling.

You must have faith, young gentlemen, in yourselves, so that when the time comes you can tell exactly how much you can do, and of how many horse power your mental and physical engine is.

Take the world as you find it, a hard and stubborn one, and do not expect to convert it in six months or any such time, but do not sneer at any genuine, hearty effort at reforming humanity.

The world is growing better. With some the time to act never comes; they devise their plans and elaborate them, but never put them into actual endeavor. You must know when the iron is hot enough, then strike.

There is a tide in the affairs of men which, taken at the flood, does lead to fortune. It was not wholly by accident that the lucky buyers of Boston back bay land become rich, or that investors in Nashville to-day may be millionaires of the future.

The civilities of life, the courtesies and deferences which dignify the true gentleman, have been the key of many a man's success. Destiny may hang on politeness.

Out of difficulties grow miracles. Misfortune crushes the weak, but nerves the strong to greater effort. To the wise man failure, instead of overwhelming him, only teaches him more carefulness in constructing his plans and conducting his experiments. The chemist detects clearly the ingredient which balked him, and makes some grand discovery through his very blunders. The Roman general was determined to "find a way or make it." Cyrus Field tried over and over again before he succeeded in laying the cable that binds the continents. Then the sarcasm of nations was changed into applause.

Whittier writes:

> Well, to suffer is sublime!
> Pass the watchword down the line,
> Pass the countersign " Endure."
> Not to him that rashly dares,
> But to him that nobly bears,
> Is the victor's garland sure.

Determination to get well has much to do with recovered health.

You can, if you try, make it as uncomfortable for the whale as the whale can make it uncomfortable for you. There will be some place where you can brace your feet against his ribs, and some large upper tooth around which you may take hold, and he will be as glad to get rid of you for a tenant as you will be glad to get rid of him for a landlord. There is always a way out if you are determined to find it.

"Truth crushed to earth shall rise again. The eternal years of God are hers." He who anticipates disaster suffers from it twice over. Cowards die many times before they taste of death; the brave man dies but once. Let us never lose heart.

> High hopes go down like stars sublime,
> Amid the heavens of freedom;
> And brave hearts perish in the time,
> We bitterliest need 'em.
> But never sit we down and say
> There's nothing left but sorrow.
> We walk the wilderness to-day,
> The promised land to-morrow.

But most of all, measure success or failure by the good you do humanity. You may not in your profession become a great authority, leading and perpetually suggesting, but as one of rank and file you will have the satisfaction of knowing that you have done something to elevate humanity. Contrast a high philanthropic life with one devoted to mean, sordid ends. There is joy in the giving of comfort and felicity in the retrospect, and enviable is he who goes down to his grave loaded with the respect of his fellow-practitioners and bewailed by the friendless.

But I want to tell my friends, the public who are here to-night, that there is in dentistry something more than the filling and extracting of teeth and the insertion of artificial ones. If an iron puddler spoils his mold, it is no great loss; or if your barber mangles your hair in cutting it, it will repair itself; but when the clergyman, through ignorance of theology, misleads an inquiring mind; when a lawyer, through mismanagement of his case, risks his client's life; when a physician, by culpable stupidity, slays while pretending to save, it becomes a crime. So with the dentist, who by his reckless hand's work inflicts useless suffering, or mars his patient, or leaves a memorial to curse him in snags and roots of bitterness. Our aim should be to discover the best methods of doing quickly and painlessly what is considered by many a slow torture.

As the medical profession has found its way from blistering, bleeding, and the administration of heroic boluses and preposterous powders and potions to the truth that the mild power cures, so our intention must be the invention of schemes which shall leave no injury and inflict no pain. It is for us to study our work upon a tooth with as much forethought and carefulness in detail and prearrangement as the diamond cutter displays in fixing upon and polishing the faces and angles of his sparkling gems. In our efforts to conserve the teeth we should remember that we are dealing with a portion of God's handiwork, a part of that creation which he pronounced " very good."

So while we study the various arts and methods whereby we cover nerves and still their throbbing, and fill its exterior with the precious metals, while we seek medications and the most efficacious process for arresting dental decay and death, let us learn reverence for our work, and how sublime that work which shall restore to primeval health and usefulness one of the Creator's gifts to man. And since the house we live in can be materially reconstructed—our eyes fitted with spectacles, our ears with trumpets, our teeth with fillings and crowns which serve the same purpose as the original parts—we learn how to replace the wreck which rottenness and death have made. We prepare the mouth for the substitutes for its late occupants and place these partial or complete dentures that are almost the counterpart of nature. Sometimes we even do better than nature herself, for some men have such miserable teeth that the poorest artificial ones would serve better.

As we find ourselves investigating the manufacture of teeth, we exclaim: " What a work for an artisan is this!" Compared with it, what is the restoration of a painting or a fresco, or the rebuilding

of a palace? These deal with lifeless stones and pigments, while we
with matter having within it the breath of life. Consider what belongs to the saving of an organ. It is in proportion as honorable as
the sacred duty of saving life itself. The decaying teeth give us our
first disagreeable impressions of our mortality; this is the first start
of that crumbling away which by and by shall attack irresistibly
the whole frame and level it with the dust. Thankful we should be
to him who can put away from us these unwelcome suggestions to
our thoughts.

Again, if there be superlative merit in the creation of any part of
the body by the first Artificer, he who interferes to prevent its destruction must share in the glory. Physicians and surgeons are co-
laborers with God in the human frame: he conceiving and creating, they preserving against untimely ruin and repairing the broken
places; and not only the saving of the organ is under consideration,
but the alleviation of pain as well. To give a quietus to the acute
suffering of toothache, to act the part of saviors to society is something to be grateful for.

We sometimes may affect to disparage the pain which proceeds
from inferior parts, we do not respect the protest of a tooth nerve
as much as the twinge of the heartstrings; but it is a thorn in the
flesh, and will make the imperious and haughty soul attend.

It is curious to notice how a little thing will destroy a great pleasure. What is the majesty or loveliness of Yosemite to him whose
back molar is calling him away from mountain and valley to be in
sympathy with its obstreperous outcries. The resources of pain
seem largely to predominate over those of pleasure; there seems to
be ten avenues for the one to one for the other; pleasure is evanescent
and pain is lingering. Blessed is the man who can allay it somewhat. What hero will requite him who, though he never groaned
in battle and is all unused to the melting moods, bows before the
power of the demons boring into the very soul of sensibility, while
he envies the boils of the patriarch who was never so afflicted.

Lives are sometimes abnormally shortened by the endurance of
excessive pain; and to-day it is not so much work as rack which
breaks down constitutions—the wear and tear and exasperation of
nerves. The patriarchs lived their hundreds of years because they
simply vegetated; a man at thirty now is a condensed Methuselah.
Well for science if she follows out what she has so well begun in the
prolongation of human life by the banishment of pain. Already the
average duration of life of the race seems on the increase, and the

over-sanguine look forward with hope to recovery of pristine longevity.

Delineate before your fancy the picture of a toothless specimen
of the *genus homo*. The sound of the grinding is low. The poet's
oft-quoted line is falsified; a thing of beauty is not a joy forever,
especially when the lips are sunken and expressionless, when
the language of Shakespeare and Macaulay degenerates into an
indistinct guttural and mumble. He becomes a mocking death's-
head, a very genius of famine. All the lines of beauty are obliterated. What would be the Venus de Medici or Beatrice de Cenci
without the warm, sensuous curves of the lips? All the force of
character departs, for the mouth lines, more than any fictitious cranial
bumps, are the proper indices of the brain within. How does the
appearance of our luckless wight prejudice him in the estimation of
the public, who judge him forceless and effeminate! Even the power of the tongue, either for kindness or malice, depends on the presence of the teeth. While it might be a blessing to humanity if some
of our effervescent talkers would be visited with a caries which
should leave their gums tenantless, where would be the eloquence of
Depew without his incisors? But this is not the worse: moroseness
sets in, and the hapless wit becomes a confirmed Puritan. Where be
their gibes now, their songs, their flashes of merriment that were
wont to set the table in a roar? Not one now to mock their own
grinning! Quite chopfallen! Bad teeth or few, with the attendant
imperfect mastication, the prelude to all the gloomy horrors of dyspepsia, wherein the sufferer's self-consciousness is only consciousness
of a stomach, give rise to all hopeless views of life and man's destiny. Even hope flags, that sentiment which is said to remain with
us to the extremest verge of life. Faith departs; the gods forsake
the sky; kind Providence is replaced by a melancholy and miserable chance; all thought and feeling are dead, and the tides of life
ebb to the lowest; the most fatalistic and atheistic views in philosophy are adopted readily; the whole universe seems bound by irrefragable chains of suffering and remorse; the stars shine with a yellowish glare to his gangrened vision; and a settled misanthropy
possesses him, taunts him ever with the unanswerable interrogation: "Is life worth living?"

Consider how all this mournful record might have been reversed
by a proper attention to our art. While happiness may not be the
end of life nor the test of virtue, yet the pursuit of it is noble and
engrosses all; and oftener than elsewhere it may be chased down

and captured in a dental chair. Here beauty is regained, character brought back, the esteem of the public repurchased, hope restored, the world clothed in beauty, life shown to be full of desire and worth, the throne of the Deity unveiled, and the silver cloud turned out. Pessimism is pushed out by a rejoicing optimism which believeth all things, hopeth all things. A tooth root may thus become a root of all evil, and its extraction may operate like the eradication of original sin. The dental profession has not, perhaps, had the credit due them as creators of English literature; but Milton could never have written "Paradise Lost," nor Tennyson have composed the reflective "In Memoriam," nor Gladstone or Stanley made Parliament and Westminster ring with eloquence had they not first well masticated and digested good English beef; requiring jaws in good order under the care of a dentist.

It comes also in our plans to discuss the questions belonging to dental pathology: the application of special medication to special ailments, whether the difficulty arise from constitutional or local causes, and the proper treatment in each particular instance; for diseases, like individuals, have a personality of their own, and it is the invariable mark of the empiric to have a stereotyped prescription for a malady when general appearances are the same. They have the same likelihood of success that a physician would have who should enter a crowded assembly and scatter his pills right and left for the healing of the people.

Neither can we pass this division without noticing somewhat elaborately that grandest discovery in modern healing, the use of anæsthetics. This great boon has given the world a foretaste of that time when there shall be no more sickness, neither shall there be any more pain, for the former things have passed away. While the most critical operations are being performed, men may have visions of the Lotus-eaters, "with half-shut eyes ever to seem, falling asleep in a half-dream," feeling that "there is no joy but calm," "resting weary limbs on beds of asphodel," and whispering: "Surely slumber is more sweet than toil." How much of the unspeakable horror of war is mitigated, and the primal curse upon Mother Eve in mercy assuaged! "I will greatly multiply thy sorrow and thy conception; in sorrow thou shalt bring forth children." While one would be writhing in agony, to be floating upon a calm and shoreless sea, oppressed only with a delicious sense of our own infinitude and ecstasy. While busy phantasy conjures us with gorgeous dreams! To sleep—and by a sleep to say we end the heartache and the thousand natural shocks that flesh is heir to—'tis a consummation devoutly to be wished. But the wished for boon is found. What King Lemuel says of wine is truer of the ethers and chloroforms and the nitrous oxides: "Let him drink and forget his poverty and remember his misery no more."

What statistical table could give the lives saved when the surgeon could work with his patient under this dreamful influence? How rapidly and unerringly can he use his implements, undistracted by the moanings of the subject or the panic of the friends! How often would the debilitated frame have been unable to endure the exhausting drain upon the nerves by operations which are now rendered safe! The man is as unconscious that his tooth is gone as is the victim of the clever Oriental executioner, not knowing himself decapitated until the latter obligingly gave him some snuff and the head therewith tumbled off. But there is enough pain in the world for which there can be no narcotic or opiate—that we bless providence for this.

And we must not pass over the correction of dental irregularities, whereby man's mouth—as overcrowded as a frontier hostelry with three in a bed and the floor stratified with slumbering humanity— is rendered commodious and presentable. When eccentric teeth, each with a predilection of a separate angle, make a line as irregular as the boundaries of counties on the Kentucky map where no two join, one imagines that the man may be one of those poets who don't know it, one of those "mute, inglorious Miltons," with heavenly visions, but no gift of language; and that his ponderous thought created an earthquake in his mouth, endeavoring to find expression. Ah! well for humanity, unprepossessing as most of it is; well for the prospects of wedlock on the part of foreordained spinsters and benedicts that cross eyes and crooked noses and jutting teeth can be brought around straight.

It is given us, then, to work with the human face divine, bringing out latent improvements and investing it with unforseen beauties. We are artists surpassing in our material and effects the production of Raphael or West. We have not to saturate canvas or chisel stiff, unyielding marble; but our matter is living flesh, and our product an animate thought, a glowing beauty! For though beauty may dwell in mountain, tree, and landscape, its highest throne is in the face of man, and we who remove the obstructions to its appearance are performing as noble a service as he who takes away the soil and muck from the beautiful statue of Apollo found deep in Roman mire.

Think well of our high prerogative in the restoration of speech when this is lost, occasioned by lesion of the palate, and so comes within our province to cure. Speech is one of the highest dowries of man; by it he is separated from the beasts that perish; for though they may give vent to emotional cries, there is in them none of that Godlike faculty which uses arbitrary sounds and symbols for the expressions of the profoundest thought in literature and philosophy. Without speech there could be little thought, for it is in language that our ideas are preserved; and as they flash through the brain they speedily vanish unless caught and imprisoned in words. Hence there could have been no development of man's brain, no civilization beyond brute knowledge. If it be a worthy thing to unstop the ears of the deaf or unseal the eyes of the blind, surely our function in bringing back a man's departed power of utterance is no less praiseworthy. The age of miracles has indeed passed, but science, by slower but yet efficacious steps, is coming to the relief of suffering mortals: the blind can often receive their sight, the dumb are taught to speak, and it is our privilege to coöperate in these achievements. Speech is said to be silver, and silence golden; but this is only said of foolish babbling, not of wise discourse. Great is our glory, therefore, if we can bring back to man the power and pleasure of communicating his conceptions to his fellow-men, of moving multitudes by his eloquence, and with the highest exercise of language of audibly praising Him who gave to man a mouth and speech and wisdom.

We should know something of the principles of chemistry, especially its relation to materia medica. We should examine with all the ardor of wise old alchemists—falsely deemed mad—the pharmacopœia of nature. Earth is indeed the nurse of man. She possesses for every sickness or wound the *vis-medicatrix*, the Gilead balm for the afflicted. Her every vein is full of life and healing. She lays her hands on the heads of her children, and the fever departs and they are healed. Into what a wonderful world has man been introduced! A world stocked with forests to give him shelter, with food to maintain life, with coal supplied to warm him—all long before he came upon the scene—an anticipation and prophecy of his lordly advent! Does it not prove a far-reaching design, this adaptation of man to his environments, this nice relation of the being to his home? Never did the bridegroom bring his bride to a house more fully furnished, and realizing before every need. But not only to dwelling and eating man, but to sick and suffering man was the earth prearranged. The soil, the air, the water, the vegetation were stocked, as the great pharmacy, ready to dispense any drug for whatever nameless malady might appear. Man was to love his earth-home.

How varied is the purpose and power of any single drug! What mind could have invested it with such multitudinous properties so nicely forecalculated for human emergency! Every day scientific explorers are bringing in new remedies from the arcana of nature, and whatever mysterious affliction—yellow fever or diphtheria—may baffle skill for the present, we never once doubt that earth has in some of her hidden vials the proper remedy, and so we do not in hopelessness remit our search. Long did the ancient alchemists seek for that one potent drug which should be a cure for every ill. But the panacea exists not in any one potion, but does exist in nature as a whole. It should create in us an ever-increasing desire to know the materials, their purposes and properties for the repair of our frames—phosphorus for the bones, phosphates for the brain and nerves, iron for the blood; and so for the proper food of every tissue, muscle, and cell we have only to eat the dust of the earth.

We should also know something of physiology, and learn "What a piece of work is man." How noble in reason! how infinite in faculties! in form and moving, how expressive and admirable! in actions, how like an angel! in apprehension, how like a God!—the beauty of the world, the paragon of animals. It is as an animal we will have to view him, yet this is sacred and ennobling. David cried: "When I consider thy heavens, the work of thy fingers, the moon and the stars which thou hast ordained, what is man that thou art mindful of him?" Yet man in his mind outweighs the universe, and even his body in its form and symmetry equals the harmony and perfection of the heavenly spheres. "The proper study of mankind is man," not only the laws of his mental and social growth, but of his physical development. Consider him only a machine, and it is enough to awe us in dumb admiration. Only recently have the sciences of physiology, histology, and embryology been written. Men confused themselves for ages over insoluble problems of will and destiny, and held a knowledge of their corporeal existence in contempt. Yet what a being! Allied on one side with the higher apes, on the other the temple of the Holy Spirit—a thought—a poem—of God, not written with pen and ink, but with blood on fleshly tablets of the heart! And is this he, who, they say, was evolved out of star-dust, a statue of clarified mud, raising itself out from the primeval ooze and slime? Can one go into the laboratory and so unite his chemicals and breathe into them the breath of life, that they shall move and think in human semblance? The stars in their courses are

Forever singing as they shine:
The hand that made us is divine.

If the " undevout astronomer is mad," no less the undevout physiologist.

We should know something of the process of digestion and the laws of hygiene. For though these subjects, with their earthly suggestions, might shock an Eastern mystic and transcendentalist, the majority of mankind are "creatures not too bright or good, for human nature's daily food." We are not yet ethereal, and cannot subsist on nectar and ambrosia. These modern times are insisting, rightly, more and more upon the necessity of properly cooked food. If a man does not intend to prescribe for others, he has the solemn responsibility of his own body to look after. Men of knowledge must preach more continually and persistently to the masses the necessity of attention to hygienic laws. The laws of cleanliness, eating, and general preservation of health must be reiterated strenuously, so that a blockade against disease may be enforced. It is the evil effect of too high a civilization that men sin constantly against their stomachs and think themselves pious, when a little of the healthful barbarism of savagism would be beneficial. The body in its health must be shown to be under as immutable laws as govern the planets. Each man has branded in him Scotland's imperial motto, *Nemo me impune lacessit*—(No man insults me with impunity). How may you add to your enjoyment of life, making it a continual ecstasy simply to be! And how many a lay sermon can you give your patients in your conversations on the care of their wonderful frames!

We should know something of the mysteries of anatomy and surgery. It may be objected by some that we have no need of a knowledge of these—that ours is a specialty. Yet the evil of the contracted specialism is seen in the belittling effects on the human mind. He who spends his whole life fixing the heads of pins becomes as small as his work. The time was, and not long ago, that tooth-pulling and blood-letting were done in shaving shops, but the profession has since attained higher dignity. The man who knows only his own specialty does not know that thoroughly. If he cares only to learn enough to make his bread, caring nothing for science for truth's sake, he is base. Dentists do not want to pull teeth as a blacksmith would a horseshoe nail. They must be more than respectable jaw-carpenters, more than mere sawbones. The good blacksmith will not stop at the hoof, but will know the whole skeleton of the horse. The good dentist will not be content with knowing only the teeth and being ignorant of all other bones and how to repair them. One man is a decent house shingler, another is the architect of St. Peter's.

Look again at the broadening and culturing effect of this study. While it lies parallel with our vocation, it is yet so deep and requires as much profound reflection that equally with the ancient languages or mathematics it enlarges the reasoning powers, cultivates acute-ness of observation, breadth of judgment, and an acumen that can be brought to bear upon every affair of life.

Finis coronat opus. Strive for an unfading crown of capability to benefit your fellow-men.

The value of education is incomparable. It is good for its own sake, for the pleasure it adds to life; besides which it will introduce you into the society of the cultivated and secure a more refined and better paying patronage. It is an inheritance that cannot be taken away—more valuable than any other form of advertising. But of what use shall the most classical culture be without the formation of good habits? We teach our children Greek roots and Latin verbs, and the number of angles in a triangle, and then they go out, steal, and get drunk. You may be ever so skillful, and yet if you are not affable and gentlemanly, cleanly in your dress and person, refined in your language, no gentleman or gentlewomen will hire you. A lady said she would not trust any prescripton of Dr. Epps's because he did not believe in the personality of the devil! Little things determine our success, just as the pebble on the Rocky Moun-tains may deflect one stream to the Atlantic and another to the Pacific. Choose a high ideal and maintain it.

CARE FOR CHILDREN'S TEETH.*

BY R. R. FREEMAN, M.D., D.D.S., NASHVILLE, TENN.

It has been said by some one that the time to commence the care of children's teeth is before they are born. To be more explicit, I would say, that to secure the ideal of perfection, the commencing

---

* Read before the Nashville Academy of Medicine.

should precede that important epoch fully three or four generations, for then we have the assurance of the freedom from visitations of the sins of fathers upon their children. If, however, we do
to-day whatsoever our hand findeth to do, and stimulate others to
follow our example, we may with the present and coming generations accomplish results that will be of lasting benefit to mankind.

We as an American nation are very jealous of our rights; but
what attention are we giving to our children? Surely, if justice is
to be done, we should concede to them the one great privilege of being *born right*. How many consider for a moment, or take any
thought of prenatal conditions and their influence transmitted to future generations, when by that magic touch the spark of a new life
is kindled? Sadly may we remember these things when we see
their mark and impress upon our children; but then it is too late.
Have not our children the right to expect better things of us than
this? To our artificial mode of life is accredited most of our ills,
and many sigh for the times of old when it was not so; but we
cannot return to those happy days, and must make the best of the
times as we find them.

Our present civilization, with its rapid movements, mental anxiety, and excessive development, may tend in a measure to depressive
reaction and nervous prostration. But I do not accept the idea of
deterioration, or degeneration in the physical structure of the Americans of to-day. We are progressing through the acquirement of
light and knowledge, but I do believe as a people we fail to appreciate our opportunities to enhance the happiness of our race,
through our neglect in the care and treatment of our children's
teeth. Heredity may and does have its tendency to interfere with
our hopes and aspirations in perfect development, but the advances
made in a knowledge of the laws of health enables us largely to sur-
mount these difficulties. Then let fathers have a care that, while
under the unusual mental excitement that they may experience in
this rapid and intricate business age, they do not, in seeking a
moment of gratification, transmit to their children temperaments
that will curse and follow them to the end of their days.

Let mothers realize the grand and noble responsibility intrusted
to their care, and instead of seeking by every way and means to
avert it, discharge with fidelity the duties of this honorable office,
so that her children may rise up and call her blessed. Everything
should be done to make this the most happy and joyous period in a
woman's life.

There are well-known and established hygienic laws which the wise would not think of disregarding in the rearing of their inferior animals. How much more important their observance as relating to our children!

If the American mothers would profit by the knowledge now attainable, they could discount the sagacity of Jacob of old, and be freed from the band: "A tooth for a child." Nor would they have to adopt the earthy salts, lime phosphates, and brown bread hobby. "Man doth not live by bread alone;" but with wholesome, luscious, delicious, nutritious, and toothsome food, comfortable clothing, easy shoes, with plenty of light, air, and judicious exercise, with mind freed from care and annoying perplexities, you might confidently expect to successfully pass the important period of birth and present a child so perfected in development as to have a fair chance to win in the great battle of life. Some will question the extent of impression to be made upon a child in utero; but when it is born, all will agree that surrounding influences soon begin to tend for its weal or woe. As relating to the truth, we have often observed that as soon as a child is born the nurse looks to the scrupulous cleansing of its mouth, and during the early months of its existence this is seldom if ever neglected, until the cutting of the first tooth, then rubbing and cleansing of the mouth ceases and the teeth are left to make their way as best they can. And here I would ask: Why at this time this general rejoicing, this universal congratulation and spreading the news from friend to friend? A custom observed in every clime, among high and low, rich and poor, from time immemorial. It is the spontaneous outburst of a natural instinct.

In the beautiful, pearly jewel of a tootsy toothsy, man recognizes the herald of the possible joys and blessings to come. We are assured of the young hopeful's being prepared to eat. But blessings perverted to a curse may turn. Alas! the tortures of a neglected tooth. For descriptive particulars we would respectively refer you to the poet Burns.

The period of dentition is an exceedingly critical one to the average American child. That which should under proper hygienic measures be a delightful physiological performance, from disproportional development, miasmatic, electrical, and other disturbing influences, becomes a perplexing and agonizing pathological procedure. Children born in the summer months have the better chance to pass safely through this trying ordeal. Cutting the first teeth in the winter or spring, there is a cessation, a period of rest, affording opportunity for general physical recuperation through the following summer, which enables them to resist the final assault which carries so many of our children to their graves. The tooth advancing through its bony walls commences a rapid eruption, and passes through the gum, which is normally void of sensation, but upon irritation is exceedingly sensitive.

Every one has noted at this stage the eagerness with which children will bite on hard substances. It is nature's way of cutting the gums—a cue which we might profitably follow.

The cutting edge of the tooth becomes a *mechanical irritant*, inducing excessive flow of blood to the part. If irritation continues, we have congestion, dull, constant pain, swelling, and pressure upon the growing tooth, which in turn induces pressure on the surrounding tissues about the growing root, stimulating the flow of blood to these parts, increasing the development of the root of the tooth. If the condition is not relieved, how can it be otherwise: inflammation of the gums, irritation of the dental pulp, resulting in loss of appetite, peevishness, fretfulness, restlessness, feverish thirst, bowel difficulties, suffering, pain, and death.

Thousands of our children have died after days of suffering who might have spent long years of useful life and pleasant association, a joy to their parents, an honor and blessing to their country, had the ever ready and efficient remedy, the oft neglected *gum lancet*, been properly and timely applied.

And here I would call attention to an old, exploded, superstitious idea, for which the doctors are largely responsible, and to which, unfortunately, many hold to-day. That is to the hardness and unyielding character of cicatricial tissue ; that should you cut the gum too soon an eschar will form, the hardness of which will retard or deflect the coming tooth, all of which is known to be untrue ; for we do know that cicatricial tissue has a very low grade of organization, and soonest yields when subjected to disorganizing influences. In perverted dentition every organ is more or less impressed, nutrition impaired, development fails to reach perfection, and therefore the teeth, after they come through, are more susceptible to the attacks of corroding agents, which can only be counteracted by the most scrupulous care.

Clean teeth do not decay. But O then, what is the difference? these teeth will all be shed. If they decay and give pain, they can be pulled out and others will come in their place, so why bother about them now? This is the usual sentiment we have expressed

5

regarding the temporary teeth, which frequently includes the six-year molar, for it is astonishing how few understand the important relation this tooth sustains. The *six-year molar*, coming as it frequently does as early as the fifth year, attracts no attention from the ease with which it is cut. Its development being impaired by surrounding unhealthy conditions, lacking therefore the physical ability to resist corroding influences, it succumbs to the inevitable. Relief from pain is sought at the hands of the dentist. Parents then, being informed of the serious aspect of the case, will exclaim: "Why, doctor, I thought that tooth had to be shed. I had no idea but that it was temporary and that another would surely take its place." Then somehow or other responsibilities get mixed: the dentist blames the parents for carelessness and inattention; the parents think: What are doctors for if not as public guardians they find ways to enlighten the people and save them from these errors of ignorance?   .

The most important epoch in the life of an individual is the first seven years of existence. This is a time when lasting impressions are made which mold and shape the being *physically* and *morally* for all time to come. The most rapid increase of the brain is said to take place before the seventh year. The crowns of the permanent teeth are in simultaneous development. How then could we expect other than devastation if we neglect the care of our children through this important stage? We are paying entirely too little attention to this important subject. Parents do not understand it, physicians say little about it, and even the dentist, to whom we should naturally look for instruction in these matters, many times fails to appreciate the great good that might accrue by a word spoken in due season. One serious obstacle to improvement in this matter is a want of a better understanding as to the *relation* that should be sustained between dentist and patient. The community are given to lying about the tortures of the dental chair, children are taught to dread the very sight of a dentist, and even the ordinarily truthful do not hesitate to draw it strong when enlightening the uninitiated. Nor are the dentists altogether blameless, for they too are sometimes given to deception, and in order to gain a little time or get the clinch upon their patient will exclaim: "It won't hurt!" These tactics might be excused on an old deceiver, but to think of practicing such on a confiding child is, in my opinion, unpardonable. "Faithful is the wound of a friend." No one will appreciate this sentiment more than a little sufferer if you gain and maintain their confidence. Truthfulness, gentleness, and firmness best control a child, and it is astonishing what suffering they will endure if they believe you are in sympathy with them, and would guard them from unnecessary pain. When I tell you that fully 95 per cent. of the children of our schools are affected with dental caries; that in these cavities almost every class of bacteria find a convenient home in which to propagate, and thence sally forth on their mission of devastation, you may begin to appreciate the importance, at least as a sanitary measure, of the necessity for the proper care of our children's teeth.

SALIVARY AND SANGUINARY CALCULUS.*

BY W. H. P. JONES, D.D.S., NASHVILLE, TENN.

SANGUINARY, I take it, means, "Of or from the blood." The blood is a nutritious fluid circulating through the tissues of all organized beings. This liquid, which is essential to life, in the plant is known as "sap;" in the animals, as "blood."

The elaborated juice constituting the former is probably simply nutritive. Blood, or "liquid flesh," as it has been called, is a nutriment and something more. It is the means by which used up materials are gather up and *probably passed to other fluids* to be removed from the system.

The characteristics of living organisms are ceaseless change and ceaseless waste. The blood becoming impaired or weakened, it cannot dispose of waste materials properly, and may therefore play an important part in these calcareous deposits. But I cannot see that it is more *conspicuous* in this than in the biliary, urinary, or salivary calculus, allowing for the variations in characteristics, according to the organ in which the deposit is found. For it is a reasonable conclusion to suppose that the blood—as general scavenger of the system—gathers up the material and passes it on to the various organs by means of the fluid passing through and peculiar to that particular organ. And therefore I beg leave in this manner to dispose of the sanguinary, and pass on to the consideration of salivary calculus.

Calculus means, literally, "a small limestone." Calculi are concretions which may form in every part of the animal body, but are most

---

* Read before the Dental Section of the Nashville Academy of Medicine.

frequently found in the organs that act as reservoirs, and in the ex-cretory canals. They are met with in the tonsils, joints, biliary ducts, salivary, spermatic, and urinary passages, and upon the teeth.

The causes which give rise to them are obscure, or not well un-derstood. Those that occur in reservoirs or ducts are supposed to be owing to the deposition of the substance which composes them, from the fluid as it passes along the duct; and those which occur in the *substance* of an organ are regarded as the product of some chronic irritation. Their general effect is to irritate, as extraneous bodies, the parts with which they are in contact, and to produce re-tention of the fluid whereof they have been formed.

The symptoms differ according to the sensibility of the organ and the importance of the particular secretion whose discharge they impede. Their *solution* is generally believed to be impracticable. *Spontaneous* expulsion, or removal by *mechanical* means is the only way of getting rid of them.

We will confine ourselves to the consideration of the kind found in the oral cavity, salivary calculus.

Salivary calculus, or tartar of the teeth, is composed of earthy salts and animal matter. According to one authority it is composed of phosphate of lime, fibrin, and animal fat. Another says that it composed of phosphate of lime, magnesia, ptyalin, and animal matter.

Dr. Dwinelle gives as a result of an analysis:

Phosphate of lime................................... 60 parts.
Carbonate of lime................................... 14 parts.
Animal matter and mucus......................... 16 parts.
Water and loss....................................... 10 parts.

The *relative* proportions of its constituents vary according as it is hard or soft, or the temperament of the individual is favorable or unfavorable to health, and therefore no two chemists give the same result.

We find the black, hard, dry tartar, which affects the teeth of those persons of good constitutions, not in very large quantities. It is hardly soluble in muriatic acid, while the dry, yellow tartar, found upon the teeth of bilious persons, dissolves more readily in it. The soft, white tartar, found on the teeth of persons of a mucous tem-perament, is scarcely at all soluble in the acids, but is readily dis-solved in alkalies.

The black tartar is the hardest, the white the softest; and the density varies as it approaches the one or the other of these colors.

There is one kind of black tartar found upon the teeth of those whose *innate* constitutions *were* good, but by disease or through intemperance and debauchery have impaired their physical powers. It is deposited in *large* quantities on the teeth opposite the mouths of the salivary ducts. It is hard and so firmly attached to the teeth that it is with the greatest difficulty it can be removed from them. It is very black, with rough uneven surface, and is covered with a glairy, viscid, and almost insufferably offensive mucus. This kind of salivary calculus is very hurtful, not only to the gums, alveolar process, and teeth, but to the general health also. The gums inflame, swell, suppurate, and recede from the necks of the teeth. The alveoli waste, the teeth loosen and frequently drop out. The secretions of the mouth are vitiated by it, and are unfit to be taken into the stomach, and so long as it *remains* on the teeth no treatment can fully restore the system to a healthy condition.

There is another kind of black tartar, but this variety rarely accumulates in large quantities, and is much less harmful to the teeth and gums. It is very hard, adheres very firmly to the teeth, and indicates a good constitution.

The dark-brown tartar is not so hard as either of those just described. It collects in large quantities on the lower front teeth, and sometimes on the first and second superior molars, and is frequently found on all the teeth, though not in such great abundance. It does not adhere so strongly to the teeth as either of the black varieties, and can therefore be more easily removed. The odor from this variety is less offensive than the first, but more fetid than the second. Those subject to this kind of tartar are of mixed temperaments; the sanguineous, however, usually predominating. Their physical organizations, though not of the strongest, may, nevertheless, be considered very good. They are more susceptible to morbid impressions than those of the more perfect constitution.

The yellow or yellowish brown variety is softer in consistence than the dark, and is generally found upon the teeth of persons of a bilious temperament. It is sometimes found on every tooth in the mouth. It contains less of the earthy salts than any of the foregoing descriptions, and owing to the vitiated mucus adhering to it, has an exceedingly offensive odor. It is so soft that it can very easily be removed.

White tartar is not often found in large quantities, generally upon the *outer* surfaces of the first and second superior molars, and the *inner* surface of the lower incisors, and frequently on all the teeth. It is almost devoid of calcareous ingredients. Fibrin, animal fat, and mucus constitute more than one-half its substance. It is quite soft, exerts but little mechanical irritation on the gums, but its acrid qualities keep up a constant morbid action in them. It vitiates the fluids of the mouth, corrodes the enamel, and causes rapid decay of the teeth. This kind of tartar affects persons of mucous habits, or those who have suffered with some disease of the mucous membranes.

Green tartar, or green stain, though commonly classed as *tartar*, is not properly a calcareous concretion. It affects the teeth of children and young persons; not often the adult. It is usually confined to the labial surface of the upper incisors and cuspidati, and bicuspids. It is exceedingly acrid, and corrodes the enamel and irritates the gums. This discoloration—it is hardly more—indicates an irritable condition of the mucous membrane, and viscidity of the fluids of the mouth.

The general effects of the deposition of tartar upon the teeth are irritation, inflammation, turgescence and suppuration of the gums, inflammation of the alveolar dental periosteum, the destruction of the sockets, and loss of the teeth. Tumors, spongy excrescences of the gums, hemorrhages, an altered condition of the fluids of the mouth, are among the local effects arising from a long-continued presence of large collections of tartar upon the teeth. The constitutional effects are hardly less pernicious, and are not well understood. Indigestion, and general derangement of the assimilative functions are among the most common.

I desire to call your attention to a calcareous deposit found on the teeth of some persons, remote from the gum margin, attached to the root at a point seemingly inaccessible to the saliva. It possesses the same characteristics as the dark-brown tartar. It is found in varying quantities, and may attack *any* tooth, but rarely have I found it on more than two or three teeth in the same mouth and at the same time. It is an irritant to the extent of producing abscess and loss of tooth.

Whence comes this deposit? Hardly from the saliva, as it seems to be beyond its reach. Can it be from the gum, itself apparently healthy? It it only of late years that I have observed this deposit. And on making inquiry I have learned that such persons were of a decided rheumatic or gouty tendency. We know that pericementitis is often directly traceable to a rheumatic condition of the system. May not this fact lead to the belief that a want of proper distribution of the acids exert a greater influence over diseases of the mouth than is generally thought? And may we not look to the adoption of systemic along with local treatment for better results in diseases of the mouth? This affection is not to be mistaken for Riggs's disease. I hope for a full discussion on this subject, for I think that there is more in it than would appear on the surface. Another point in this connection. It is the peculiar and destructive effects of tartar upon the teeth of *some mouths*, teeth that have been filled, notably on the neck of a tooth, along the gingival border. You find the gum turgid and spongy, discharging constantly a poisonous fluid—serum, perhaps—fearfully destructive to enamel and dentine alike.

How shall we alter these conditions and bring about a healthy action in the parts, and stop the destruction going on around the fillings under this *moist, gummy, disgusting* deposit? And this brings us to the treatment of tartar.

The thorough removal of *every particle* of tartar is the first step, and of the first importance. One, two, three, *five* sittings, if necessary, until you do get it, every particle. For this purpose, variously constructed chisels, hoes, hatchets, scalers, scrapers, turned at every conceivable angle—in fact, anything of any shape in the way of an instrument best adapted to getting it off is what you want. Operators will differ in their selection of instruments for this purpose. So it is almost useless to recommend any particular set or make of instruments.

Having removed the tartar, ordinarily you will need very little medicaments beyond a simple astringent and stimulative tonic, such as nutgalls and cinchona, soda, sage, and honey, tincture of white oak bark and honey, alum and cinchona, among the old, with a world of new solutions and mouth washes. In some cases you may find the inflammatory action so great as to call for general constitutional treatment in addition to the local, in which you will be governed by the indications, and select your remedies to suit the peculiarities of each particular case. Manifestly it would make this paper too long to attempt a detailed description of the treatment of the varied cases. In my opinion there is a great deal in this subject. I mean in the causes that lead to this deposit. And I regret my inability to present this paper to you in better shape.

EXOSTOSIS: ITS INFLUENCE AND PREVENTION.*

BY E. F. HICKMAN, D.D.S., NASHVILLE, TENN.

By the caption of my paper you will see that osteomata in general are included as well as the form which we as dentists are called
upon to treat—namely, exostosis of the teeth; or, more properly
speaking, hypercementosis.

The osteomata are tumors consisting of bone, either compact or
cancellous. They are divided, like all other tumors, into two great
classes—namely, homologous and heterologous—according as they
resemble or differ from the tissue from which they spring.

Osteomata are the result of the ossification of newly formed connective tissues, which is not a product of inflammation, and are to be
distinguished from those of simple ossification of normal connective
tissue, and also from that derived from an inflamed tissue. They
must also be separated from calcareous degeneration.

As above stated, there are two principal kinds of osteomata; the
first of which, the homologous, is subdivided into exostosis and enos-
tosis, according as they project *from* the surface or into the medulla-
ry canal of a bone.

Exostoses are divided, according to the density of the bone of
which they consist, into two kinds: 1. The compact ivory or ebur-
nated. 2. The cancellous or spongy.

The ivory exostosis grows from periosteum. It occurs most frequently on the external or internal surfaces of the skull. The orbit is a place where it is more frequently observed than in any other part of the body. It is met also on the scapula, pelvis and upper
and lower jaws. In the last-named situation it also grows from the
dental periosteum.

These growths are wide based, and are covered by periosteum,
which is connected with that of the old bone from which they grow.
They are small and rounded, and on section they are throughout of
ivorylike density, and are readily distinguished from the adjacent
tissue.

The spongy or cancellous exostosis is an ossifying condroma. It
grows from cartilage usually near the line of junction of an epiphy-
sis of a long bone with the shaft, especially at the lower end of the
femur and the upper of the tibia and humerus. Its outlines are
less regular than that of the ivory growths. It is covered by a cap

---

* Read before the Dental Section of the Nashville Academy of Medicine.

of cartilage while growing, and when this ossifies its growth ceases. On section the mass consists of spongy bone directly continuous with the cancellous tissue of the bone from which it springs.

So far we have treated the subject mainly in a general way, but we will now take hypercementosis or osteomata from a dental standpoint.

This abnormality consists in excessive development of the cemental tissues of the roots of teeth. While the condition cannot be called a common one, it is met with too frequently to be considered rare. This variety may be further divided into "circumscribed" and "diffused."

Circumscribed hypercementosis includes those cases where the enlargement is confined to a limited area, and is usually small in extent. It may appear in rounded masses at the apex or on the sides of the root; or it may have the form of a hooded or domelike covering of the end of the root.

The nodular form is generally found on the bicuspids and molars, more frequently on the former; and aside from any local trouble it may give rise to, it offers one of the most serious obstacles to its removal, when this becomes necessary.

The diffused variety may involve one-half or more of the length of the root; or the entire root from apex to neck may be involved. It may be larger at the apex, or midway between the apex and neck, and in some cases it is found the largest at the neck. This diffusion has been known to be so great as to include all the roots of the teeth, and combine them into one solid mass.

The secondary deposit is the product of the peridental membrane, and is formed by a process differing in no way from the formation of normal cementum.

The pericementum is that birdnestlike tissue which lies between the roots of the teeth and the alveola wall, and is the agent by which both of these hard tissues are produced. The side lying next to the alveolus produces bone, while the opposite side produces cementum. The fact that this membrane produces two hard tissues —in most respects alike, yet differing enough to prevent their union —has caused this to be described as a double membrane; but more recent authorities show it to be but a single membrane performing the functions of *two*, by being provided upon the one side with osteoblasts and on the other with cementoblasts.

The normal amount of cementum having been deposited, the peridental membrane becomes inactive, and so remains in this dormant condition, as it were, until it is again stimulated into activity by some pathological influence, when it again resumes to all intents and purposes its formative activity, and new tissue similar to the first is added to that already formed.

This affection may be confined to a very small area, or it may involve a considerable territory of the roots' surface, from which it is known as localized and general hypercementosis. When the last-named condition prevails, the growth may assume enormous proportions, not being confined to a single tooth, but may progress to such an extent as to involve the roots of its neighbors, resulting often in their coalescence. When this takes place, some of the roots are enlarged to such a degree as to result in great distension of the pericementum, bringing about the absorption of the intervening septum of bone, the dissolving of the membrane itself, and the final fusion of the roots into one solid mass.

It is generally conceded that this abnormality is brought about by some of the forms of irritation of the peridental membrane. The most common among these are malocclusion of the teeth, calcareous deposits, protrusion of fillings at or under the gum margins, and localized irritation of constitutional origin. The pathological results brought about by this abnormal development make the study of it a very important one to the dentist.

The diagnosis is at all times very difficult, and in a majority of cases it is impossible, as there is usually no enlargement or bulging of the tissue over the affected part, and its presence can only be determined by the symptoms; and if there are no symptoms to warrant us in diagnosing the case as exostosis, we would have to fall back upon the oft-resorted method of guessing at the cause of the trouble. If upon the other hand we were satisfied from the symptoms that a genuine case of hypercementosis was the cause of the trouble, we would be warranted in treating it in one of the following ways: It has been suggested that large doses of iodide of potassium be given; but as to the efficiency of the remedy the profession is considerably in doubt. Again, where the teeth were living, the pulps have been devitalized, and the roots filled in the hope of obtaining relief. Again, a more efficient remedy, where the trouble is brought about by an improper adjustment, is to properly occlude the teeth. Last, and by no means least, at least in one thing—to wit, giving relief—we may mention extraction.

DISEASES OF THE ANTRUM, AND TREATMENT.*

BY HENRY W. MORGAN, M.D., D.D.S., NASHVILLE, TENN.

THE subject selected for me to-night, "Diseases of the Antrum, and Treatment," is one of so much importance and magnitude, and one with which I have such a limited personal acquaintance, and have had such a short time in which to make preparation, that I must confine myself to a few brief sentences, with a hope of arousing an interest which will more properly develop the subject in the discussion which is to follow.

All writers on pathology and therapeutics recognize the frequency with which antral diseases are met and the formidable and dangerous character of these diseases; yet we are impressed with the scarcity of the literature on the subject. All of these are not, however, of dangerous character, but some very simple and readily yield to treatment.

In sixteen years the writer has been confronted with but four cases, only one of which proved fatal, yet if neglected in the earlier stages the others might have assumed such a form as to have bid defiance to the skill of the surgeon and his remedies.

"The form which the disease puts on," says one writer, "is determined by the state of the constitutional health, or some specific tendency of the general system; and we readily imagine that a cause which in one person would give rise to simple inflammation of the lining membrane, or mucous engorgement of the sinus, would in another, produce an ill-conditioned ulcer, fungus hæmatodes, or osteo-sarcoma."

The necessity of early diagnosing and administering the proper surgical and therapeutic treatment is very apparent, and it may be said that the greater danger lies in neglect than the necessarily fatal character of the disease. A variety of theories have been offered as causes of disease of the antrum. Briefly stated, we may group them under three general heads: (1) Results from injury; (2) the extension from a diseased tooth; and (3) from the extension of the inflammatory process of the nasal mucous membrane produced by continuity of tissue.

One writer holds : " My experience with inflammatory diseases of the maxillary sinus is that they mostly follow pathological proc-

---

* Read before the Dental Section of the Nashville Academy of Medicine.

esses of the nasal mucous membrane, and accordingly the soft parts of the nasal and maxillary cavities are generally diseased together." Many coincide with this opinion, but view it from the standpoint of the rhinologists. On the other hand there are many who hold that a very few cases can be traced to this cause. Dr. Carr, of New York, in considering this subject, quotes Dr. Bosworth, of London, as claiming that inflammation of the mucous membrane shows but slight tendency to extend from one anatomical region to another, and observes: " We have all noticed that an inflammatory condition of the mucous membrane of the oral cavity, although severe, seldom extends to other anatomical regions." Dr. Bosworth states: " But few patients suffer from antral diseases compared with the great number who are affected with chronic rhinitis." He, however, admits that the hypertrophic rhinitis does produce antral disease, not by extension, but because the hypertrophic process in some manner causes occlusion of the ostium maxillare.

Dr. Louis McClane Tiffiny, in the American System, holds that inflammation of the antrum is not met with as an idiopathic affection; yet if any cause should operate to close the sinus so as to exclude the air, and prevent the escape of that confined within it, a condition would soon exist dependent on no other affection.

The first and second molars are the teeth more frequently to act as the exciting cause, as their fangs project into the floor of the antrum.

Symptoms.—The usual symptoms of inflammation are not always present, and the onset of the disease is apt to be insidious, unless the nasal opening is closed. A catarrhal inflammation of the lining membrane expresses the pathological condition—a sense of weight which is relieved by the discharge of pus. Pus will flow on lying on one side and will be noticed in the throat and nose; a bad taste and smell may be complained of. The amount of pus passing into throat and swallowed may be so great as to produce nausea or ill-health. If the opening into the nose is closed, the symptoms of retained pus become evident: pain, fever, rapid pulse, distension of the antrum and absorption of the bony wall, perforation and sudden discharge of pus, with immediate subsidence of all symptoms. An eye may protrude greatly from an antral distension, and the inflammation may extend and involve the meninges.

There is not often difficulty in diagnosing a case. I have not time to consider this, but with the symptoms laid down above, it would seem impossible to confound a diseased antrum with any other affection. I cannot do better than to quote Dr. Carr on the treatment, which I do in full, with the report of a very interesting case:

"After diagnosis has been established, the treatment of antral suppuration is simple in its character. If caused by an obstruction of the ostium maxillares, the obstruction should immediately be removed, and an effort made to effect a cure through the nares. Should this be unsuccessful, the question will arise: Where shall an opening be made for evacuating the sinus of the accumulated secretions, and for subsequent treatment? Various operations have been suggested, and, owing to a natural hesitancy to remove sound teeth, it has been suggested that an opening be made in the meatus below the inferior turbinated bone. He thought this operation unjustifiable, because, though the opening could be made without difficulty, it would be repugnant to the patient, seeming more formidable than it really is. Nor would it afford the same facilities for thorough cleansing as would an opening through the alveolus. He would make an opening through the alveolus even at the sacrifice of a sound tooth, and, in such a case, in the choice of the tooth to be extracted he would be governed by the character of the tooth, always selecting the weakest. The second molar would be chosen usually, as it makes an opening nearer the center of the floor of the antrum; but if either the first or the third molar or the second bicuspid is defective, this should be sacrificed in preference. Should it seem best to extract the second bicuspid, the opening should be drilled upward and slightly backward, as at this point the floor of the antrum is thicker than at the other points mentioned. In case the teeth on the affected side are all missing, the point corresponding to the second molar should be chosen. If the trouble arises from decayed teeth, and necrosis is present, all necrosed bone should be removed. Before commencing either to remove necrosed bone or to make an opening into the antrum, a ten per cent. solution of cocaine should be applied to the gums three or four times, in order to produce local anæsthesia of the mucous membrane. The operation is then performed with but slight inconvenience or pain to the patient. The opening being established, the antrum should be thoroughly cleansed with tepid distilled water, to which a little salt has been added, until all traces of pus have disappeared. If the discharge is offensive, the cavity should be syringed with a solution of permanganate of potash and dressed with a stimulating solution composed of carbolic acid, one drachm; glycerine, one ounce; and distilled water, seven ounces; or, if it is preferred, Dobell's solution or Listerine may be used. The cavity should be cleansed twice daily with salt water, followed by either of these stimulating solutions, until the discharge diminishes, when once daily will be sufficient. Should the orifice in the meatus be closed by inspissated mucus, the patient should be instructed to use either of the above stimulants several times daily, by means of a nasal spray, until the obstruction is removed.

"Some practitioners insert a silver drainage tube through the orifice, kept in place by ligatures. He sees no advantage to be derived from this practice, as thorough cleansing twice daily is all that is required. Besides, there is always the possibility of food being pressed through the tube into the sinus, or of becoming clogged at the entrance, so that the object sought is not obtained. The better method is suggested by Dr. Abbot: Closing the opening by means of a broom straw, serrated and wrapped with carbolized cotton, then pressed firmly into place, and retained by means of ligatures to the teeth. If there are no contiguous teeth, a plate may be inserted with a projection one-half the size of the orifice, the projection wrapped with carbolized cotton to fill the space. This is sufficient to exclude all foreign substances. When the discharge ceases, the cotton should be lessened in quantity daily; thus the orifice will be permitted to close gradually."

This treatment Dr. Carr finds all that is necessary, and in his hands it is uniformly successful. He gave the following history of a case which he treated.

"Mr. M., aged thirty-three, who had until two years before been in perfect health. He then noticed an offensive discharge from the right nostril, which usually disturbed him greatly upon retiring, causing violent coughing when lying on the left side. Also, upon arising in the morning, he experienced nausea, which continued until the nasal cavity had been entirely cleaned. He supposed that he was suffering from catarrh, for which he sought and received treatment at intervals for twenty-two months, when the following additional symptoms were manifested: At intervals of three or four days he experienced attacks of vertigo, followed by severe otalgia and great tenderness of the teeth. For these symptoms he was treated by his family physician for three months, who finally advised him to consult me regarding what he supposed to be an alveolar abcess. The right side of his face was then greatly swollen. Upon examination the full number of teeth were found, but the second bicuspid had been filled, and was pulpless. My diagnosis was not that of simple alveolar abscess, but suppuration of the antrum. The extraction of the second bicuspid was followed by a slight flow of pus. A further examination showed the alveolar process greatly necrosed, but there was no visible opening into the antrum. After an application of cocaine all necrosed bone was removed, and an opening made into the antrum, when a great quantity of offensive matter escaped. The lining membrane of the antrum had thickened to at least ten times its normal thickness. This pathological condition I have found in all chronic cases upon which I have operated. Then proceeding to syringe the sinus, I failed to establish an outlet through the opening into the nares. The sinus was first cleansed with salt water, then with permanganate of potash, after which the orifice was closed in the manner already described. The patient was then directed to use Dobell's solution, by means of a nasal spray, in order to remove any secretion from the nares. The following day the pledget of straw and cotton was removed, when the discharge seemed greater than on the previous day. The opening through the meatus had then been established, and the cavity was thoroughly syringed with warm water until all traces of pus disappeared. Then it was cleansed with a stimulating solution, and the pledget renewed. This treatment was continued daily for three months, when the patient was dismissed cured. I have seen him since at intervals during the past three years, and there are no signs of recurrence of the disease."

School Commencements. · · · ·

VANDERBILT UNIVERSITY.
ANNUAL COMMENCEMENT EXERCISES OF THE DEPARTMENT OF DENTISTRY.

THE Thirteenth Annual Commencement of the Department of Dentistry was held at the Theater Vendome on the evening of the 24th of February. A large and representative assembly of Nashville's cultured citizens were gathered to witness the graduation of seventy-one young gentlemen who had chosen the Vanderbilt as their *Alma Mater*.

The stage was tastefully decorated with potted flowers and evergreens; and seated thereon, besides the Faculty and friends of the the department, were the President of the University, Bishop Hargrove; Chancellor Garland, Judge E. H. East, of the Board of Trust; and the distinguished guest and speaker of the occasion, Dr. C. S. Stockton, of Newark, N. J., who had been chosen to deliver the address to the graduates; also Dr. Collins Denny, of the Vanderbilt; and Dr. Clarence J. Washington, the class valedictorian.

The exercises were opened by an overture from the orchestra, after which the Rev. Dr. Collins Denny offered an appropriate and inspiring prayer.

Prof. W. H. Morgan, Dean of the department, next proceeded to give his report of the progress and work accomplished during the past session. He made a most interesting address, which was listened to with marked attention by the audience, whom he succeeded in convincing that the facilities offered by this department for the successful acquirement of the art and science of dentistry were not surpassed by any similar institution in America. The superior clinical advantages afforded had attracted students from other schools all over the country to the Vanderbilt University.

Dr. Morgan called the roll of graduates, and one by one they came upon the stage, gradually forming a semicircle. Each one as he came into view was an object of eager interest on the part of the audience. The honor men were liberally applauded.

·

Chancellor L. C. Garland's honored figure, bent with the weight of years, arose after the class had ranged on the stage, and in well-chosen words of advice and wisdom he conferred the degree of "Doctor of Dental Surgery." He complimented the students on their high standard of right doing. He said that the Vanderbilt University took peculiar pride in the Dental Department—the youngest department but one—and found great gratification in the wonderful strides it had made, elevating dentistry from an art to a science.

The diplomas were awarded, and the dentists, whose names are appended, filed off the stage: V. U. Alexander, New York; C. K. Adams, Mississippi; C. S. Alred, Alabama; G. M. Brown, Michigan; L. Bland, Louisiana; J. M. Ashburn, Tennessee; A. E. Brown, Texas; F. Bartel, Illinois; T. K. Barefield, J. A. Beaver, Alabama; J. R. Beach, Tennessee; J. S. Brown, Mississippi; J. P. Corley, Alabama; R. Z. Chapman, Alabama; J. J. Cook, Michigan; D. P. Cook, Kentucky; S. C. Cawthon, Florida; R. H. Carrette, Iowa; W. J. Dillard, Texas; J. S. Dalton, Missouri; S. K. Davidson, Kentucky; E. H. Denison, Connecticut; O. C. Evans, Illinois; Charles Eshleman, Iowa; T. A. Fayett, Alabama; F. B. Gaither, North Carolina; C. B. Graham, South Carolina; W. I. Hale, Alabama; W. L. Hansbro, Tennessee; A. C. Jones, Mississippi; R. A. Jones, Jr., Alabama; E. L. Kendrick, Alabama; E. C. Kidd, Alabama; O. G. Mingledorff, South Carolina; T. W. McKell, Mississippi; M. B. McCrary, Tennessee; J. M. Millen, Tennessee; G. Minnich, Illinois; A. J. Newcomer, Illinois; J. B. Penney, Missouri; J. M. Murphree, Alabama; W. J. Johnson, Alabama; C. W. Mathison, Alabama; J. H. Palm, Germany; W. H. Powell, Louisiana; A. L. Pedigo, Texas; M. D. Steele, Louisiana; C. A. Sevier, Tennessee; R. Sanderson, Alabama; T. W. Simons, Texas; W. K. Slater, Tennessee; C. C. Sims, Arkansas; N. W. Sherman, Tennessee; M. O. Sallee, Kentucky; H. E. Spencer, Mississippi; W. S. Taylor, Kentucky; R. E. Thornton, Georgia; F. O. H. Thiele, Germany; C. J. Washington, Tennessee; V. B. Warrenfells, Virginia; J. D. Wise, Alabama; F. P. Ward, Alabama; N. F. Weatherby, Mississippi; H. W. Walker, Georgia; V. H. Ward, Mississippi; V. A. Williams, California; H. Wiggins, Texas; A. Walker, Georgia; C. M. Walton, Tennessee; W. L. Weathersby, Mississippi.

The next feature of the programme was the class valedictory, for the performance of which important and delicate task Clarence J. Washington, of Tennessee, appeared. Mr. Washington's address was decidedly unique, showing great versatility, evincing dramat-

6

ical power, and sparkling with humor and occasional witticisms. His attempt at originality was decidedly successful. He was frequently interrupted by enthusiastic expressions of approval.

Prof. Henry Morgan, in a few well-chosen words, gracefully introduced Dr. C. S. Stockton, of Newark, N. J., as having been selected to deliver the Faculty address to the graduates. "Dr. Stockton, who is a forcibly graceful speaker, got at once into the good graces of his audience by remarking that he came from a State that always went Democratic. He said that when he first received the invitation to come to Nashville he thought that the distance was too far; but several influences had drawn him despite that obstacle. He said that he had desired to see the famous Vanderbilt University, to see Nashville, and to once more shake the hand of Dr. W. H. Morgan, than whom a nobler man did not exist." Dr. Stockton's able and scholarly address will be found fully reported elsewhere in this issue.

Dr. Collins Denny, of the Theological Faculty, in a happy speech, in which he fully sustained the reputation of being one of the best platform speakers of the South, awarded the class honors as follows: First honor, Founder's Medal for best general average, to R. A. Jones, Jr., of Alabama; second honor, Morrison Brothers' Medal, to O. G. Mingledorff, of South Carolina; third honor, Honorable Mention, to Charles Eshleman, of Iowa; Prof. H. W. Morgan's Medal, for best gold filling, to A. E. Brown, of Texas; Prof. Ambrose Morrison's Medal, in anatomy and physiology, to R. A. Jones, Jr., of Alabama (C. A. Sevier, of Tennessee, it was announced, was beaten for this prize by only one-sixth of a point in a possible one hundred); Dr. S. S. Crockett's Medal, for best examination of first course students in anatomy and physiology, to J. K. Campbell, of Mississippi. (*American.*)

UNIVERSITY OF TENNESSEE.

COMMENCEMENT EXERCISES DEPARTMENT OF DENTISTRY.

THE Sixteenth Annual Exercises of this department were held in conjunction with those of the Medical at the Theater Vendome on the evening of the 23d of February. A large audience was drawn together and much interest manifested in the occasion.

Seated upon the stage, which was handsomely decorated for the occasion, were the members of the Medical and Dental Faculties, supplemented by Chas. W. Dabney, Jr., LL.D., President of the University, and Gen. H. H. Norman, who represented Gov. Buchanan.

Dr. J. Berrien Lindsley offered prayer, after which Robert B. Harris, of Tennessee, the valedictorian of the dental class, was introduced. He spoke eloquently on the subject of the "Unsolved Problem of Life." His address evinced careful preparation and was well received by the audience, who greeted his effort with frequent and timely applause.

Dr. Charles W. Dabney, Jr., as President of the University, then conferred the degrees of D.D.S. on twelve dental graduates. Each came forward as his name was called, and received his diploma. The following is a list of the graduates: H. D. Broaddus, Missouri; E. A. Cowles, Minnesota; Marcus A. Coykendall, Michigan; Greenwood W. Crow, Pennsylvania; Andrew J. Galbreath, Colorado; Ore J. Harrelson, Mississippi; Robert B. Harris, Tennessee; Charles W. Marvin, Florida; J. W. Middleton, Fred Moss, Missouri; Ed M. Roberts, Alabama; Albert S. Swett, Mississippi.

The charge to the graduates was delivered by T. Hilliard Wood, M.D., professor of physiology. His address was designed for both medical and dental graduates, and we regret that lack of space prevents us from giving more than a bare synopsis. Among other things he said:

Remember 'tis human life you handle, and if the confidence of friends, the groans of the suffering, and the gasp of the dying do not stimulate you to study, then, in the interest of humanity, step aside and give place to others.

As a safe guide for your conduct, I recommend to your study and adoption the "Code of Ethics of the American Medical Association." This code, born of the best brains and indorsed by the best men of your profession, explaining, as it does, the duty of physician to patient, of patient to physician, and of physician to physician, should be your invariable rule of practice You will see upon every line the impress of the golden rule: "As ye would that men should do unto you, do ye even so unto them."

Medical men as a class are said to avoid politics. Possibly this is true to too great an extent; but if it be an error, I am disposed to believe it an error in the right direction. While I would warn you against an excessive indulgence of your political tastes, yet any legislation looking toward the betterment of science and the protection of the masses against fraud and incompetence should receive your earnest support.

In this direction, under the influence of the medical profession, the State of Tennessee has made progress. Already empiricism has received a check and charlatanry is beginning to look for other fields.

A review of the progress made by medicine in modern times is most enchanting. In no way can this be better illustrated than by comparing the methods and results of the present with those of the past. The progress is truly astonishing. Antiseptic surgery is of modern development. Within the last half century dentistry has lifted itself from an ordinary vocation or trade to the full recognition of the medical profession as an important department of the healing art. The aggregate amount in the improvement of dentistry, from a surgical, mechanical, therapeutic, and hygienic standpoint, is almost without a parallel in history. I congratulate the Department of Dentistry on its achievements, and on the opportunities that lie before it for the advancement of dietetics and hygiene.

The retina of a living eye was first seen in 1851. The interior of that most wonderful of all musical instruments, the living larynx, was first seen by Garcia in 1854; and with the aid of the laryngoscope the play of the vocal chords, and the production of musical notes can be as easily studied and as thoroughly explained as the mechanism of any musical instrument.

Time forbids that I should speak of the discovery of the various anæsthetics, both local and general; of the numerous beneficial operations which surgery has devised; of the innumerable instruments, modifications, and apparatus which have been invented; and of the investigation into the causes and modification in the treatment of the many ills to which flesh is heir. And in studying the modern history of our profession you will notice that these improvements have not been made by any single nation; but in them you see represented medical men from every civilized country, showing that there is a general enthusiasm which pervades the entire medical world. The medical men of the civilized world stand shoulder to shoulder, and, although separated by oceans, the discoveries of the one are promptly communicated to the others, who in turn take up and continue the investigation.

It is during the devastating reign of fatal epidemics, when the people of whole cities and districts are frenzied with fear, when disease and death are borne on every breeze—it is then the physician's most noble, most self-sacrificing qualities are developed; it is there his honor shines with valor ne'er surpassed on any tented field. What is a grander scene than the medical profession standing shoulder to shoulder, defending the public against invading death?

The following medals were awarded by Prof. J. P. Gray, D.D.S.: Robert Russell Faculty Medal, A. J. Galbreath, Colorado; Faculty second honor (Morrison Bros.), M. A. Coykendall, Michigan; Faculty third honor, E. A. Cowles, Minnesota.

Dr. J. Bunyan Stephens pronounced the benediction. (*American*.)

Selections. . . . .

ARSENIOUS ACID IN DENTISTRY.

THE fact that the world of thought moves in circles is nowhere more apparent than in dentistry. This is, perhaps, inevitable, from the fact that each generation brings with it men cultured, it may be, in the scientific thought of the time, but not familiar with the past history of their profession, as they should be, nor conversant, as they certainly ought to be, with the treatment of pathological conditions in past epochs.

The extraordinary rush of apparent intelligence into what may be called the startling in therapeutics is without reason except that individuals may regard this as an open door to fame or notoriety; the difference between these two states may not be clear to some minds, and to them the terms are possibly interchangeable.

When Spooner, in 1836, introduced arsenic for the devitalization of pulps, he probably had no idea that fifty-four years thereafter it would be recommended for filling root-canals; but such has recently been done, and the matter gravely discussed by a learned society.*

Arsenious acid, from its peculiar action, is fitted for the purpose of devitalizing pulps as no other known agent. It is not necessary to call the attention of our readers to any extent to its peculiar properties. These are well understood, and were it not for the fact that some seem to forget that in using it they are not only dealing with an escharotic, but one of the most powerful and insidious of poisons, it would not be necessary to allude to it here.

While there is still much to learn in regard to the action of this agent in its local, general, and cerebral effects, it is well understood that it does not destroy the tissue as some other escharotics, but the destruction of life is through a more complicated process, and has no immediate effect upon the continuity of the pulp. The action of arsenious acid is to first render the sensory nerves torpid, for Sklarek, quoted by Ringer, says: "Arsenic, therefore, paralyzes first

---

* Mississippi Valley Association of Dental Surgeons, March, 1891.

sensation and reflex action, and some time afterward voluntary power." Ringer further writes: "My own experiments, conducted with Dr. Murrell, confirm this statement; but they show also that arsenious acid is a paralyzer of the motor and sensory nerves and of the muscles."

Probably no class of professional men have had this drug more closely under observation than dentists. For over fifty years it has been their one agent for the destruction of pulps, and when properly used is always reliable. Hence those who have observed closely have long since arrived at the conclusions quoted, and that the paralysis observed was a gradual process requiring a definite time for its completion, and its extent was only limited by the amount of arsenic applied. The torpidity of the sensory nerves of the pulp means a retardation of the blood supply, and eventually death.

The action of this agent shows also a clearly defined progress, and an examination of a removed pulp will exhibit a distinct line of demarcation between the affected and the nonaffected portion, and this is of deep interest in a macroscopic as well as in a microscopic sense. It will have been noticed, in practice, that the antiseptic property of arsenious acid is of no value. This, in the discussion alluded to, was dwelt upon as the main reason for its use. While the agent fails to break up the tissue, it leaves the pulp and, consequently, all its connections in a condition to rapidly decompose, which takes place in a varying period of from ten days to two weeks.

The evidences are ample to demonstrate that the use of arsenious acid should be confined to that purpose which led to its introduction, and that even here it will fail and be productive of serious results in the hands of the careless or ignorant.

In the annual address alluded to the following formula was offered for "root-filling:"

> R Arsenious acid, gr. ii ;
> Precipitated chalk, ℥ i ;
> Glycerin, q.s.
> Sig.—To be made into a paste.

He says: "The dose of the arsenic is from $\frac{1}{20}$ to $\frac{1}{40}$ of a grain. The $\frac{1}{40}$ of a grain is probably all that is applied in this use of the drug, and this is *not in contact with soft tissue*. Even should it be forced through the foramen, *its effect would only be beneficial*." (Italics ours). It is not necessary to enter into any lengthy discussion of this quotation, as it probably carries with it its own antidote.

His mode of filling the canals is as follows: Take "one-half a grain of the mixture. equal in size to half a grain of wheat. Mix with a single drop of water on a slab. Introduce into the canal, which should first be repeatedly washed with alcohol and thoroughly dried. Fill the canal perfectly with the compound. Then dry again, and in the majority of cases fill the tooth at once."

The supposition that there are no avenues for the agent to reach the pericementum is a fallacious one. Not only is the foramen present as an outlet, but it has been demonstrated that not infrequently other openings are found to expose a greater surface to the irritant. It has, further, never been proven that a continuous chain of canals do not exist. connecting the tubuli with the pericementum. On the contrary, there is much evidence to show that solutions can be transmitted from the pulp canal to the periphery of the tooth.

Those who have had much experience appreciate the difficulty in preventing that much to be dreaded pericementitis from arsenious acid. And so far from its being true that, "even if it did pass it could do no harm," it is certain to arouse periosteal inflammation of a serious character. If it fails to affect this, it is simply because the amount has been too minute to produce results beyond a limited area of tissue.

If this agent destroys the life of the pulp, in the manner previously outlined, it will have a similar effect on all tissues. This is well understood, and the sloughing of cervical margins is an apt illustration, and demonstrates what takes place when the pericementum is saturated with the solution. The result is death of this important organ, necrosis of the cemental tissue, and necessarily destruction of the corresponding alveolar plate. That any other view should be accepted is strange; and it is, further, a matter of surprise that any organization should seriously consider the subject, or conceive, as one present expressed it, that "there may be more in this method than we are now able to conceive, after all."

In the May number of the *Journal* is a paper read before one of our own most prominent societies, in which arsenious acid and cocaine are recommended for sensitive dentine, although the writer acknowledges that "in private practice it is not desirable to try it as an obtunder, except in certain teeth like bicuspids, or others that are to be removed for regulating, for it might endanger the vitality of the pulp." While this is a saving clause, it would seem that the whole process should then and there have been denounced.

This mode of treating the hyperæsthetic conditions of dentine

with arsenious acid has been repudiated by the dental profession
for many years. A certain very prominent operator in this city
(Philadelphia) applied it for this purpose more than a quarter of a
century ago, and took one step in advance of the writer of the
paper, for he regarded it possible to use it as an aid in the extraction of teeth.

It is only recently that several of the dental journals, including
the *International*, felt called upon to expose a preparation that had
deceived some of the most intelligent dentists, and we are certain to
have a flood of nostrums of a similar character if those who are the
leaders of thought are not constantly guarding the best interests of
their profession.

Dentistry has suffered enough reproach from the careless use of
this drug, and no one having its welfare at heart dare pass by any
reckless application of it. It has only one legitimate place in our
work, and that is the one proposed by Spooner. (Editorial in *International Dental Journal*.)

ETHER has long been recognized as one of the most effective anæsthetics, but its debilitating effects upon the system and the slow-
ness with which it could be eliminated from the blood have caused
physicians to be very chary of its use. It is now reported that Dr.
Charles McBurney, of the New York College of Physicians and
Surgeons, has devised a method of administering it which does
away with its harmful effects. In the experiment which Dr. Mc-
Burney performed before a number of medical students on Saturday, the legs and arms of the patient were tightly bound at the
thighs and shoulders, so that the circulation in these members was
stopped. When the ether was applied to the mouth and nostrils it
saturated only one-third of the system's blood, and unconsciousness
was produced with about one-quarter of the quantity ordinarily required. When the operation was finished, the bandages were removed and the untainted blood, which they had confined, rushed
back to the heart and throughout the body, producing a revivifying
effect, and almost immediately consciousness. The patient spoke
intelligently within three minutes. The small quantity of ether
used, and the rapidity with which the stupor is dispelled, render this
method peculiarly adapted to weak or sickly persons. (Exchange.)

Extracts. • • • • • • • • • •

AN OLD IDEA USEFUL.

WE believe that it was Dr. W. H. Eames who said: "If you wish to remove a deciduous tooth, and through fear the child will not permit it, slip a piece of rubber tubing over the crown down to the neck of the tooth, and in a few days the tooth will be so loose that it can be extracted with the fingers." If you have such a case, try it and see the exact result. (Editor *Review.*)

NITRATE OF AMYL FOR HICCOUGH.

MISS ALICE WOODWARD, of Shelton, Conn., who hiccoughed herself to the point of death, is now out of danger. In consequence of the circulation given the case remedies were sent from all parts of the country. Many were tried, and that suggested by Frank W. Mack, of the Associated Press, New York, was effective. The remedy is nitrate of amyl, a few drops to be inhaled from a handkerchief.

COPPER AMALGAM.

COPPER amalgam in itself may not be injurious to the general health, but when combined with oxygen an oxide is formed, for which acetic acid has great affinity, and subacetate of copper is formed, which is a very active poison. Let us not follow the course of many who, in their efforts to restore the health of their patients, do so at the sacrifice of the teeth. On the contrary let us not destroy the health of our patients to save the teeth. (Dr. Miller, in Rochester Dental Society.)

A NEW METHOD OF BANDING A ROOT.

DR. E. PARMLY BROWN, of New York, has a new method in making a band, using platinum, gauge 30 to 36. The band is made wider than necessary. After it has been fitted and soldered with pure gold, the portion projecting beyond the end of the root is clipped and bent over so as to cover the end, being neatly malleted with a plugger, to properly fit the root. It can then be soldered to any pintooth by simply pressing the pin through the platina, or it can also be used without soldering it to the tooth, as the cement will hold it in place.

PAIN: ITS DIAGNOSTIC VALUE.

PAINS coming on in paroxysms are neuroses, and usually involve nerve troubles. Rhythmical pains always belong to nonstriated tissues, and are usually associated with some one of the hollow organs.

Coldness, chilliness, heat, burning, itching, creeping, crawling, and similar pains are peculiar to the skin and mucous membranes.

Sticking, darting, stabbing pains are always found connected with serous membranes or connective tissue.

Throbbing, tearing, aching, and pressing pains indicate that the sensory divisions of the cerebro-spinal nerves are involved. (Dr. Owens.)

ANTISEPTIC TOOTH WASH.

IT is now universally admitted that the only way to prevent decay of the teeth is to use frequently a good antiseptic lotion. The following is amongst the best:

Phenic acid ....gr. xv.
Boric acid ....dr. vij.
Thymol...gr. viij.
Essence of peppermint..............................gtt. xx.
Tincture of anise.... ....dr. iiss.
Water................................... ....O ij.
M.

Rinse the mouth and brush the teeth with an equal portion of this lotion and water, night and morning, or after each meal. (*Popular Medical Monthly.*)

AMALGAM.

IT is surprising to me that any one who has had such success in the use of amalgam as Dr. Douglas should ever use anything else. While it may be true that amalgam is a safe and reliable filling material in Dr. Douglas's hands, it has woefully fallen short of giving satisfactory results in the hands of most operators. One great difficulty with this material is that it encourages faulty and careless manipulation. There is nothing in its manipulation to call out the highest ability of the operator, and the very ease with which it is manipulated is a source of evil in more ways than one. The material itself has inherent, harmful, and bad qualities which ought to debar it from use. Of its practical use perhaps I ought not to say anything, for I have never yet inserted an amalgam filling in any patient's tooth; but from what I have seen of those made by other operators I don't think that I shall begin now. (Dr. J. Taft, before the Michigan Dental Association.)

COCAINE POISONING.

In the *Medical and Surgical Reporter* Dr. J. B. Mattison cites numerous instances of the lethal and dangerous effects of cocaine, from which he draws the following conclusions:

1. Cocaine may be toxic.
2. This effect is not rare.
3. There is a lethal dose of cocaine.
4. The lethal dose is uncertain.
5. Dangerous or deadly results may follow doses usually deemed safe.
6. Toxic effects may be the sequence of doses large or small, in patients young or old, the feeble or the strong.
7. The danger, near or remote, is greatest when given under the skin.
8. Cardiac or renal weakness increases the risk.
9. Purity of drug will not exempt from ill result.
10. Caution is needful under all conditions.
11. Réclus method, Corning's device, or Esmarch's bandage should be used when injecting.
12. Nitrate of amyl, hypodermic morphia, hypodermic ether, alcohol, ammonia, and caffeine should be at command. (*Medical Brief.*)

Correspondence. . . . . . .

A NOTABLE ANNIVERSARY.

To the DENTAL HEADLIGHT.

A notable anniversary meeting of the First District Dental Society of New York took me out of my office in January. While the account may not prove of great professional value to your readers, still it is well for them to see what is going on in our world of dentistry outside of the narrow limits of home. This, we trust, will be ample apology for daring to take a limited portion of your valued space.

It is well known, doubtless, to many of your readers, that this First District Society of New York is one of the two great city associations of the metropolis. During a season comprising a large portion of the year it has monthly meetings. Once in two years the January meeting is enlarged into a kind of professional, fraternal celebration, where not only the members themselves contribute, but professional brothers also from a distance are called upon to engage. Owing to the specially enthusiastic nature of the Executive Committee for this year (official), the idea was planned upon even larger conceptions than usual. The two days' meeting of past years was extended to three days this, with two sessions of the society each day. A programme was thoroughly matured and the essayists and clinical exhibitors of varied kinds and interests conferred with and put in practical order.

It would possibly be an imposition to ask space enough to present this programme here. Suffice it to say, essays were presented by Drs. C. N. Pierce, C. S. Case, N. W. Kingsley, D. R. Stubblefield, and J. H. Marshall, and the special discussions were participated in by such men as Drs. J. S. Dodge, George Allan, Charles Heitzman, James Truman, C. J. Essig, J. Y. Crawford, S. C. G. Watkins, S. H. Guilford, J. N. Farrar, as well as some of the essayists themselves.

These essays and discussions were supplemented by clinics and exhibitions of new and interesting matters pertaining to progressive dentistry. If any fault could be found with this feature, it was that so much was placed in so little space that it was almost impossible to obtain a clear idea of anything except your neighbor's elbows in your ribs or lumbar region. Upon the whole, however, it was a memorable occasion creditable to even the First District Dental Society, so famous for its contributions to our professional advancement.

In concluding this very imperfect though well-meant review of the annual meeting, tribute must be paid to the elegant and entirely welcome banquet given to the visitors by the members. The actions on that occasion spoke even more fulsomely than the words that were themselves cordially uttered of appreciation and splendid enjoyment. Too much praise cannot be given to (1) the idea rendered practical by this society of inviting professional brothers from other parts to contribute to the work. It engenders vigor and a better understanding of the questions professionally of the hour. And (2) the scheme of preparing matured discussions. It is indubitable that few, if any, men can do themselves or the subject justice speaking impromptu. This valued feature should be taken hold of and carefully applied to all meetings.

I promise myself the pleasure of other agreeable anniversaries.

Yours very truly,   D. R. STUBBLEFIELD.

DR. S. E. BEST, a dentist of 224 Lenox Avenue, was killed in a collision on the New York Central railroad, near Sing Sing, Thursday night, December 24. (The *Doctors' Weekly*.)

THE death of Mr. George Claudius Ash occurred in London on January 17, in the seventy-eighth year of his age, from an attack of influenza. Mr. Ash was for many years senior partner of the firm of C. Ash & Sons, Limited, of London, manufacturers of and dealers in dental supplies. He was originally apprenticed to the late Mr. Thompson, of London, with whom he studied dentistry and afterward entered into practice as a dental surgeon, but subsequently retired to devote himself exclusively to the business established by his father, the late Claudius Ash. He was for a long time in charge of the manufacturing department of the concern, to which he devoted great care and attention. Mr. Ash was highly respected, both as an employer and in all the relations of life. His loss will be keenly felt by all who were associated with him. (*Dental Cosmos*.)

Editorial. . . . .

TEMPORARY LICENSE AND JUNIOR EXAMINATIONS.

BOARDS of Dental Examiners should require of students asking temporary license that they should hold certificates of having passed the junior examination.

If the colleges are to demand this examination and it is to be worth anything, the Boards should uphold it and make it at least obligatory that the student should remain at college until it is held. The standing of the student his second year is largely determined by the vigor with which he pursues his studies the first year; and if there is nothing dependent upon the junior examination, the incentive for hard work is small with those who are inclined to indolence.

Let the Boards heartily second the colleges in anything that will have a tendency to make students more industrious and ambitious. The goal of most students is to get to practice with a diploma and a license with as little hard work as possible. There must be behind them as many forces as possible to arouse their energies to close attendance on college duties if the demands for higher and better education are met at all.

No one doubts for a moment that great progress has been made in the matter of education—not alone in dentistry, but in all professions and trades as well. Training schools are the order of the day. Better facilities exist to-day for obtaining knowledge than ever before in the history of the world. The demonstrator of to-day is a very different order of man to the one of a few years ago, and his duties are, when honestly discharged, as onerous as that of the professor. The results of his work over the chair and workbench largely determine the future as practitioners of those who are under him, and as this fact has been recognized and the need supplied, better operators and workmen are the result, and State Boards should everywhere join hands with those colleges who are thus honestly endeavoring to elevate dentistry.

RESIGNATION AND ELECTION.

DR. SIDNEY S. CROCKETT, who for several years past has occupied the position of Demonstrator of Anatomy in the Department of Dentistry of Vanderbilt University, has tendered his resignation, having been unanimously elected Demonstrator of Anatomy in the Department of Medicine. His many friends in the dental profession will join us in expressions of regret for his withdrawal, and wish for him great success in his enlarged field of labor and usefulness.

It is with much pleasure we announce the election of his successor, Clifton R. Atchison, M.D., the talented son of our esteemed colleague, Prof. Thomas A. Atchison, M.D.

MEHARRY SCHOOL OF DENTISTRY.

The Sixth Annual Commencement Exercises of the Dental Department of Meharry Medical College were held in connection with that of the Medical and Pharmaceutical, February 18, at Nashville, Tenn. President J. Braden conferred the degree of Doctor of Dental Surgery on J. B. Singleton, of South Carolina.

George W. Miller, of the medical class, gave the address of welcome, and J. W. Holmes delivered the valedictory. Robert Tyler, of Alabama, represented the pharmaceutical classes.

The charge to the graduates was given by R. F. Boyd, M.D., D.D.S., Professor of Physiology, Hygiene, and Clinical Medicine. Rev. John Pierson, D.D., of Cincinnati, was present and gave an earnest and appropriate address to the graduating class.

During the past session seven students have been enrolled in the Dental Department.

THE WORLD'S COLUMBIAN EXPOSITION.

Send 50 cents to Bond & Co., 576 Rookery, Chicago, and you will receive, postpaid, a four hundred page advance guide to the Exposition, with elegant engravings of the grounds and buildings, portraits of its leading spirits, and a map of the city of Chicago; all of the rules governing the Exposition and exhibitors, and all information which can be given out in advance of its opening. Also other engravings and printed information will be sent you as published. It will be a very valuable book, and every person should secure a copy.

NEW BOOKS.

Catching's Compendium of Practical Dentistry for 1891. By B. H. Catching, D.D.S. Constitution Publishing Company, Atlanta, Ga. Price ——.

The wonderful strides which are almost daily made in the realm of dentistry, the introduction of new and improved methods, appliances, and treatments, appearing monthly in various dental periodicals, defy the efforts of the busy practitioners to collect, sift, and preserve those of practical importance and value, and require more leisure than it is possible to command. Therefore to have them annually presented in a well-bound and well-printed book, judiciously selected, is most desirable, and the demand is well met by Dr. Catching. He has with numerous cuts handsomely illustrated the volume. It should be found in every dentist's library.

Man, the Masterpiece; or, Plain Truths Plainly Told About Boyhood, Youth, and Manhood. By J. H. Kellogg, M.D., M.B.S.A., etc. Published by Health Publishing Company, of Battle Creek, Mich.; also publishers of the monthly magazine *Good Health*.

The above publications are the crystallized hygienic and sanitary teachings of Dr. Kellogg, who is at the head of the largest and most popular sanitarium in America. He is considered one of the ablest physicians and medical writers in this country, and certainly well qualified by both practice and large experience to give advice in all matters relating to hygiene and sanitation. Every physician and dentist should procure these publications, which are also admirably adapted to the needs of the laity, and should be read by all classes of the community.

Chart of Typical Forms of Constitutional Irregularities of the Teeth. By Eugene S. Talbot, M.D., D.D.S., of Chicago. Published by Wilmington Dental Manufacturing Company, Philadelphia. Price $2.50.

A collection of sixteen beautiful plates made from drawings and models of the author and reproduced in colors by the Shober & Carqueville Lithographing Company, Chicago. The author's object is to "illustrate the typical forms of constitutional irregularities of the jaws and teeth so that the teacher and student may readily comprehend the various positions which the jaws and teeth may assume," which are produced by excessive or arrested development of the maxillary bones or possible local causes. The forms are selected from a collection of three thousand models, and must prove of much interest to the advanced student. Plates enlarged from four to six times the size of these would be ornamental framed and hung in the lecture room, where they would be seen by all students, proving much more instructive and exciting much inquiry and research. Dr. Talbot is devoting much time and earnest work to orthodontia, and has by the results of this work, which he has stamped with an honesty of purpose highly commendable, won a reputation as an authority on the subject which places him in the front rank of the profession in this country and abroad.

Associations.

AMERICAN DENTAL ASSOCIATION.

W. W. Walker, New York, N. Y., President; J. D. Patterson, Kansas City, Mo., First Vice President; S. C. G. Watkins, St. Clair, N. J., Second Vice President; Fred J. Levy, Orange, N. J., Corresponding Secretary; George H. Cushing, Chicago, Ill., Recording Secretary; A. H. Fuller, St. Louis, Mo., Treasurer.

The next meeting will be held at Niagara Falls, N. Y., on August 2, 1892.

SOUTHERN DENTAL ASSOCIATION.

Gordon White, Nashville, Tenn., President; E. L. Hunter, Enfield, N. C., First Vice President; J. T. Calvert, Spartanburg, S. C., Second Vice President; W. H. Marshall, Oxford, Miss., Third Vice President; D. R. Stubblefield, Nashville, Tenn., Corresponding Secretary; H. C. Herring, Concord, N. C., Recording Secretary; Henry E. Beach, Clarksville, Tenn., Treasurer.

The last Tuesday in July, 1892, Lookout Mountain, Tenn., was selected as the time and place of next meeting.

ALABAMA DENTAL ASSOCIATION.

George Eubank, President; C. L. Boyd, First Vice President; J. H. Hall, Second Vice President; J. H. Allen, Recording Secretary; G. M. Rousseau, Treasurer.

Place of meeting, 1892, Montgomery, second Tuesday in April.

FLORIDA DENTAL ASSOCIATION.

L. M. Frink, President; W. A. McQuarg, First Vice President; W. A. Snead, Second Vice President; L. F. Frink, Corresponding Secretary; C. P. Barrs, Recording Secretary; C. P. Carver, Treasurer. Executive Committee: B. B. Smith, W. McL. Dancy, J. H. Crossland, James Chase, and T. B. Hannah.

The ninth meeting will be held in Jacksonville, Fla., in May, 1892.

GEORGIA DENTAL ASSOCIATION.

W. G. Browne, President; S. M. Roach, First Vice President; W. W. Hill, Second Vice President; L. D. Carpenter, Corresponding

6½

Secretary; S. H. McKee. Recording Secretary; H. A. Lawrence, Treasurer.

The Board of Examiners consists of J. H. Coyle, A. G. Bouton, G. W. McElhaney, William C. Wardlaw, and D. D. Atkinson.

The Association is to meet in Rome, Ga., at such time as the Executive Committee may elect.

KENTUCKY DENTAL ASSOCIATION.

H. B. Tileston, Louisville, President; M. W. Steen, Augusta, Vice President; J. F. Canine, Louisville, Treasurer; J. H. Baldwin, Louisville, Secretary.

MISSISSIPPI DENTAL ASSOCIATION.

D. P. McHenry, Grenada, President; A. A. Dillehay. Meridian, First Vice President; J. B. Rembert. Jackson, Second Vice President; A. A. Wofford, Columbus, Third Vice President; W. E. Walker, Bay St. Louis, Recording Secretary; P. H. Wright. Senatobia, Corresponding Secretary; L. A. Smith, Port Gibson, Treasurer.

TENNESSEE DENTAL ASSOCIATION.

D. R. Stubblefield, Nashville, President; S. B. Cook, Chattanooga, First Vice President; W. W. Jones, Murfreesboro, Second Vice President; P. D. Houston, Lewisburg, Recording Secretary; J. L. Newborn, Memphis, Corresponding Secretary; H. E. Beach, Clarksville, Treasurer.

Place of meeting, Lookout Mountain, Tenn., last Tuesday in July, 1892.

THE POSTGRADUATE DENTAL ASSOCIATION OF THE UNITED STATES

WILL hold its Annual Meeting on April 29 and 30 next, at the Leland Hotel, Chicago, Ill.

Dr. W. C. Barrett, of Buffalo, N. Y., Drs. T. W. Brophy, Louis Ottofy, and others, of Chicago, will present essays and addresses. An interesting programme has been prepared, and a good attendance is expected. All members of the profession are invited.

Graduates of recognized dental colleges may become members by paying membership fee of $1, and dues for one year in advance, $1.

R. B. TULLER, *President;*

96 State Street, Chicago. L. S. TENNY, *Secretary.*

MEMBERSHIP IN THE AMERICAN MEDICAL ASSOCIATION.

This is obtainable, at any time, by a member of any State or Local Medical Society which is entitled to send delegates to the Association. All that is necessary is for the applicant to write to the Treasurer of the Association, Dr. Richard J. Dunglison, Lock Box 1274, Philadelphia, Pa., sending him a certificate or statement that he is in good standing in his own Society, signed by the President and Secretary of said Society, with five dollars for annual dues. Attendance as a delegate at an annual meeting of the Association is not necessary in order to obtain membership. On receipt of the above amount the weekly journal of the Association will be forwarded regularly.

ALABAMA STATE DENTAL ASSOCIATION.

The following announcement and programme of the twenty-third annual meeting of the Alabama State Dental Association, which has been furnished for publication by its worthy Secretary, J. H. Allen, D.D.S., is so comprehensive that it will hardly be necessary for us to add to it, except to heartily second his appeal to every dentist throughout the entire State to be present and participate in its deliberations. To those who have had that privilege in the past, it would seem superfluous for us to speak of the great benefit to be derived from their regular attendance. One entire day will be devoted to clinics; and prominent dentists who are members of the Association, as well as quite a number from other States, will give clinics, and this feature alone should call forth every dentist in Alabama. In this age of rapid advancement in the science of dentistry no dentist who desires to keep apace with its progress can afford to stay at home, and thus deprive himself and his patients of the valuable knowledge to be gleaned from this source. "Here a little and there a little" and each one who attends will either give or receive some information that will more than repay him for the time and money expended.

It is with pleasure that we present in this issue the photograph of the Association's popular young President, George Eubank, D.M.D., of Birmingham, Ala., a dentist of recognized ability and well known to the profession throughout the South. With him in the President's chair, aided by his well-selected corps of officers, the success of the meeting is already assured.   J. C. M.

ANNOUNCEMENT.

The Twenty-third Annual Meeting of the Alabama State Dental Association will be held at McDonald's Opera House in Montgomery, Ala., April 12, 1892. If you are a member, it is your duty to be present, if in your power to do so; if you are not a member, you are cordially invited to attend.

Yours truly,   J. H. ALLEN, *Secretary.*

Special Notice.—The Board of Dental Examiners will meet on Monday, April 11, at the same place. All those desiring to obtain license to practice in this State must be on hand. All must have license—none are exempt.

Hotel Rates.—The Exchange Hotel, $2.50 per day; Windsor Hotel, $2 per day; Merchants' Hotel, $2 per day.

Railroad Rates.—Be sure and purchase tickets on the L. and N. system, which is the only one to give us rates, which is one and one-third fare round trip. Be sure and get your certificates signed properly.

PROGRAMME.

FIRST DAY.

Meeting called to order by the President at 10 A.M.
Prayer.
Address of welcome.
Response on behalf of the Association, R. C. Young, D.D.S.
Annual address by the President.
Calling of the roll and collection of dues.

SECOND DAY.

Report of committees and discussion. Afternoon, papers on special topics. Incidents of office practice.

THIRD DAY.

Clinics will be given by prominent dentists in the different branches, with the latest methods and appliances. The entire day will be devoted to clinics. Those expecting to operate will bring their own instruments with them.

FOURTH DAY.

Finishing reports of Standing Committees. All unfinished business of the Association must be attended to. Election of Officers. Adjournment.

COMMITTEES.

Executive Committee.—George Eubank, D.M.D., Birmingham, Ala.; E. Wagner, D.D.S., Montgomery, Ala.; G. M. Rousseau, D.D.S., Montgomery, Ala.; J. C. Wilkerson, D.D.S., Selma, Ala.; J. H. Allen, Birmingham, Ala.

Committee of Ethics.—J. A. Hall, Chairman; T. S. Jordan, H. D. Barr, W. F. Slaughter.

Histology and Physiology.—E. S. Chisholm, Chairman; R. Y. Jones, T. P. Whitby, G. H. Hudson, A. A. Pearson, F. H. McAnaley, J. C. Chisholm.

Chemistry and Therapeutics.—J. A. Hall, Chairman; M. J. Lunquist, T. J. Jones, O. Y. Garnett, J. P. Allgood, J. S. Hill, G. W. Slaughter.

Pathology and Surgery.—J. C. Wilkerson, Chairman; J. E. Frazier, J. M. Clark, G. H. Taylor, W. F. Brown, S. W. Foster, T. M. Allen, S. W. Yarbrough.

Operative Dentistry.—A. Eubank, Chairman; E. Wagner, Charles A. Merrill, R. C. Duboise, S. G. Robertson, R. A. Rush, R. E. Watkins, R. G. Ragan.

Prosthetic Dentistry.—G. M. Rousseau, Chairman; C. P. Robinson, R. L. Allen, G. S. Cobb, W. M. Corley, T. L. Adams, G. M. Orr.

Dental Education and Literature.—A. A. Pearson, Chairman; W. A. Patrick, W. E. Proctor, R. C. Young, J. S. Hill, J. A. Frazier.

Clinics.—E. S. Chisholm, Chairman; J. A. Hall, T. S. Jordan, R. L. Allen.

Committee of Arrangements.—E. Wagner, Chairman; George Eubank, A. A. Pearson, G. M. Rousseau, J. H. Allen.

MISSISSIPPI STATE DENTAL ASSOCIATION.

We desire to call special attention to the Mississippi Association and to urge, through the columns of the HEADLIGHT, a full attendance on the part of the profession throughout the State. No dentist in the State can afford to absent himself from this meeting, which gives promise of being one of the most interesting and instructive in the history of the Association. A glance at the list of officers and efficient committeemen bespeaks in advance a profitable and enjoyable meeting. Add to these the name of Westmoreland as Chairman of the Committee of Arrangements, and no better guarantee of an enjoyable time, with lots of fun, could be given. Columbus is so well and widely known that it is needless to sound its praise. Suffice it to say that no city in the South, of its size, can excel it in beauty, climate, and whole-souled Southern hospitality. The following letters, from Messrs. Walker and Hilzim, will give some idea of the programme and the determination of the managers to insure a successful meeting.   J. C. M.

To the Editors Dental Headlight.

Dear Sirs: The Eighteenth Annual Meeting of the Mississippi State Dental Association will be held at Columbus, Miss, May 3, 4, and 5.

Officers.—Dr. D. B. McHenry, Grenada, President; Dr. A. A. Dillehay, Meridian, First Vice President; Dr. I. B. Rembert, Jackson, Second Vice President; Dr. A. A. Woffard, Columbus, Third Vice President; Dr. W. E. Walker, Bay St. Louis, Recording Secretary; Dr. P. H. Wright, Senatobia, Corresponding Secretary; Dr. L. A. Smith, Port Gibson, Treasurer.

Executive Committee.—Drs. W. W. Westmoreland and A. A. Wofford, Columbus, and P. H. Wright, Senatobia.

Clinics.—Dr. A. H. Hilzim, Chairman of Clinic Committee, is making every effort to have the clinics a special feature of the meeting, and with the aid of the other members of the committee, whose names we give below, insures the success of this important item. A first-class line of clinics will be presented both on the part of members of the Association and by prominent members of the profession from other States, who will demonstrate the latest methods now in vogue in dentistry.

Clinic Committee. —Dr. R. K. Luckie, Holly Springs; Dr. W. H. Marshall, Oxford; Dr. Morgan Adams, Sardis; Dr. W. E. Walker, Bay St. Louis; Dr. J. B. Asken, Vicksburg; Dr. J. B. Broadstreet, Grenada; Dr. T. O. Payne, Vicksburg; Dr. George B. Rembert, Macon; Dr. W. T. Martin, Yazoo City; Dr. George B. Rembert, Natchez; Dr. L. G. Nesbit, Aberdeen; Dr. T. C. West, Natchez; Dr. J. A. Warriner, Corinth; Dr. J. D. Killian, Meridian; Dr. J. Hack Rush, De Kalb.

I have not yet learned what the Executive Committe are doing in the way of local arrangements, but with Dr. Westmoreland as Chairman feel secure of having a good time generally.

Yours truly, W. E. WALKER, *Rec. Sec.*

To the Editors Dental Headlight.

Dear Sirs: The Eighteenth Annual Meeting of the Mississippi State Dental Association will be held at Columbus, Miss., May 3, 4, and 5, 1892.

The Board of Dental Examiners will meet to examine applicants on Monday, May 2.

Every thing possible will be done to make the meeting interesting and profitable to those who attend. The attendance promises to be large, as is usual with our gatherings.

A good clinic list is in course of preparation, which will be added to between now and May 3.

The usual reductions on the railroads and at the hotels will be given.

The following gentlemen so far have been heard from in regard to clinics:

Dr. R. L. Allen, Birmingham, Ala.: Crown and bridge work, demonstrating the use of a new instrument for measuring necks of teeth preparatory to crowning.

Dr. S. W. Allen, Birmingham, Ala.: Seamless knuckled gold crowns.

Dr. W. H. Morgan, Nashville, Tenn., expects to be present, and clinic.

Dr. George B. Clement, Macon, Miss.: (1) Treating and filling pulpless teeth, using material to suit cases presented. (2) Melting gold, silver, etc., with oil-burning furnace.

Dr. W. H. Marshall, Oxford, Miss.: (1) Will demonstrate the use of Marshall's matrix in filling. (2) Will show how the crowns and bands are made for his anchor dentures, and simplify the construction of all gold crowns.

Dr. Morgan Adams, Sardis, Miss.: Will do contour gold filling.

Dr. W. T. Martin, Yazoo City, Miss.: Glass inlay fillings.

Dr. R. K. Luckie, Holly Springs, Miss.: Gold filling contours.

Dr. George W. Rembert, Natchez, Miss.: Will do anything but bridge work, with a mouth blowpipe.

Dr. W. E. Walker, Bay St. Louis, Miss.: (1) flexible rubber plate (Morris process); (2) treatment and filling root canals; (3) extreme case of orthodontia, exhibition of models and bandages in connection with the same.

Dr. J. P. Broadstreet, Grenada, Miss.: Contour filling of soft and cohesive foil.

Dr. J. Hack Rush, De Kalb, Miss.: Contour gold fillings.

Yours truly, A. H. HILZIM,
Jackson, Miss. Chairman Clinic Committee.

D. B. McHENRY, D.D.S.,
PRESIDENT MISSISSIPPI DENTAL ASSOCIATION.
1891-92.

THE DENTAL HEADLIGHT.

VOL. 13.　　　　NASHVILLE, TENN., JULY, 1892.　　　　No. 3.

Original Communications. • • • •

TREATMENT OF SENSITIVE DENTINE.*

J. P. GRAY, M.D., D.D.S.

THE treatment of sensitive dentine is a subject of great importance to the dentist. In fact, I believe I can state without fear of contradiction that it is of more importance than all others, because we come in contact with it every day. It is necessary for us to understand the anatomy and physiology of tooth structure, to some extent, that we may the better understand the treatment of this subject. The entire human anatomy is susceptible of demonstration. For instance, the anatomist can take two portions of the brain, and by the use of his microscope can tell you to what portions of the brain they belong. The circulatory system is easy of demonstration, and the heart's every action is thoroughly understood. So is it with the liver and stomach. Their functions are easy of demonstration, and we have a perfect knowledge of their every action, and by the aid of the opthalmoscope we are enabled to demonstrate and to observe the inner working of the eye into the optic nerve, and by similar instruments the ear, with all its intricacies, is easy of exploration and demonstration. But it is not so with the organs of mastication. When the microscopist or anatomist attempts to demonstrate tooth structure, they are just about as much at sea and demonstrate with just as much certainty as the astronomer who points his telescope to one of the planets and says that it is inhabited, and when we ask him how he knows it is inhabited, he tells us that because it is surrounded by atmosphere sufficient to sustain animal life he knows it is inhabited. It is the same with the microscopist in his demonstration upon tooth structure. He tells us that a tooth is composed of certain elements; that it is an osseous structure, which I do not believe is altogether so. I believe that it is more of a nervous structure or a nerve growth than it

---

* Read before the Section of Oral and Dental Surgery, Nashville Academy of Medicine.

7

is an osseous growth. Therefore we should treat it as such. He will also tell you that it is composed of a nerve, dentine, cementum, and enamel, and that within the dentine we have odontoblastic cells, dental tubuli, or dental fibrillæ, and within these tubuli or fibrillæ are contained a fluid, and that he does not thoroughly understand what it is.

One writer, I have forgotten who it is, says that it is protoplasm. Now, gentlemen, we have a little outline of the anatomical arrangement of tooth structure. I do not believe, gentlemen, the profession has made any very great progress in the treatment of this disease. I have read to a considerable extent, and at length upon the subject, and find there has been practically no advance in the last decade or two. It seems to me that our treatment has been handed down to us as it were, from generation to generation, or from preceptor to student, until we have about the same treatment now that our fathers used forty or fifty years ago. In fact the subject has become so common, and every day the same thing, that we give it little or no attention, thinking there is nothing to be done. I believe, gentlemen, that if the same amount of scientific thought and investigation was given to the treatment of sensitive dentine that has been given to the treatment of pyorrhea alveolaris, or the scientific investigation that has been given to exostosis and similar diseases, we would have accomplished and obtained some remedy or instrument that would have been a complete pain obtundent. I do not know that I can give you anything new upon the subject, but will simply give you a few thoughts that you may be better enabled to discuss the subject more fully. The treatment has been a failure, the application of remedies have not been made in a scientific manner. Take, for instance, the application of carbolic acid and some of those well-known remedies, such as carbolic acid and the essential oils. Carbolic acid, as is well known, is an escharotic in its effect as well as an anæsthetic in its application. We have the medicine coming in contact with the dental tubuli. When these dental tubuli are filled with protoplasm, and when the drug is passed into the tooth, we have pain, for protoplasmic bodies are sensitive to mechanical and chemical action. The pain is produced by the drug or foreign substance, causing it to dispel the food, or the building material, and therefore rushing back upon its neighboring cell, and producing pain. The carbolic acid, when entering the cavity, produces the coagulation by its escharotic effect, completely sealing the mouth of the tubuli, there preventing the absorption or further taking up of the drug. Therefore, when we procede to excavate, we immediately enter sensitive dentine. I believe that it would be better in the case where we find it so sensitive that it is impossible to excavate without treatment, that we should seal the cavity without medication, excluding the air and secretions, and thereby giving nature a chance to heal and relieve itself. In the use of the essential oils, such as the oil of cloves, which is the standard remedy, we find that it is a little bit more compatible, and that these little protoplasmic bodies will receive it with less pain, but it does not serve the purpose for which we use it to any great extent. Why? Because it simply drives these little bodies back into the odontoblastic cells, and before the drug has time to enter the tubuli or penetrate the structure nature rallies and sends forth a troop of well-loaded protoplasmic bodies to fight against the enemy and throws up breastworks, as it were; or, in other words, these little protoplasmic bodies will throw over a cap. Like the bee, when its cell is broken, it immediately builds a new cap to protect itself and honey. Therefore, when we remove the treatment and procede to excavate, we immediately enter into sensitive material. You will ask me if that is not a success? How would you proceed to treat sensitive dentine? There are two methods whereby we can obtain complete relief; one which I will designate as the dry method, and the other by vaporizing. I do not believe we have yet produced an instrument constructed upon scientific principles for the proper treatment of sensitive dentine. The instruments we have are clumsy and are inadequate for the treatment. In treating by the dry method we simply have an air syringe, with the bulb of metal that we hold over the flame and heat, and then pump into the cavity the hot air, and fairly make the patient howl with excruciating pain— pain that is far more excruciating than that produced by sharp instruments. I do not believe that that kind of treatment is a bit better than the old carbolic acid theory or oil of cloves, but if we had a scientifically constructed instrument, we could, with heat, produce a complete obtunding of the pain. Take, for instance, a cylinder filled with compressed air, and as the air passes out of this cylinder into a tube, have it heated and hold the nozzle of the instrument away from the tooth and begin with the heat about normal, and gently heat the tooth, coaxing these little protoplasmic bodies back into the odontoblastic cells, and then turn the heat on until you get a high temperature, completely demoralizing, as it were, the protoplasmic bodies, or drying them up, and then you may cut without pain. If we use the vapor method, we may have an instrument constructed upon the same basis, and I believe there is an instrument constructed after this method, but I would use it a little differently. I would first dry the cavity in the manner in which I spoke of, and then begin with the vapor just at normal temperature, holding it gently away from the cavity, and put closer and closer until these little protoplasmic bodies expel the food, and are forced back into the odontoblastic cells, and thereby heating the vapor hotter and hotter all the time, until finally we bring the mouth of the instrument into the cavity, and upon that have the sealing plug, that the cavity may be sealed, and the force of the air drive this vapor into the tubuli, completely filling the tube with the remedy. By this manner I believe, gentlemen, we can produce a complete obtunding of the pain.

ARRANGEMENT OF ARTIFICIAL TEETH.

BY R. M. WALKER, D.D.S., CHICAGO.

In view of the many inartistic efforts at the arrangement of teeth which are brought to our notice, a few points on this important branch of dental art may not be out of place. A set of artificial teeth should not only feel comfortable in the mouth, and be of service in masticating the food, but should be so arranged as to hide their artificial nature, and this, above all other things, requires the close attention and artistic ability of the dentist. Unfortunately, too many of our brethren dislike the work, and say they wish they were never called upon to make a plate. If they really mean what they say, they must either be poor workmen or take no interest in the work; in either instance, the patient must suffer. One great mistake is in trying to have a fixed rule for the arrangement of teeth. There are, comparatively speaking, no fixed rules in dentistry. Each case must stand or fall upon its own merits, and each should be judged without any regard to any other.

The first point to which I shall call attention is the selection of teeth. We will speak first of what we have most commonly to deal with—upper sets with natural lower ones—the teeth should be the same color as the natural ones, of a size to form what we call a good articulation, the point of the superior cuspid fitting between that of the lower and first bicuspid; next they should be the same shape and *general appearance* of those remaining. A little space judiciously applied will, in a majority of cases, tend to a more natural appearance, and in this way a good articulation may be obtained with teeth much smaller that would be required were they set close together. The most space should generally be left between the centrals and laterals; the latter should be a trifle shorter than the former, and just a trifle out of line; not enough to be called an irregularity. Unless the case demands it, the cuspids should be made prominent, particularly at the necks, thus hiding the bicuspids, which should scarcely show from the front. We see so many cases where the rule has been reversed, and apparently an effort made to make the bicuspids more prominent than any of the others. These points must all be modified to suit each individual case. This can only be intelligently done by trying in the mouth on wax or other temporary plate; then the little peculiarities of the case may be studied and changes made to accommodate them. The bicuspids and molars must of necessity form a good occlusion with the opposing ones, but even here a little judicious grinding may make a great improvement. The teeth must, of course, be the proper length; the lip, when at rest, is usually a safe guide. If they are too short, the jaws close together too much, giving the appearance of nose and chin trying to meet; if too long, of course they are just as bad.

Another point which is often overlooked is giving the teeth the proper slope from the gum margin to the cutting edge. They are too often made nearly perpendicular, when the points will generally stand a slight outward slant. This must be governed by the shape of the mouth and position of the lower teeth. Avoid what is called the horse shoe arch, the widest point where the bicuspids would come, and narrowest at the second molar. By just reversing the order you can come much nearer the desired arrangement. It is sometimes necessary to place the six anterior teeth nearly straight across in order to restore the contour of the lip, and this regardless of the shape of the ridge, which, though often a safe guide, is not infallible.

In arranging upper and lower sets more liberty may be taken in some directions and less in others; a little variation in shade is not so noticable, but little, white teeth in the mouth of a large, dark, swarthy person would not suggest a very close study of nature. More care is required in selecting the size. Not having any to go by, you must have each tooth occupy the place which you are led to believe the natural one did, spacing or lapping as the case may

require. Not having any natural teeth to match gives you more
liberty with the bicuspids and molars. The former may be placed
in so as not to show, the latter turned outward and upward. In
some cases the upper teeth must protrude in order to hold the upper lip in its normal position without forcing the lower to an abnormal one, while others will require being set nearly or quite "on end,"
as we say, for comparatively the same reason.

I have written this paper with a view to the exclusive use of
plain teeth, as it is with them only the proper variations of arrangement can be obtained; and were a little more skill used in this respect, it would do more to overcome the prejudice in some minds
against them. There are very few cases where the teeth are properly arranged, in which the artificial gum shows enough to be noticeable, and as this is the only advantage the gum teeth have, I think
in view of the better occlusion and more perfect results obtainable
by using the plain, it is our duty to throw our artistic skill into
this branch, and thus overcome this foolish prejudice. I do not
wish to be understood as making the fit and working qualities of a
set of teeth subservient to the appearance, but I will say that many
times were the patient perfectly pleased with the appearance first
the fit would be a foregone conclusion.

FITTING COLLARS TO TEETH.*

W. K. SLATER, D.D.S.

IN order to properly treat the subject under consideration, at least
a fair knowledge of the various forms of the roots or rather necks
of teeth are necessary.

Strange it seems, but to within the last year or two there was not
in general circulation a complete anatomy of the teeth; but recently
Dr. Black has written a very complete work upon this subject, to
which I refer you for a detailed description of them.

To attempt to describe the different methods of making collars
would be an arduous task indeed, as well as an uninstructive one to
you. And so if there are any better methods for the particular
cases herein described we would like to know them.

---

* Read before the Section of Oral and Dental Surgery, Nashville Academy of
Medicine.

Of course a paper upon a subject like this is necessarily to many a mere repetition of oft-practiced methods.

I suppose the first in order would be one of the single-rooted teeth, to which say a porcelain-faced crown is to be attached. We would, before attempting to trim the root any, shorten it almost as much as necessary, and then with safe-sided files, sandpaper, disks, corundum points, and the various forms of scrapers give the root the desired shape. With No. 34 soft iron wire now take an accurate measure of the root. From this a band can be made, and if we have remembered the shape of the measure as it came from the tooth, it can be made to approximate the desired shape with slender pliers. This can now be forced upon the stump, and such portions as would go too far under the gum can be cut away, so that it extends beneath the gingivus at every point alike; and if the root has been properly shaped, the extreme edge of the crown will fit the closest. The stump and band may now both be ground down together, using a stone that will cut the proximal sides as well as through the center of the tooth. The reason for not shortening the stump to the necessary height at the start is that the difficulty of taking an accurate measure would be greatly increased.

The procedure in fitting the band for an all gold crown is necessarily very much the same as described, except that in most cases it is well to contour the band until there is sufficient knuckle to touch the adjoining teeth after the crown is finished, thereby preserving the interdental space.

One word more about the preparation of a certain class of teeth for gold crowns. I refer to those which need very little or no trimming, as they taper from gum to grinding surface. In such cases something more than the thickness of the collar should be cut from the buccal surface of the tooth, and after the collar is properly fitted, small v-shaped pieces may be clipped from the end that is to be closed, and the edges of the gaps brought together so as to touch. These will easily solder up as the grinding surface is being attached. In this way glaring gold, which is conspicuous enough at best, may be made to appear less so.

The method of taking a piece of gold and pinching it around a tooth, and depending on this for a fit, will reward us with a fit about one time in ten, or maybe not quite so often.

In casts of molar teeth, where there is enough recession to expose the depression between the roots, a groove may be cut from depression to grinding surface that will permit us to bend in the collar

with pliers out of the mouth, and then put it in place. If the teeth
are too sensitive for this, 24 K. gold may be used for the band, as
this can the more easily be burnished into the depressions as the
crown is being set.

Would that I could always make such accurate fits, like some
whom I have heard discuss this subject, that a piece of silk would
not catch upon the edge of the collar. There are a great many
bands that go too far under the gum, and in my short experience it
has been clearly demonstrated that if we err either way it would
be better for the tooth if the collar stopped just a little short than
to go too far. The farther we get beneath the gum the harder it
is, as a general thing, to obtain good adaptation, and the more liable
are we to have the band standing off at some point, thereby becoming a constant source of irritation, and as a consequence, in a short
time there is the swelling and tenderness characteristic of the pinched
gum pedicle, and we being unable to relieve this permanently without
removing the cause, our work is likely to prove to the patient to be
anything but a joy forever.

In all cases where collars or partial crowns are used for the support of bridge work, the less of the tooth that is covered, in order
to secure the necessary strength, the better it is for that tooth.

It seems to me that too much cannot be said in condemnation of
letting such collars extend beneath the gum like ordinary crowns,
to become, in a short time, a source of irritation by pumping up and
down upon the teeth. For instance, a typical case would be this;
A gap upon the upper jaw, extending from the first molar to a
sound cuspid. Upon the attachment to the molar all are agreed,
but when it comes to the cuspid there is a diversity of opinion. It
seems to me that a perfect-fitting collar, slightly broader upon its
lingual than its labial surface, extending just to the largest portion
of the tooth, is the thing indicated in almost every case. These
may be made by burnishing a piece of gold directly upon the tooth
which has been previously pinched around it by flat-nosed pliers,
taking out small pieces where it is inclined to pucker. But what
looks to be a better way is to make a perfect die of the tooth, and
around this shape the collar to fit.

Associations. . . . .

PROCEEDINGS OF THE EIGHTEENTH ANNUAL MEETING, COLUMBUS, MAY 3–5, 1892.

THE Mississippi State Dental Association convened in eighteenth annual session Tuesday, May 3, 1892. The meetings were held in Concordia Hall.

The meeting was called to order by Dr. W. W. Westmoreland, Chairman of the Executive Committee, and opened with prayer by Rev. J. L. Johnson, pastor of the Baptist Church of Columbus.

Dr. Westmoreland then addressed the Association briefly, saying that feeling doubtful of the ability of the resident dentists of Columbus to fittingly welcome the State Association, he had deputed that pleasant duty to Col. L. M. Meek, a prominent and eloquent lawyer of Columbus, whom he had the honor and pleasure of presenting to the members of the Association. Col. Meek then delivered the following

ADDRESS OF WELCOME.

Mr. President and Gentlemen of the Mississippi Dental Association: I was no little astonished, a few days ago, when I, a lawyer, was called upon by one of the most distinguished dentists of your Association with the announcement that I had been selected to deliver the welcoming address to your body upon this occasion. But nothing daunted, and ever feeling a lively interest in whatever adds to the culture, advancement, and elevation of any one of the learned professions, it is with sincere pleasure and unalloyed gratification that I undertake the task and now extend to you, one and all, individually and collectively, a warm and hearty welcome to the hearts and homes and firesides of the residents of our beautiful city. It is no "scandal in disguise" when I speak in words of laudation of the beautiful little city within whose portals you sit to-day, for such words are not "undeserved." In all the appointments necessary for the upbuilding of a large and prosperous city Columbus has been bountifully blessed and gifted by Providence; and her men and women too are every day becoming more actively alive to the cultivation and advancement of those enterprises which tend to make her worthy of the historic name she bears, and to render her people great, prosperous, and happy. To a climate as genial and delightful as any portion of our country, to a healthfulness unsurpassed (the medical statistics showing that it is the second healthiest city of its population in the entire Union), to an active, energetic, and thriving population, its men being cultured, refined, and gentlemanly in all the relations of life, and its women gentle, fascinating, and beautiful, it adds the finest and most thorough institutions of learning in the State, manufacturing establishments of no mean grade, together with houses of worship of all denominations pointing their spires to the skies, manifesting the elevated Christian character of her people. In exterior beauty, too, its private residences fringed with flower gardens that laden the air with perfumes, its ornamental and evergreen forest trees waving their wings of beauty as they are kissed by the springtime breezes and bathed in the mellow light of this May day sun. It might be well said, in the language of an eminent scholar while roaming around our city, his mind in a poetic mood, "Truly she looks like a little city that had gone on a picnic and camped for a time amid forest trees and flowers." This is the character of the people, and this the little city which deputes me to extend to you a hearty welcome to-day. When you leave us and shake the dust from your feet, if perchance they catch any, we doubt not that you will bid us good-bye with a sigh of regret and feel like coming to see us again. But "something too much of this."

But what brings you here to-day? Is it simply to enjoy the good things that may be set before you, to shake hands with our stalwart men and bask beneath the smiles and listen to the honeyed words of our fascinating women? These are all well enough, but as I understand it you are here for higher, nobler, and more elevating purposes: to associate yourselves together as professional brothers, to give the experiences of the past, to consult and advise, exchange and interchange sentiments and views upon the scientific principles that lay at the root (for your profession has to deal with many roots) of your profession; in short, to improve yourselves and to lay down rules and enunciate principles for the benefit of the junior members of your calling. If these be your purposes, they are "laudable ones," which those who come after you may well follow you as their guides and "fear no evil."

Associated effort in any occupation, calling, or profession is the great lever and hand-guide for the accomplishment of good. There is scarcely any enterprise on earth, which is considered worthy of the interest of a free people, but lifts its drooping crest and rises to a lofty height of prosperity beneath the genial influence of associated effort. The sacred councils of wisdom and of peace which are to soothe the expiring bosom and to raise the emancipated spirit to realms of eternal light are propagated and planted under the benign wing of associated effort. Do you wish to establish a road, to cut a canal, or open some stream which is to bear upon its palpitating bosom the precious fruits of the farmer's annual toil, to build a railroad or a dummy line, as we have recently done, the laudable undertaking must be accomplished by associated effort. Yes, the chief magistrates of 60,000 of freemen, our own grand and matchless government, are elevated to their high places and hurled thence again; candidates for popular favor and distinction are raised to general acceptance and effectually put down; the causes of temperance and religion, with their benign and peaceful influences, and of every other cause within the expanded range of human thought and human care, are both advanced and depressed by the efficacious influence of public meetings, conventions, and associations. Is it to be expected then, that a profession based upon the loftiest principles of art and science is to form an exception to the extended catalogue of human interests and force its way to solid strength and permanent prosperity without the combined power of associated effort? Your annual convention, then, gentlemen, held in different sections of the State, is to be highly commended, and from the periodic association and combined efforts and experience of its members, where the thought of your ablest minds may come in contact with each other, explaining and eliminating their individual experiences, is sure to advance your profession to the highest point of respectability, distinction, and honor.

As deplorable as it is, gentlemen, it is nevertheless true that dentistry has not until within comparatively a recent period been recognized by the world as deserving to rank among the learned professions. The world long looked upon its practitioners as a set of charlatans and pretenders, and sneeringly called them "bone thumpers and tooth carpenters." As bloodletting was formerly beneath the dignity of the scientific physician, and barbers performed this operation, thus originating the barber's pole, their sign, with three colors—the blue line representing the veins, the white the skin, and the red the blood—so the extractors or fillers of teeth were looked upon by the world and by the members of that learned profession as bone carpenters and tinkers. But thanks to the better spirit of a better age, and to the learning and ability of your predecessors, dentistry is now universally recognized as occupying the position which its merits and usefulness deserve. In truth, I have been astonished myself by the reading and investigation which I have given it in the last few days to find the important pinnacle to which it has risen.

From the earliest periods of which history gives any account, attention was directed to the means of preserving and improving the beauty of the teeth. The ancient Hebrews appreciated their importance in giving form and expression to the countenance, as when Jacob, blessing Judah, says, "His teeth shall be white with milk," and Solomon compares a fine set of teeth to a flock of sheep even shorn. In the time of Heroditus the art of dentistry was practiced in Egypt as a distinct branch of surgery. History tells us that in the ancient Egyptian tombs artificial teeth of ivory or wood were found by Belzoni and others, some of which were fastened upon plates of gold. It is also stated that the teeth of mummies have been found filled with gold. Artificial teeth are alluded to by several of the Greek and Latin poets, as Ovid, Horace, Martial, et al. From an article in the American Cyclopedia I find that from advertisements in the newspapers of 1703, the practice of the art, as it was called, of making teeth and cleaning them appears to have been in the hands of silversmiths and jewelers. In 1826, a work called, "The Principles of Dental Surgery," by Dr. Leonard Koecker, who had practiced dentistry from 1807 to 1822 in Baltimore and Philadelphia, appeared in London and fully established the claims of the art to take rank as a distinct science. Many able works rapidly appeared in Europe, and year after year attracted the attention of the *litterateur* and the scholar. The practice of dentistry was introduced into the United States by Le Mair, of the French forces which joined our army under Lafayette during the Revolutionary War. An Englishman named Whitlock also commenced the practice soon after the arrival of La Mair, and in 1788 John Greenwood established himself in New York, the first American of this profession. In 1790, and again in 1795, he carved in ivory an entire set of teeth for George Washington, our first President. In 1820 the number of practitioners in the United States was probably a little over one hundred. Ten years later there were about three hundred, and in 1858 there were four thousand. The "American Journal and Library of Dental Science" was the first important work established in this country at Baltimore in 1839, and from its issuance and circulation dentistry was fully recognized as one of the learned professions. The impetus thus given to this noble and indispensable profession has moved on with unexampled rapidity, until to-day its members are almost countless; its magazines, periodicals, journals, and other publications are being weekly, if not daily, issued from the press; dental schools, and colleges are springing up in nearly all the states of the Union, laws have been passed for the direction, regulation, and control of those desiring to enter its ranks, and the prospects before it are gratifying and encouraging.

By an act of the Legislature of our state in 1882, prepared no doubt by some of the leading men of your profession, safeguards are abundantly thrown around the profession, and the public cannot be imposed upon by incompetent men if the "Board of Examiners" perform their duty faithfully and well. Blacksmiths, silversmiths, and jewelers must stick to their callings and not tamper and tinker with that of which they know nothing. Quacks, pretenders, and charlatans must and will stand aside if this Board, appointed by the Governor, of men eminent for their learning and ability, meet the demands of a confiding public.

You will pardon me, gentlemen of the Association, for this hasty and imperfect glance at the rise and progress of your profession, as it is intended more for the benefit and information of outsiders than for yourselves. You are familiar with all that I have said in relation thereto and more.

I have read your admirable Constitution, which seems to come up to all the demands and requirements of your Association.

But I must bring this rambling address to a close, as I have already detained you, I fear, too long. In conclusion allow me to say that in extending to you this welcome on behalf of the citizens of Columbus, may this the eighteenth annual meeting of the Dental Association of Mississippi prove a source of gratification, information, and pleasure to you individually and collectively! and may the welcome which I now extend prove to be so warm and cordial, and your stay among us so pleasant, that at no distant period you may meet beneath the shade of our trees again!

At the conclusion of this noble tribute to the rise, progress, and scientific standing of the dental profession, Dr. Westmoreland introduced Dr. R. K. Luckie, of Holly Springs, the retiring President of the Association, who, on the part of the members of the Association, made the following

Response.

Mr. President, Ladies, and Gentlemen: The State Dental Association has been cordially, eloquently welcomed to this charming city, the " Queen City " of our eastern border. We have listened with delight to these assuring words of greeting, and we feel that by them a new obligation has been imposed upon our noble profession to march to the front rank of science and mechanical appliances to meet the advancing wants of civilization.

To be selected to represent so important an Association as ours at such a meeting, and in response to the welcome from the fluent representative of such a cultured city, is both an honor and an embarrassment. In behalf of the Mississippi State Dental Association I return to you our heartfelt thanks for your cordial greeting. At the time Columbus was selected as the place for our next annual meeting, there was general rejoicing and hand shaking; the vote was unanimous—not a dissenting voice—for we knew to come to Columbus meant a royal time for us all, and I can assure you that we are very glad to be here and to accept the hospitalities of your city. Whenever or wherever we meet we try to conduct ourselves in such a way that when we depart for our homes, it is a source of regret to the people among whom we have met, and I trust such will be the case at Columbus this year.

It is a fact that all true Mississippians are proud of Mississippi—and why should they *not* be, with all of her vast undeveloped domain inviting the agriculturist with promise unequaled, with a climate almost matchless, with a soil as fertile as the valley of the Nile, with millions of acres of virgin forests, with schools and colleges, male and female, churches, Sunday schools, and missions all in a prosperous condition? It is also a fact that all true Mississippians are proud of Columbus, one of the gem cities of our great commonwealth. This is a beautiful city to look upon ; a city of fragrant flowers, of taste and refinement; a city of beautiful homes, a city of great men and noble Christian women ; a city of historic associations and traditions. It stands in perpetual memory of the greatest, most adventurous navigator that ever directed a vessel over the ocean's waves. Christopher Columbus was an heroic man; a man of scientific, philosophic, religious faith ; a man of indomitable perseverance under intimidating difficulties, who could stand before kings and advocate a project that was derided as visionary, and in pursuance of it stand amid shrieking, curdling, roaring waves and mutinous sailors. This city, resting so beautifully and quietly here on the banks of the Tombigbee, is a fit memorial of such a life. There are many things here to impress the observing visitor. Your broad, beautiful, graded, well-drained streets tell us that sanitary science has an advocate here. You are now building a dummy line. Your public schools and machine shops are in a prosperous condition. In your open yards *Camelia Japonicas* grow and evergreen oaks mark your lovely sidewalks. Your church spires, pointing heavenward, show that you are a God-loving and a God-serving people. Your unceasing artesian wells of purest, sparkling water are inexhaustible in supply, and their value is beyond estimation. Their influence can be seen in the clear complexion of your citizens, and we seem to hear their ripple in the merry laughs of your children and see their purifying influence in the clear eyes and peach-blossom cheeks of your lovely girls. Time would fail me to tell of all the enterprises here which evince the spirit and progress of this people. I will mention one other. The South is and has been the great producer of raw cotton, but the manufacture of it has for generations past been left to Old England and New England. Thereby we have left in the hands of others the largest profits that are to be realized on our own staple. Here in our own midst are all the powers of nature—water, steam, air, human hands—all that is needful to its proper transformation to meet the varied demands of society, all that is needful of skill to develop articles of taste and ingenuity; but as a rule we have not employed them, and have lost millions annually. I am gratified that in this city you are setting the example of manufacturing cotton goods, and I am gratified especially to know that your mills are making a splendid success as well as an enviable reputation for high-grade products. May your worthy example be followed by scores of places till the whirl of cotton machinery shall be heard in every available place and the smile of prosperity overspread our state.

This town is the home of the I. I. & C., one of the state's brightest ornaments. It is said that Mississippi has in her soil no gold, no silver, no copper, not much iron ore, no rubies, no diamonds, no sapphires nor emeralds, but no state of the Union, no country in the world, can show purer jewels, jewels of more priceless value than we have in our pure and lovely girls. What a privilege that you have in your midst the institution established by the state for polishing these jewels, who will go forth to beautify Christian homes and brighten human hopes! With what knightly chivalry you will defend this noble institution! and with what enlightened zeal you will foster its glorious purposes! It is a pleasure as well as a privilege to hold one of our meetings in such a city as this.

We feel proud of our State Association, for we realize that it is one of the best and most active dental societies in the South. On the 21st of April, 1875, its star arose above the horizon, and has gradually ascended till now its influence is seen and felt throughout the entire state and its fame gone to the four quarters of the continent. It has stood the test for eighteen long years, and is honored because of its active membership and for the good work it has done. It was through the influence of this Association that wholesome dental laws were promulgated, thereby protecting the people from the ravages of ignorance and empiricism. Dental education has been encouraged, our members stimulated to renewed efforts in scientific investigations as well as introducing new methods of work in the different departments. The association of men together in an honest cause is something grand and noble. We invariably feel stimulated and benefited by these annual gatherings, and it is not unreasonable to say that those who patronize us are likewise greatly benefited.

The dental profession is moving on grandly in the various departments, and the progressive dentist of to-day is prouder of his calling than he has ever been. Advances are being made all along the line: scientific investigations are more thorough; new discoveries are being made; new instruments, materials, and appliances are being introduced: the standard of dental education has been elevated; and there is a constant desire for higher, nobler, and better things.

It has been said by the highest medical authority that the teeth and jaws from childhood up are as important a part of the human organization as the eyes, the ear, or any other part of it, and that a knowledge of how to treat their diseases, their accidents, and their injuries depends on the same application of scientific principles as in all other departments of medicine; and to-day the dental profession stands on the same plane with pathology, laryngology, ophthalmology or any other ology that may be named as a part of the great, broad field of medicine. We are getting up higher; good things have come; better things are *yet* to come—better not only for the profession, but for the public at large—and I can truthfully say that no one can foretell the future of our loved and honored profession. Again I thank you for the warm welcome you have given us. .

Dr. Westmoreland introduced the President of the Association, Dr. D. B. McHenry, of Grenada, who delivered the customary

ANNUAL ADDRESS.

Gentlemen of the Mississippi State Dental Association : Another year has been added to the past, whether for weal or woe to the individual or collectively, time only can demonstrate. We are apt to look adversely upon the old, whilst hope with all of its alluring arts beckons us on to new fields of thought and labor. Advanced thoughts and practical ideas have each their followers, and yet the ideal is never reached. Criticisms and critics have their place in life's history, yet they too are doomed to vanish like their predecessors when their work has ended.

Out of the labyrinth of ideas propounded, the good must be separated from the bad and individually applied, as the success of one may prove abortive in the hands of another. So all along the line of mechanics and operative, he who is most successful should fill the place of the teacher and impart to others less endowed those difficult problems of application and success, laying aside selfishness and individuality.

If we as advanced thinkers and workers would free our minds of criticisms and look to our interest as a whole, many seemingly crude ideas of the individual could be useful to the many, and often lay the foundation of what might prove some great and useful idea.

The rapid advance of our profession stands second to none, and he who would keep pace must take hold of the problems himself, and wait not for his neighbor.

I would call your attention to the great oversight we are laboring under at our annual meetings, and take some concert of action looking toward a change, adopting the rules of some of our sister associations which have worked out to the good and interest of their meetings. I refer to the reading and discussing of papers. It is often the case that questions arise in the presenting of these papers which would prove of great interest to the many if they had had time to examine the questions at issue and thereby intelligently discuss them. I note that papers to be read are handed in several weeks before the meeting and printed copies forwarded to the different members, which gives them time to examine more fully the matter therein and come prepared to make the paper of more interest to all.

I have in the past noticed many good papers read which should have met with a different reception but for the want of proper time on the part of members to intelligently discuss them. I hope that during this meeting some action will be taken looking toward a material change in that direction.

I would also call the attention of the members to an unprofessional practice which I am sorry has been the means of alienating many of our profession and causing ill feeling where brotherly love should predominate. I refer to the disposition on the part of a few to lower the profession by cutting rates and thereby dragging the honored calling down on a par with the day-laborer. It is a well-known fact that we are our own architects, and if we would enjoy the praise and respect of the community in which we are laboring, it is with ourselves to elevate or depreciate our usefulness. If there is anything unprofessional and at once stamping its author with shame, it is to offer to do work at a less figure than a fair remuneration will allow; and I believe as an Association we have it within our power to regulate between our members the practice whereby this demoralization of fees can be regulated.

At the last meeting the Association authorized your President to appoint in each judicial district one of its members to see that the law governing the practice of dentistry was enforced. This I did soon thereafter, and hope it met the approbation of the individuals so appointed, and that they will continue to exercise a watchful care over the interests of our profession.

It is my opinion that during this meeting the law governing the practice of dentistry passed by the last Legislature, which antagonized our Association and our Board of Dental Examiners, should be fully discussed, and if possible such changes brought about in the law at some future time that will harmonize more fully the two bodies.

Finally, gentlemen, I am informed that at the coming "World's Fair," to be held at Chicago next year, it is contemplated by the dental profession of the United States to show to the visitors the rapid strides that our profession has made in the Western Hemisphere during the last decade, and they ask that every State Dental Association appoint a special committee to assist thereby in the great undertaking, both mechanical and operative. I do not think that our own state can be laggard in this great work, and will during this meeting, if it meets the approval of this body, appoint a committee to coöperate and look after the interest of our profession in this state.

At the conclusion, Dr. McHenry took the chair and called the meeting to order.

On motion of Dr. Westmoreland, the Lowndes County Medical Society, now in session in Columbus, was invited to seats in the Dental Association, with privilege of discussions.

On motion of Dr. Hilzim, the roll was called, and the following members responded to their names and paid their dues: D. B. McHenry, Grenada. President; A. N. Dillehay, Meridian, First Vice President; A. A. Wofford, Columbus, Third Vice President; W. E. Walker, Bay St. Louis, Recording Secretary; P. H. Wright, Senatobia, Corresponding Secretary; W. T. Allen, Amory; C. C.

Crowder, Kosciusko; George B. Clement, Macon; J. O. Frilick, Meridian; A. H. Hilzim, Jackson; R. K. Luckie, Holly Springs; W. T. Martin, Yazoo City; W. H. Marshall, Oxford; K. S. Moffat, West Point; J. H. Magruder, Jackson; L. G. Nisbet, Aberdeen; T. O. Payne, Vicksburg; C. R. Rencher, Meridian; J. Hack Rush, De Kalb; H. T. Segrist, Brandywine; E. E. Spinks, Meridian; J. A. Suber, Winona; J. A. Warriner, Corinth; Charles E. Ward, Shubuta.

The following members were absent, but sent their dues: J. P. Broadstreet, Grenada; O. B. Hilzim, Vicksburg; J. D. Payne, Vicksburg; I. B. Rembert, Jackson; L. A. Smith, Port Gibson.

In the absence of the Treasurer, Dr. L. A. Smith, Port Gibson, the Recording Secretary was made Acting Treasurer.

Dr. A. H. Hilzim announced clinics at 2 P.M., and the next session at 8 P.M.

Dr. McHenry introduced Prof. Francis Peabody, of the Louisville College of Dental Surgery, an honorary member of the Association, who proceeded to make some remarks explanatory and descriptive of an instrument or appliance he expected to use in clinics, in the treatment of root canals. It consists of a small glass cylinder inclosed in a metal tube, with a rubber bulb attached as in the ordinary syringe. The cylinder is filled with crystals of iodoform and held over a lamp to sublime the iodoform. A vapor is formed which, by pressure on the bulb, is forced through the nozzle into the root canal of a tooth. This application of the fumes of iodoform accomplishes two effects—the application of heat and of an antiseptic, by which all forms and modes of germ life are destroyed. One great objection to its use is the unpleasant odor of the drug, with the fumes of which the office is filled. Another objection is the damage done to instruments. In the Infirmary of the Louisville College of Dental Surgery it has been used to great advantage in from three to four hundred cases of teeth which would otherwise have been consigned to the forceps. He had subsequently extracted teeth into which three or four puffs of the vapor had been introduced, and on grinding them down on the lathe, had found from one-half to two-thirds of the tooth substance perfectly permeated with iodoform, through the tubuli to the peridental membrane. By precipitation of crystals from the vapor, the root of a tooth may be filled to the very apex with solid iodoform, which being soluble only in ether or chloroform, must make a more reliable filling than the plastic materials now in use. In one

8

case, a tooth which was very sore, very loose, with a fistulous opening, was filled after one treatment in this manner, with perfect satisfaction. He has had only one failure in two years.

On motion, the Association adjourned to 8 P.M.

NIGHT SESSION.

The Association was called to order by the President at 8:30 P.M. Dr. Clement reported a paper from Dr. W. E. Walker.

The reading of the minutes of the last meeting was dispensed with, as they had been published and distributed. The report of the Section of Operative Dentistry was called for; Dr. George B. Clement, Macon, Chairman.

Dr. Clement stated that he had prepared some diagrams illustrating some defects in tooth form and structure as primary causes of decay, on which he would make some remarks, though he had not prepared any paper on the subject. He said that the subject of operative dentistry had been treated until threadbare. There is really nothing new or interesting to be said about it. The mouth and its organs should be kept pure and clean, because of its importance as the entrance to the alimentary canal. The teeth are the hardest parts of the human organism, or of any other organism, and yet they are the first to decay.

The brain, which is the softest of all the tissues, is the last to decay, resisting after even the most extreme emaciation of the muscular tissues. This is perhaps due to the fact that it is mechanically better protected. It is rare that we find a perfect set of human teeth. There may be thirty-two in the jaws, but they will seldom all be perfect. As dentists we are called upon to repair their defects and restore their utility. If the laity understood this importance, they would be more careful of them. Patients come to us for the first time at the age of ten, twelve, and fourteen years. It is then too late, and the blame lays at the door of the parents. If the teeth were properly cared for from infancy, parents would be put to much less expense and their children would experience much less discomfort.

The first chart displayed by Dr. Clement represented the lingual surface of the central incisors. When do they decay? In the center of the lingual surface there is a fissure where decay usually begins. This may be caused by germs or by acids; the two theories are really one and the same. This fissure is a natural defect in enamel structure, and it should be cut out and filled before decay sets in. We don't want to wait until it is decayed and aching. The agents to which decay is usually attributed are not its real cause of decay. It is due to these natural defects. In another chart showing the labial surfaces, there is seen a defect or crevice between the cementum and the enamel. This will prove a cause of decay. Why do the approximal surfaces decay? Because of abrasion by pressure and the lateral movement of the teeth, causing a defect which invites decay. In proof of this statement Dr. Clement passed around three bicuspids embedded in plaster; one having a large cavity, one slightly decayed, the third having only a fissure. If the first or the second had been filled when in the condition of the third, the ravages of decay would have been forestalled. From this primary cause, the natural defect, secondary agents get in their work. Filling a fissure is an aseptic operation; a natural defect is not a septic lesion. If the operation is deferred, then we must employ antiseptics. In the molars we have the same lateral movement, producing these lesions and inviting the secondary agents, fermenting food and acid formations. In bicuspids there is the fissure and the two approximal points of attack. In the molar fissures there are from three to five tiny points which invite decay. Cutting out and filling these fissures is now an accepted operation. In the incisors, if the teeth touch at full length, as shown in a diagram of straight-sided square teeth, decay will be found all along the line of contact; if pressed apart with rubber, this will readily be seen. This is a tooth form. A perfect tooth represents a perfect organism; and when we find perfect teeth, we will find all the other tissues more than usually perfect. Another chart showed incisor teeth with a very short root and touching only at the incisor edge like an inverted truncated cone. (*Reporter.*) From grinding together with a lateral motion a spot of black decay will be found at these points of contact. With this class of teeth, with short roots and broad at the cutting edge, we always find low vitality. Another chart showed four incisors, broad at the base, with narrow incisor edge, touching only at the cervical margins, where we will always find decay, though food will lodge much higher up. In the germ theory it is immaterial whether we consider the germs as the result of acids, or the acids as the result of germ life, the germ theory is correct in either case, and the work is done in these crevices. In the soft tissues alkaloid ptomaines invade the territory with poisonous results. Whenever continuity is interrupted, decay is invited. We may introduce the hand into septic matter and no harm will be done if the surface is sound; but if there is an abrasion, there will be infection, and it is the same with the tooth. Microörganisms are present waiting for an opening, when they may make an attack, and they seize upon the slightest defect, so minute often that we are not able to detect it. They do not eat up the tooth structure; they do not feed upon living tissue, nor upon decomposed tissue, but upon dead matter. Serum is the pabulum upon which they feed—not putrid matter, not living tissue, but dead matter. They in turn die and break down, and furnish the products which destroy tooth structure. These organisms form the ptomaines, which weaken the vasomotor nerves. Blood flows there and oozes out in thin serum.

In a perfectly formed tooth the external shape indicates the form of the pulp cavity. In the narrow-base tooth we are very liable to strike the pulp, and in excavating we must be very careful to avoid the horns of the pulp. Whenever we find these slight defects, pits, or fissures, we should excavate at once, and fill as perfectly as possible with gold. If it is a mere abrasion, it may be sufficient to separate and polish.

The discussion was opened by Prof. William Crenshaw, of Atlanta, Ga.

Prof. Crenshaw said that he had been much interested in the lecture of Dr. Clement. He considered his conclusions correct in regard to the shape of the teeth and location of the pulp, and also correct in theory. He said: "Over our way we are too busy to investigate causes; we only have time to stop up the results." The square teeth with parallel sides were a very unusual form except as the result of slashing and filing. If ever found in nature, it would be the ideal shape for a filling. By studying types of tooth form we are able to make more satisfactory fillings, and especially, we learn how to avoid the horns of the nerves.

Dr. Westmoreland, the last speaker, said he had no time to examine into causes; that he was too busy. Dr. Westmoreland thought we ought to pay *more* attention to causes. It should be our endeavor to have our children safeguarded, even in embryo, and have them come into the world prepared to have good teeth. He considers eruptive diseases a great source of decay. We often find the marks they have left upon the teeth, in these imperfections of structure and labial defects, which are caused in nine out of ten cases by eruptive diseases suffered by the child just when the teeth were in process of formation. It is too late when decay has already set in and the child is suffering pain, at eight, ten, or twelve years of age. But we must do the best to remedy what it is too late to prevent.

Dr. R. R. Freeman, Nashville, Tenn., said that if he could talk as pleasantly and as agreeably as he felt in being with the Association, we would have a good time together. He desired to indorse the remarks of Drs. Clement and Westmoreland as to the early care of the teeth. All understand the importance of the early care of children as to refinement, purity, and healthfulness, but all do not see that this also involves the early care of the teeth. It is the want of care and cleanliness that causes the defects mentioned. If all particles of food are kept thoroughly cleaned out, we will not have chemical abrasion at the points of contact, where acids are formed and retained by capillary attraction. If we could prevent our children suffering from eruptive diseases until after the eruption of the teeth, they would have a better development. He believes so thoroughly in the effects of cleanliness upon the teeth that he puts across one corner of his billheads the maxim, neither original nor new, but always true: "Clean teeth don't decay." It is true we can't keep them perfectly clean, but we must do the best we can toward it. To insure good teeth it is not enough to commence at three years, nor at six months, but, as has been said, the work should have been begun three generations back. It is true that children forget to brush their teeth, but not more so than to wash their faces or to brush their hair; parents should look after the one as much as the other. In the little child the secretions are acid, as evidenced by the green stain on the enamel, but this must be prevented and kept away. It is not the persons of most robust development who live the longest, but those who take the best care of themselves, of what they eat and how they live. And it is so with the teeth. In spite of defects in development, if the best of care is taken of what we have, we may carry our teeth to the grave, and live to a good old age. Establish good habits in these respects in the earliest years of a child's life, and the work is done.

The paper of the Section being called for, Dr. W. E. Walker, of Bay St. Louis, responded.

PRACTICAL ANÆSTHETICS.

Mr. President, Ladies, and Gentlemen: Having been requested by the Chairman of the Committee on Operative Dentistry to prepare a paper for that Section, I have selected the subject of "Anæsthetics" as being a topic of vital importance to us, as well as of great interest.

Our profession is eminently humanitarian. We are so frequently called upon to relieve pain, or to give prophylactic treatment. When, as it often happens in the course of our operations, we foresee resultant pain, it is our duty to use the preventive means placed in our power by the judicious use of anæsthetics.

The first agent I will mention, though personally I have had no experience with it, is "hypnotism." If the half that is told be true, we would seem to have in hypnotism a valuable adjunct to dental practice. If without any interference with pulse or respiration such complete insensibility to pain can be induced by the mere will of the operator, that the severest and most painful tests can be applied to the unconscious patient, why not extract teeth, or operate upon the dental pulp?

The *Dental Review*, in July, 1890, republished from the *Medical News* the report of an interesting trial of hypnotism as an anæsthetic agent, employed by Milne Bramwell, in the presence of upward of sixty medical men and dentists. Among the many operations performed, I will only mention the painless extraction of three teeth for a woman of twenty-five years of age, and the removal of sixteen stumps for a servant girl nineteen years old, during the hypnotic sleep. The latter awoke smiling, having felt no pain, not even experiencing any subsequent soreness of the mouth. A number of other severe dental and surgical operations were performed on patients of different ages and both sexes, with unvarying success.

At a meeting of the Chicago Anæsthetic Club, in October, 1890, Prof. Anderson, of Denmark, lectured on this subject to a large audience of physicians and dentists, applying many severe tests. At a more recent meeting of the same club, Prof. Norman J. Roberts gave a clinic, with hypnotic demonstrations, before a large number of prominent physicians and dentists, among the latter Drs. Haskell, John S. Marshall, J. J. R. Patrick, and others well known to you. Among other tests, needles were put through the flesh of hands and faces of different individuals, no pain being felt, the flesh apparently insensible; with respiration 16 and pulse normal, as tested by a physician present.

According to the writer of the report, a number of these wide-awake scientific men were placed in a semicircle around the apparatus devised by Prof. Roberts for inducing the hypnotic sleep, and in less than fifteen minutes many of them were asleep and snoring. At his command they awoke simultaneously, rubbing their eyes, dazed and ashamed of their skepticism.

Prof. Roberts predicts that hypnotism will, within a few years, take the place of ether and chloroform, specialists being employed to hypnotize patients for all operations where anæsthesia is desired.

It may be well, though, to give a word of caution as to the introduction of this agent into our practice unless our field is sufficiently large for us to make it a specialty. A case was brought to my notice, during the past year, where a member of our profession has injured his practice very seriously by the employment of hypnotism, patients fearing that they might be rendered unconscious against their will, or without their knowledge.

I will at this time confine myself principally to the consideration of local anæsthetics.

First, because the operations we are most frequently called upon to perform are not such as would require or justify the use of general anæsthesia, not only because of the risk and inconvenience attending their exhibition, but also on account of the frequent impracticability of securing proper antecedent conditions of the patient. For instance, when a patient has nerved herself up to the point of having a tooth extracted (an operation which must be performed sometimes, and which is always much dreaded even when an anæsthetic is promised), that is the time to do the work; for if we send her home to return with looser clothing, an empty stomach, and a lady friend (precautions which must be taken in the use of general anæsthetics), the golden opportunity is lost, and weeks if not months of both physical and mental suffering may be endured before she again acquires sufficient fortitude.

By the employment simply of a local anæsthetic all this may be obviated, as also the possible risks, or the disagreeable after effects which may follow the administration of general anæsthetics.

As dentists, unfamiliar as we are liable to be with the constitutional idiosyncrasies, or organic peculiarities of our patient, we find in local anæsthetics all that is requisite for the minor operations of our circumscribed field.

In thus limiting my paper we will also avoid the unprofitable discussion of legal points involved in the use of general anæsthetics. Even with this restriction, however, we have a long list to select from, though the ideal local anæsthetic has not yet been found.

Probably the most frequent call we have for a local anæsthetic is in the treatment of sensitive dentine.

My own mainstay has been, and is, warm air gently applied, with encouraging, sympathetic words and gentle firmness.

"Herbst's Obtundent," a saturated solution of cocaine hydrochlorate in chemically pure sulphuric acid, to which solution sulphuric ether is added to the point of saturation (all excess of ether floating upon the surface and evaporating) will often be found to give very satisfactory results.

The application of ether spray, for obtunding sensitive dentine, I find usually causes about as severe pain as the excavation of the cavity without the application. Dr. George F. Eames says: "Put on the dam, secure perfect dryness by using warm air, and apply cocaine dissolved in chloroform in a 10 per cent. solution." Dr. Littig, of Philadelphia, uses a saturated solution of cocaine in glycerine.

The removal of the pulp can frequently be accomplished with little or no pain, by the use of a local anæsthetic. I have successfully used a preparation recommended by Prof. A. W. Harlan, a solution of 10 grains of hydrochlorate of cocaine in 90 minims of sulphuric ether. This should be left in contact with the pulp for five minutes.

An operation which very frequently calls for the use of a local anæsthetic is the extraction of teeth, for which purpose I have found cocaine, alone or in combination with other agents, to best serve the purpose.

From the introduction of cocaine into surgical practice, up to the year 1891, I had not sufficient confidence in myself, or in the knowledge then possessed of the drug, or in the stability and pureness of the preparations

then used, to use it by injections, though for slight operations I found the
aqueous solutions quite serviceable applied externally. Finding the drug
store preparations so very unstable, I adopted the plan of having ready a
definite quantity by weight of the hydrochlorate crystals in a set of little
vials so marked with a file that when a patient presented it was only necessary to fill the vial with water even with the file-mark to have a fresh 4
per cent. solution for each patient, without delay or uncertainty. This, however, I discontinued on the introduction of Parke, Davis & Co.'s hydrochlorate hypodermic tablets. For a single injection, one tablet dissolved
in ten drops of water makes as much of the 4 per cent. solution as it is
advisable to inject "above the shoulders."

Prof. Wölfler (in the *Medical Bulletin*, March, 1891) points out that the
cases in which unpleasant results have followed the injection of cocaine are
chiefly those in which the drug has been employed about the head, it being
most dangerous when inflicted in proximity to the brain. He says that
while 16 minims of a 5 per cent. solution may be safely injected in other
parts of the body, about the head the same amount of a 2 per cent. solution
is sufficient.

My own experience with cocaine has led me to believe that while we get
some anæsthesia from very weak solutions, yet the 4 per cent. solution is
more satisfactory, the results being more uniform.

Dr. Delbose (in a report found in the *Boston Medical and Surgical Journal*,
November 20, 1890), after very careful investigation of the published his-
tories of accidents following the use of cocaine, concludes that the few well-
authenticated cases of death were due to the enormous doses administered.

In experimenting to ascertain how weak a solution of cocaine could cause
anæsthesia, a German physician has announced the remarkable fact that
pure water, when injected under the skin has an anæsthetic effect, causing
insensibility of the tissues for several minutes. He states that the effect of
the water is to create a slight swelling, resembling that caused by the sting
of a gnat. The space marked by the swelling is insensible to pain, so that
incisions can be made without causing even discomfort. This calls to mind
an incident in my own practice.

A lady, very nervous and weak, called to have a root extracted. She
called for cocaine injection. In view of her nervous condition, I endeavored
to persuade her that it was not necessary, but did not like to say anything
of the danger of the injection, as she was accompanied by another patient
with whom I had a later engagement to extract several teeth with cocaine,
there being no objection to its administration in her case. I therefore made
a pretense of preparing the cocaine solution, in place of which I used pure
water only. She was delighted with the results, and assured us that cocaine
was a splendid thing, as it had saved her from all pain!

From the archives of dentistry I have copied the following prescription
for local anæsthesia, which impresses me favorably, theoretically, though I
have not yet tried it:

Chloroform.. 10 parts.
Ether.. 15 parts.
Menthol.... .... 1 part.

Combined and used in a hand atomizer. It is claimed for this combination that such a degree of anæsthesia is produced after one minute's application of the spray that incisions can be made for the removal of growths, opening a felon or an abscess, without causing pain.

The chilling sprays are often of service, but for operations in the lower jaw, when one hand is necessary to keep the parts dry, having the other hand to both work the bulb and direct the nozzle, it is difficult to manage successfully unless an assistant is employed, and I find this to be one place where the judgment of an assistant does not usually concur with my own. It is well, however, always to have the ether at hand for spraying with the atomizer, and also chloride of ethyl in the glass tubes as furnished by the dental depots, as cocaine is not permissible in all cases.

While with the simple cocaine solution it is unsafe as a rule to inject more than from one-eighth to one-fourth of a grain, which in a 4 per cent. solution is from five to ten drops, a quantity sufficient only for the extraction of a single tooth, yet when cocaine is combined with other drugs which modify and restrict its systemic action, we can use a much larger quantity of a higher percentage, and successfully anæsthetize a larger area.

The following formula was given to the dental profession by Dr. L. C. Wasson, of Topeka, Kans., through the Kansas State Dental Association, about a year ago. It was given to him by Dr. Guibor, of that city, a specialist in the treatment of diseases of the nose and throat.

Cocaine hydrochlorate	gr. 20
Sulphate of atropia	gr. 1-10
Carbolic acid crystals	gr. 10
Chloral hydrate	gr. 5
Aqua pura, add	oz. 1

As we should be very cautious in adopting new preparations, merely because Dr. So-and-so says it is a good thing, I studied this formula carefully before using it. I found that what Dr. Wasson considers the maximum dose, two syringefuls, or 60 minims, contains (omitting decimals in thousandths for brevity) :

Cocaine hydrochl.	2 3-10 gr.	Hypodermic dose.... $\frac{1}{4}$-1 gr.
Sulphate of atropia.	1-100 gr.	Hypodermic dose....1-120 gr.
Phenol crystals	1 16-100 gr.	Hypodermic dose.... $\frac{1}{4}$-$\frac{1}{2}$ gr.
Chloral hydrate....	58-100 gr.	Hypodermic dose.... 10 gr.
Distilled water....	55 91-100 gr.	

The whole making a 6$\frac{1}{4}$ per cent. solution of cocaine combined with antidotal and localizing agents.

This combination makes the use of cocaine much less objectionable, as Dr. Wasson thus points out: "(1) Atropia, in small doses, as given in hypodermic injections from this preparation, is a cardiac, respiratory, and spinal stimulant, which tends to counteract the toxic effect of the cocaine; (2) carbolic acid aids the chloral in localizing the anæsthesia; and (3) both tend to increase the anæsthetic properties of the cocaine and localize the effects; and both aid in the preservation of the solution, which is of itself quite desirable, as the ordinary cocaine mixture is almost worthle-s at the end of a week, while this preparation is good for months."

I have used this preparation very extensively for hypodermic injections in the gum tissue, for the past eight months, and find it more satisfactory than anything I have ever u-ed, especially for tooth extraction.

Though I was exceedingly cautious in using cocaine alone, never to inject more than ¼ gr., yet I frequently found it necessary to administer brandy and ammonia, and to use nitrite of amyl, while with this injection of Dr. Guibor's I have not once required anything of the kind, though I have frequently injected 30 minims, containing from 1 to 1¼ grs. of cocaine, a quantity we would not dare to use about the head except as combined with the antidotal components of this preparation.

In the eight months I have used this formula I have had but three patients complain of systemic effects, which I will describe to illustrate how trivial the disturbance has been.

The first case was that of a young man of twenty-three, of hereditary hemorrhagic diathesis, so that although he had quite a number of broken-down teeth to be removed, I did not care to extract many at a sitting. I therefore used this injection for him a number of times. He made no mention to me of any peculiar sensations, but a little later, when his younger brother came to me with his mouth in a similar condition, at the second sitting, while I was waiting for the cocaine to take effect, he exclaimed: "Edward was right, for when I went home the other day, I told him that I felt very funny, and he told me that he felt the same way the first time, but that I would not notice anything the next time; and it is true, for I feel all right this time!"

This patient returned twice after that, and I made a point of asking him about his sensations, but he assured me there was nothing noticeable. At each sitting I injected about 15 minims.

While it is true that no untoward *systemic* effects follow the use of this anæsthetic injection, I cannot close my paper without a word of caution as to one unpleasant feature attending its use, in some cases, unless proper care is taken to keep within the limit as to quantity.

In my early experience with it, I had such perfect confidence in it that in three cases where the gum tissue was very dense and rigid, I injected more than the proper quantity, which resulted in gangrene of the soft tissues—confined, however, very exactly to the area subjected to too tense pressure by the forcible injection of so much fluid. The soft tissues assumed a grayish-black hue, appeared fibrous in character, and sloughed. Of the three cases, one caused me for a time considerable anxiety. This was not a case of extraction. I had occasion to use a Perry's separator between the superior incisors, which were so badly decayed that it was necessary to press the separator upon the gum. This caused so much pain that I resorted to the injection, but which served its purpose only too well. The operation was rendered painless, but a slough supervened, which being posterior to the incisors, on the median line, caused me to fear rupture of the anterior palatine vessels, with severe hemorrhage. Fortunately, it did not go deep enough for that.

This was my first case with such result. The other two were cases of extraction where the gum tissue was too dense and rigid to accommodate the amount injected, though the same amount in soft, flabby gum tissue would have been absorbed without ill effect.

I have already trespassed too long upon your kindness, and will not weary you with a list of formulas given in the different journals, and vouched for by men of high standing in our profession. I cannot close my paper, however, without mentioning *cocaine phenate*, which I had hoped to test before coming before you. I have been disappointed in this expectation, but if the claims made for it prove valid, it must take its place as a valuable local anæsthetic. According to "Merck's Bulletin," its practical insolubility in water is a physical property most favorable to its therapeutic action, as it cannot be readily washed away from its site of application by the lymph currents. Its local action is therefore more intense and prolonged. In consequence of this insolubility very small doses have effect, while on the other hand, excessively large doses (15½ grs.) are not prone to produce toxic symptoms. This makes it of value in topical applications, as by the brush, when the dosage cannot be regulated with exactness.*

This paper was discussed by Dr. Brownrigg (physician, Columbus), Profs. Freeman and Peabody, Drs. W. H. Marshall, Westmoreland, and Walker.

Dr. Brownrigg said that he felt a delicacy in speaking on such a specialty as dentistry, a specialty which was doing so much to add to the longevity of the human race, but that in speaking of anæsthetics they met on common ground. In the able and interesting paper to which he had listened he noted one important omission in the list of serviceable anæsthetics, and he wished to enter a plea in behalf of sulphuric ether before this learned body. He hoped that its merits would be investigated and that it would be given a fair trial. Chloroform is very objectionable, and it has been repudiated by many of the great hospitals; but sulphuric ether is harmless and may be used with perfect safety. Statistics are all in its favor. In minor operations full anæsthesia is not necessary; analgesia is all that is required. The operations of dental surgery are of such an *awakening* nature that the patient recovers readily and there is no trouble from blood in the mouth. In administering ether he always places the patient in a reclining position, pouring ether upon a handkerchief for inhalation. He tells the patient to hold his hand up;

---

* In a private letter received from Merck's laboratory, since reading the above, I am officially informed that the percentage of cocaine alkaloid in cocaine phenate is about seventy-five; that it is soluble in alcohol of from thirty to fifty per cent., the solution having a faint odor of carbolic acid. That, and hypodermically as an anæsthetic in dental operations, it produces complete topical anæsthesia without subsequent derangement of the general well-being. An alcoholic solution containing 1 part of the drug in 12.50 of alcohol was the form employed.—W. E. W.

and when it drops, the tooth can be pulled. He may feel the *pulling sensation*. but he will feel no pain. A doctor's time is not valuable, but to the dentist, whose every hour is occupied. the saving of time is a matter of great importance. The time required to reach the stage of analgesia with ether is very short, and this alone should give it popularity with the dental profession. He has used it largely for twenty-seven years, and almost exclusively for nineteen years, and he could safely say that it is perfectly harmless.

Dr. Walker said that the gentleman had failed to recognize the fact that he had limited his paper to the use of *local anæsthetics*, and therefore he did not include ether except as a spray.

Dr. Westmoreland stated that one of the physicians present, Dr. E. P. James, was possessed of the power of hypnotizing, and would, if desired, give a clinic in hypnotism.

The offer was accepted with thanks, and 8:30 A.M. appointed as the hour.

Dr. W. H. Marshall, Oxford, had used chloride of ethyl, but found that it would not keep at a temperature above sixty· degrees. About two months ago he had a very nervous patient for whom it was impossible to excavate without anæsthesia, but with the aid of chloride of ethyl he made the excavations with perfect ease. It produces perfect insensibility of the dentine. For extracting, one-half a tube is sufficient.

Dr. R. R. Freeman said that he must take exception to the statement in the young brother's paper that dentists are not acquainted with the constitutional idiosyncrasies of patients. We must bear in mind that the introduction of anæsthesia was made by dentists— we should uphold the honor of our profession. He was glad to hear what the physician had said of ether. He uses it himself and likes it, but he would not say it was harmless. Nature gives pain as a protection, and he wanted tooth-pulling to hurt; but when it is *necessary* to extract a tooth, we are justified in reducing the pain to its lowest degree. We should not administer ether merely because it will save our valuable time, though it is true that sufficient for the extraction of a tooth can be given in a very short time. He does not know that he employs the subtle force of hypnotism, but he does believe that the dentist may show by force of the *ego* such command of character that the patient will yield to him. He can say to them, "Do as I tell you, and you will suffer less pain," in such a manner that they will believe him, and they *will* suffer less. By the exercise of *rapid breathing*, or rather *blowing*, a patient may so anæsthetize himself that after a few inhalations of a small amount of ether the patient will fall away, and four or five teeth may be extracted. But always impress upon your patients that we are taking into our hands an agent that requires great care. Beware how you handle it. I am glad to hear one physician say that he will not use chloroform. I believe I would withstand the temptation of all the gold of Ophir rather than permit the use of chloroform in my office. There is no trouble with the rapid breathing followed by a few inhalations of ether, but first get the lungs in free swing. As to the use of nitrous oxide gas, I regard it as similar to holding a man's head down in a bucket of water, and then extracting his teeth; the only difference is that you don't have to pump out the water. The appearance in both cases is equally repulsive. Ether is a reliable agent. But in not one out of a thousand cases of extraction is any thing required. Put it plainly before them: "Which do you prefer, the continuous pain of the ache or the brief pain of the removal?" From my own experience with cocaine, I don't want it! it excites hysteria and other evil influences. Pure water, or even "make believe" will suffice, provided we have control of the mind of our patient.

Dr. Walker said that he was far from being one to belittle our profession. He had the highest opinion of the scientific attainments of the educated dentist, and he did not deny to them the right and privilege of using a general anæsthetic when it is required, but for that there are too many preliminary requirements. We would often have to send them home to return with an empty stomach and looser garments. It is generally conceded that we must have a third person present, and this is not always convenient, especially for a transient case of extraction. As to not understanding the constitution of our patients—those whom we have had opportunities to know well, and who are under our regular care, and not, as a rule, those for whom we extract teeth—we *save* the teeth of our regular patients. Those for whom we do the most extracting are "transients," whom we perhaps have never seen before, and never expect to see again.

Dr. Peabody said that, although he had declined to speak in opening the discussion, some of the remarks made him desire to say a few words. He wished to go upon record as saying that there was no systemic anæsthetic known to-day which is absolutely free from danger, whether it be ether or chloroform or nitrous oxide gas or cocaine—they all have their possible dangers. If Prof. Freeman is capable of producing anæsthesia by rapid breathing and inhalation of ether in one or in three minutes, he must have strong enough hypnotic powers to make them *think* the operation painless. To use these things, a man must be a good judge of human nature, of idiosyncrasies, of temperaments, and then he is not always safe. Prof. Peabody said that he had taken thirty minutes in administering one-half pint of sulphuric ether without producing anæsthesia. He had seen patients take two ounces of chloroform without anæsthetic effect. There is no general law, but different effects are produced upon different people. No man can take a perfect stranger and predict what the effect will be. The most skilled physicians cannot do it. With the known death rate of one in one thousand from chloroform, one in ten thousand from ether, one in one hundred thousand from nitrous oxide, we are not justified in using these agents on trivial occasions.

It has been said here that we should be able to judge of the fitness of the patient.

A physician administered chloroform for the seventeenth time to a patient, with whose idiosyncrasies he was presumably well acquainted, as she had been under his care for fifteen years, and had previously been safely anæsthetized by him sixteen times, and yet the seventeenth time she died in less than five minutes. In justice it must be added that sixteen times she was laid prone upon a table, while on the fatal seventeenth occasion she was reclining in a chair. No sum of money would now induce me to allow the administration of chloroform in my office. Dr. Bonwill, who originated the "rapid breathing" method, says "one hundred times in a minute." He thought it doubtful if any one can breathe one hundred times in a minute, and that it would prove one of the most trying operations ever attempted. Nitrous oxide gas is simply asphyxiating, cutting off the nerves of sensation. If the patient will draw in a long breath four or five times, filling the lungs to their fullest capacity, and then hold the breath as long as possible, allowing it to escape very gradually, the most sensitive cavities can be excavated. For extraction it is not so servicable, as at the critical moment an involuntary "ough" will expel the air and the effect is lost.

In the fanaticism of so-called "Christian Science," there is nothing Christian and no science.

The subject of anæsthesia is one of the greatest interest to us. We should know that, notwithstanding the Haidarbad Commission with chloroform, we have paralysis of the cardiac muscles; sulphuric ether affects both the heart and respiration; nitrous oxide asphyxiates. Cocaine should never be used unless ammonia and nitrate of amyl are at hand. In some countries of Europe the use of these agents is by law denied to dental surgeons.

Dr. Walker, in closing the discussion, said that he was much pleased with the criticisms and the discussion which his paper had called forth, especially as it had been said that "it takes a good paper to call out a good discussion."

As general anæsthesia is liable to prove fatal, even when administered by skilled physicians to patients whom they have brought into the world, and who have been under their charge all their lives, the risk is even greater with us, as those for whom we are most frequently called upon to extract teeth are strangers whom we perhaps have never seen before and may never see again. It is true that we can give to even these the necessary examination, but with the local anæsthetic, which answers all purposes, there is less risk and less inconvenience and less time required.

On motion of Dr. Luckie, the subject was passed, and the Association adjourned to 8 P.M. Wednesday, May 4.

Wednesday was devoted to clinics, when the following report of the Clinic Committee was made by Dr. A. H. Hilzim, Chairman:

Dr. Francis Peabody, of the Louisville College, demonstrated the use of Blair's vaporizer for treating roots of diseased teeth preparatory to filling. By the use of iodoform, vaporized in his instruments over the flame of a spirit lamp, his experiment was quite a success.

Dr. William Crenshaw, of the Southern Dental College, of Atlanta, Ga., at the request of some of the members, demonstrated his mode of setting the Logan crown, using Blair's vaporizer in the treatment of the root.

Dr. W. H. Marshall, of Oxford, demonstrated the use of the Marshall matrix by filling several teeth with amalgam. Also showed manner of making and adjusting gold crowns and bands for his anchor plates.

Dr. J. Hack Rush, of De Kalb, made contour gold filling.

Dr. Frank Smith, of Grenada, also made handsome contour gold filling in central incisor, using the Bonwill engine mallet.

Dr. D. B. McHenry, of Grenada, demonstrated the manner of making his new lock plates.

Dr. George B. Clement, of Macon, demonstrated the use of the injecter coal oil furnace for melting metals; also filled tooth with Williams's crystalloid gold, finishing with cylinders.

Dr. T. O. Payne, of Vicksburg, made contour gold fillings, with Nickold's gold.

SECOND DAY.—NIGHT SESSION.

The members of the Association met as per adjournment at Concordia Hall. Finding the Concordia Hall occupied by the Concordia Glee Club, those present adjourned to the large room occupied by Morrison Brothers' Dental Exhibit of the Nashville Dental Depot, where they witnessed a very interesting exhibition of hypnotism, through the courtesy of Dr. James.

They then adjourned to the reading room of the Gilmer Hotel, the headquarters of the Association, when the meeting was called to order at 9:30 P.M. by the President.

The minutes of the preceding sessions were read and approved.

The following applications for membership, having been approved by the committee, on motion of Dr. A. H. Hilzim, the rules were suspended and the Secretary authorized to cast the ballot in their favor, and they were declared duly elected: B. F. Worsham, Ripley; V. B. Watts, Brookhaven; A. G. Tillman, Vicksburg; O. J. Harrelson, Sylvarena; J. D. Wise, West Point; and J. D. Sugg, Aberdeen.

The Secretary read a letter from the President of the Southern Dental Association, as follows:

NASHVILLE, TENN., May 2, 1892.

To Dr. D. B. McHenry, President of the Mississippi Dental Association.

Dear Doctor: Please present to the Mississippi Dental Association an invitation to be present at the meeting of the Southern Dental Association on July 26, 1892, at Lookout Mountain. We feel assured the meeting will be interesting, and we hope your membership will attend. Wishing you a successful meeting,

I am most sincerely,   GORDON WHITE.

President Southern Dental Association.

On motion of Dr. Westmoreland, Dr. R. R. Freeman, of Nashville, Tenn., was elected an honorary member of the Association.

Dr. Marshall stated that the Board had been reliably informed that a number of dentists, whether or not members of the Association was not stated, had already applied to the Governor for positions on the new Board. He thought that the interests of the profession demanded action on the part of the Association, and felt assured that if proper steps were taken the Governor of the state would consider the best interests in the matter.

After a lengthy discussion by Drs. Luckie, Marshall, Clement, Dillehay, Hilzim, and Moffat, the following resolution, offered by Dr. A. H. Hilzim, was carried:

Resolved, That we proceed to elect five members of this Association, whose names shall be presented to the Governor as the choice of this Association for appointment as a new Board of Dental Examiners.

Dr. Dillehay moved that this be made the special order of business for Thursday morning, 10 o'clock. Carried.

Dr. Westmoreland introduced Dr. B. A. Vaughn, of Columbus, who addressed the Association on a matter of public concern and of national interest, two bills now pending before the Congress of the United States, for which he asked the consideration of the Association, and their official indorsement if they saw fit to give it. One, the bill to create a Bureau of Public Health, the chief official of which shall be a member of the cabinet. This bill only requires a little push in putting its merits properly before our representatives in Congress. Dr. Vaughn recommended the passage of a resolution approving the objects and urging its adoption, a step that will undoubtedly be taken by the Medical Association. The other measure referred to is the bill designed to secure pure foods and pure drugs, for the indorsement of which a similar resolution should be passed. These are public measures, and as an Association working for the public benefit, such indorsement would undoubtedly have weight with our representatives in Congress. At the conclusion of Dr. Vaughn's remarks Drs. Westmoreland and Clement were appointed a committee in consultation with Dr. Vaughn to prepare suitable resolutions.

There being no other business before the Association, the President called for the report of the Section on Prosthetic Dentistry.

As there was no paper to be read, Dr. Marshall suggested the discussion of the subject of Prosthetic Dentistry, and invited Dr. R. R. Freeman, Professor of Mechanical and Corrective Dentistry in Vanderbilt University, to open the discussion.

Dr. Freeman said that before coming to the meeting he had not heard of Dr. McHenry's device, the skeleton lock plate, but when his attention was called to it in the clinic rooms it had struck him most favorably, and he had examined it with great interest. He was so inspired with its obvious merits that he would not hesitate to give his indorsement of its utility as a most ingenious and useful method of constructing a partial artificial denture, relieving it of the cumbersomeness of the old plate. It makes a substantial plate for the most difficult conditions. He said that he felt fully repaid for his trip merely in seeing this device, feeling that we have in it the means of relieving our patients from the inconveniences of the old methods. We never have been able to accomplish as readily and easily what this device will do. It is worthy the attention of every dentist. It is less expensive and more satisfactory for many cases than anything we have had in the way of crown and bridge work methods. He desired to thank its inventor for presenting it to the

9

profession. Though slow to pick up new devices, he should certainly profit by the lock plate of Dr. McHenry.

The telescope attachment of Dr. Marshall's anchor plate had also struck him very favorably as one of the things needful. He has had some experience with it, and knows that it has bridged over serious difficulties. It is perfectly simple, without any complication, and accomplishes all that its inventor claims for it. He said that he was glad to have seen the man who has presented it to the Association and to the profession. In some cases one and in other cases the other of these two methods will be found *par excellence*.

Dr. L. G. Nisbet, Aberdeen. said that he had a patient who could not wear a suction plate, but the anchor plate which was given to the Association last year gives perfect satisfaction. Another patient wanted a bridge, having worn one for one tooth and liked it. He suggested a trial of an anchor plate, which had given great satisfac-

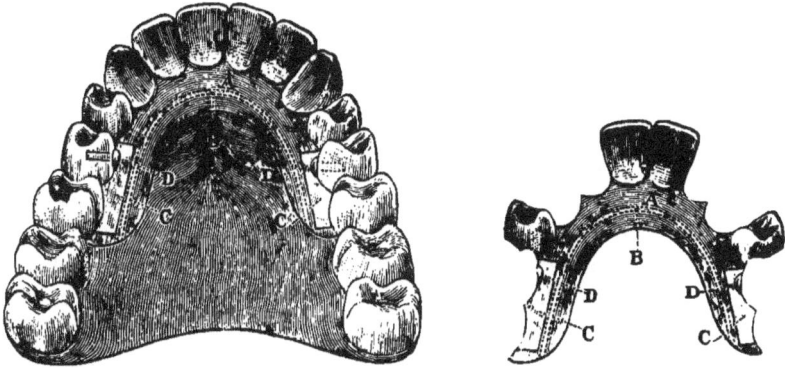

THE McHENRY SKELETON LOCK PLATE.

tion. He had constructed a full upper plate anchored to a canine for another patient, but this had not proved satisfactory. Absorption set in and in three months the tooth got very loose. Within the next twelve months he will have to make a full suction plate.

Dr. Marshall suggested that probably the plate had not been accurately adapted, or the articulation was not perfect. He had not as yet seen a sound, firm tooth get loose from serving for anchorage. There must have been some lack of adjustment.

Dr. A. H. Hilzim had made precisely the same piece—a full upper set anchored to a cuspid tooth, which had given entire satisfaction. It has been worn for six months. He expects to use this method whenever he has a suitable case.

Dr. E. E. Spinks, Meridian. has used the anchor plate successfully in two cases where the suction plate was a failure. In one case he had crowned two teeth; in the other, one. The patients pronounced them perfect. In another case a lady had been paralyzed and was still partially so when the plate was made. She could barely walk, but could not talk well. There was no power of suction. He crowned the upper molars and put in an anchor plate, which she wears with perfect comfort. It was impossible for her to wear a suction plate because of the paralysis. He had put in two more anchor plates just before leaving home. Both seemed perfectly satisfactory. From ten plates he had not had a single word of complaint.

Dr. Marshall said that he liked Dr. McHenry's plate. It was perfect for the cases for which it was designed, but instead of putting a pin in the lingual side of a natural tooth he would prefer to crown it and have a button on the lingual side. Then there would be no wear on the side of the tooth where the plate comes in contact with it. He thinks this would be more permanent, as otherwise there might be wear and tenderness produced. He wished to ask Dr. McHenry his experience in this regard.

Dr. McHenry said that the simplicity of his plate was its recommendation. The crowns and caps would complicate it. There was no necessity for protection against wear if the plate was accurately fitted. From the natural lateral movement it would adjust itself—there was no undue pressure—it was only necessary that it should fit snugly around the necks of the teeth. He had not as yet had any complaint, except in one case where there was a slight pressure, this was relieved by a little dressing of the plate, which had since been worn for six months without any complaint. He cited the case of a man who had been an ardent flute player. But while wearing a suction plate for sixteen years he had been obliged to give up his flute entirely. After he got a lock plate he found he could play with as perfect ease as ever, and this was a source of great joy to him.

Dr. Westmoreland said that he was not favorably impressed with the idea of drilling into a sound tooth for the insertion of a pin.

Dr. McHenry replied that no force was used in inserting the pin; it would be put in with the fingers.

Dr. Westmoreland asked if there was not a liability to decay around the pin.

Dr. Freeman thought it better to break the enamel of one tooth to save all the rest than to lose all. There was an old superstition about breaking the enamel, but such plates are not needed in a mouth where the teeth are all sound. There would be no trouble at all with the pin, compared with the grinding down necessary for a telescoping crown of bridge work. A sound tooth with such a pin inserted would last quite as long as a tooth with a cap or crown. The only precaution necessary was not to touch the pulp.

Dr. Westmoreland said that his question had been misunderstood· He wanted to know why a contour filling would not be better than the pin?

Dr. Freeman said: "Does not Dr. Westmoreland put in gold posts to secure a large filling? A pin or a post can be equally well adjusted. There is no chance for decay any more than around a contour or other filling. The right-sized pin, in a well-cut hole made with a corresponding drill, will always prove satisfactory."

Dr. Marshall said he was perfectly satisfied that Dr. McHenry's plate was simplicity itself. The crown would add both to the pain and the expense, and would not be needed except in extreme cases, not in one case out of a thousand.

Dr. McHenry said that the thin cement around the pin would never melt out.

Dr. Westmoreland thought that in some mouths cement would not stand.

On motion of Dr. Nesbit, the subject was passed.

Dr. Hilzim, Chairman of Clinic Committee, announced that he would complete his contour filling early the next morning; that Dr. William Crenshaw also had a clinic; and that Dr. Walker would exhibit a case of extreme absorption, with the plates worn for restoration of the same.

On motion, the Association adjourned to 8:30 A.M., Thursday. Clinics at 7:45 A.M.

THIRD DAY.—MORNING SESSION.

Meeting called to order at 8:30 A.M., the President in the chair.

On motion, reading the minutes was dispensed with.

The Secretary read the Treasurer's Report.

A lengthy discussion of incidents of office practice consumed most of the morning hour.

The hour for the special order of business having arrived, the subject was laid over, and the resolution of Dr. Hilzim put before the Association.

Dr. T. O. Payne, a member of the present State Board of Dental Examiners, rose to say that before coming to the meeting he knew no more of the new law passed than the Queen of England. He had had no hand in it, and was in no way responsible for it.

Dr. A. H. Hilzim said that he desired to make the same statement. He was surprised to hear that Dr. Payne had been accused of it. It was possible that it might be supposed that he too had had a hand in it. He had watched the papers daily during the session of the Legislature. When they reached that portion of the Code, he had made a written copy to send to the Board, and had requested some one to see that the November term be stricken out. He had asked prominent men where they had got their ideas of dental legislation, and they said they merely wished to simplify the law, and were acting on their own original ideas. That was all he knew about it.

The resolution passed at the preceding night session was read for the benefit of those not present at that time:

Resolved, That we proceed to elect five members of this Association whose names shall be presented to the Governor as the choice of this Association for appointment as a new Board of Dental Examiners.

The following gentlemen were elected to constitute the new Board: Dr. W. E. Walker, Bay St Louis; Dr. George B. Clements, Macon; Dr. R. K. Luckie, Holly Springs; Dr. George W. Rembert, Natchez; Dr. J. A. Warrener, Corinth.

Dr. Luckie begged to be excused; and Dr. P. H. Wright, of Senatobia, was selected in his stead.

The Secretary was instructed to prepare a copy of the resolution, with the names of the five members elected, and forward the same to the Governor of the State.

Dr. R. K. Luckie addressed the Association in behalf of the World's Dental Congress to be held in Chicago next year in connection with the Columbian Exposition. He said that up to this time many committees had been appointed for different lines of work. Among others, a Committee of Conference with state and local societies, of three or more persons to stir up an interest in the objects of the Congress and secure the coöperation of the dental profession. There are many things for this committee to do. They had expected a full meeting at this time, but Drs. Adams and Miles being absent, they had not been able to organize their lines of work, but he would distribute to the members present copies of the questions prepared by the General Committee to be answered by dentists in every state and territory of the Union. The answers to those questions will form the basis for a "History of Dentistry in the United States." He hoped that no one would throw the sheet aside as a trivial matter, but give it prompt and full consideration.

Dr. Dillehay said he hoped the Association would take a deep interest in the matter, and not let the State of Mississippi fall behind in this matter. Each member should consider himself a committee of one and get to work at once.

Dr. Peabody said he was a member of the Kentucky state committee and felt the deepest interest in the matter. He had found that members in the interior who were not well posted in the matter were afraid there was some trap laid; that members of the profession who were not graduates might be prevented from practicing, or something of that kind. We must do all that we can to drive out such ideas. There is a grand work to be performed, and each individual must do all in his power. The meeting at Chicago will be a World's Congress, where we will ascertain the *status* of the profession throughout the world.

Dr. Freeman said that he felt deeply interested in the matter, and that they were endeavoring to interest all the active members of the profession in his state. At the meeting of the Southern Dental Association in July next, at Lookout Mountain, in Tennessee, the various state committees would have a meeting, and it was expected that much information would have been obtained by that time. By conferring together they would give and receive suggestions by which all will be assisted in the work. The State of Tennessee had extended offers of hospitality to all the Southern State Committees and Associations. One and all were most cordially invited to be present and bring wives and daughters. He said that he took pleasure in promising that they would all be well cared for. He hoped and trusted that it would be such a meeting that when its history was recorded those who had been absent would consider themselves as unfortunate indeed.

Dr. Clement, Chairman of the Committee on Resolutions, offered the following resolutions, which were adopted:

1. *Resolved*, That we approve of and urge the passage of the bill creating a "Department of Public Health."

2. That we approve of the bill "Pure Food," and urge its passage.

GEORGE B. CLEMENT,

W. W. WESTMORELAND,

Committee.

B. A. VAUGHN,

Consulting Physician.

Dr. Vaughn said that the Pure Food and Drug bill has passed the Senate and been unanimously recommended by the House Committee, but some effort outside of Congress is being made to defeat it. If this action of the Association be supplemented by every member writing to the Congressman from his own district, a postal card, saying, "I am in favor of the Pure Food bill and urge its passage," great good will follow. The opposition is chiefly from manipulators and adulterers of foods and drugs.

There being no voluntary essays, the election of officers was the next order of business, and the following were elected for the ensuing year: Dr. A. A. Dillehay, Meridian, President; Dr. A. A, Wofford, Columbus, First Vice President; Dr. L. G. Nisbet, Aberdeen, Second Vice President; Dr. W. T. Allen, Amory, Third Vice President; Dr. W. E. Walker, Bay St. Louis, Recording Secretary; Dr. Frank H. Smith, Grenada, Corresponding Secretary; Dr. C. C. Crowder, Kosciusko, Treasurer.

Dr. A. H. Hilzim stated that the Morrison Brothers, of the Nashville Dental Depot, offered to publish the proceedings of the Association in a special issue of the DENTAL HEADLIGHT, sending a copy to each member of the Association, and printing as many copies as might be wanted. This liberal offer was accepted, and the DENTAL HEADLIGHT made the official organ of the Association. Dr. Hilzim was authorized to request the publishers to furnish twenty-five extra copies to the Secretary for distribution outside of the membership at his discretion.

On motion, the Association adjourned to 3 P.M.

AFTERNOON SESSION.

The meeting was called to order at 3:30 P.M., the President elect, Dr. A. A. Dillehay, in the chair.

The President announced that all bills duly approved would be paid on presentation to the Recording Secretary (Acting Treasurer).

The following bill was presented and paid: Concordia Club, per Samuel Straus, Treasurer, for rent of hall, $5.

Jackson was selected as the next place of meeting, and Wednesday following the first Tuesday in April fixed as the time for the meeting in 1892. Monday and Tuesday preceding, the Board of Examiners will meet.

The Committee on Resolutions of Thanks presented the following series of resolutions, which were adopted:

1. *Resolved*, **That** we, the Mississippi State Dental Association, return

thanks to Drs. Westmoreland and Wofford for the courtesy and kindness shown us during this session, and the efforts they have made to make us comfortable.

2. That we return thanks to Dr. Frazer for the pleasure and privilege of visiting the I. I. & C.

3. That we return thanks to the citizens of Columbus for all courtesies.

4. That we return thanks to Morrison Brothers, of Nashville, Tenn.; to Stuart & Adams, of New Orleans; to the S. S. White Dental Manufacturing Company; and to R. I. Pearson, of Memphis, for favors extended the members of this Association.

5. That we return thanks to Mrs. J. M. Walker for her presence and assistance.

6. That we return thanks to the M. & O. railroad, G. P. railroad, I. C. railroad, and hotels for reduced rates, etc.   G. B. CLEMENT,

C. C. CROWDER,

R. K. LUCKIE.

Dr. B. F. Worsham, of Ripley, begged to present the following preamble and resolutions for adoption by the Association:

Whereas the law of the State gives to his Excellency, the Governor, the prerogative of appointing gentlemen to constitute a State Board of Dental Examiners; and whereas we, the members of the Dental Association of the State of Mississippi, being now in session at a regular meeting, having the utmost confidence in the integrity and purpose of the Governor to advance as far as his position enables him the best interest of the whole people; therefore be it

Resolved by this Association, that Gov. Stone be and he is hereby most respectfully asked to appoint as said Board of Dental Examiners the five members of the State Association elected by them, and whose names are herewith forwarded to him.

The resolution was adopted, and the Secretary instructed to forward a copy of the same to the Governor.

The Association, on motion, adjourned to meet at Jackson, Miss., on the Wednesday after the first Tuesday in April, 1893.

MRS. J. M. WALKER,   W. E. WALKER,

Reporter of Discussions.   *Recording Secretary.*

STANDING COMMITTEES, 1892–93.

Committee on Membership and Finance.—The three Vice Presidents.

Committee on Publication.—Recording and Corresponding Secretary and Treasurer.

Committee on Appliances.—E. E. Spinks, Meridian; and other dealers in dental goods that attend the Association.

Executive Committee.—F. Smith, Grenada; A. H. Hilzim, Jackson; J. H. McGruder, Jackson.

Clinic.—R. K. Luckie, Holly Springs.

Section 1. Surgical Dentistry.—J. B. Askew, Vicksburg; J. P. Frazier, Canton; G. B. Dasey, Vicksburg.

Sec. 2. Prosthetic Dentistry.—J. O. Frilick, Meridian; A. H. Hilzim, Jackson; T. C. West, Natchez.

Sec. 3. Physiology and Dental Histology.—R. K. Luckie, Holly Springs; L. S. Nisbet, Aberdeen; J. D. Killion, Indianola.

Sec. 4. Pathology and Dental Therapeutics.—M. Adams, Sardis; J. T. Wise, West Point; J. H. Magruder, Jackson.

Sec. 5. Dental Materia Medica.—W. H. Marshal, Oxford; W. R. Walker, Bay St. Louis; P. H. Wright, Senatobia.

Sec. 6. Metallurgy.—D. B. McHenry, Grenada; J. A. Watson, Lexington; T. O. Payne, Vicksburg.

Sec. 7. Dental Chemistry.—G. B. Clements, Macon; J. D. Miles, Vicksburg; T. B. Birdsong, Hazlehurst.

Sec. 8. Dental Education and Literature.—W. T. Mastin, Yazoo City; E. B. Robins, Vicksburg; W. T. Allen, Amory.

Sec. 9. Incidents in Office Practice.—A. H. Wofford, Columbus; J. A. Warriner, Corinth; I. B. Rembert, Jackson.

Sec. 10. Voluntary Papers.—W. W. Westmoreland. Columbus; E. E. Spinks, Meridian; K. S. Moffat, West Point.

NORTH CAROLINA STATE DENTAL SOCIETY.

The eighteenth annual meeting of this Society was held at Winston, N. C., May 24–26, Dr. C. L. Alexander, President, presiding. The attendance was large. The State Board of Examiners was in session at the same time, with about twenty applicants for license. Many interesting papers were read, two of which deserve particular mention. Dr. J. H. Durham, of Wilmington, essayist for the occasion, read a valuable paper, the subject of which was "The North Carolina State Dental Society," which was a most creditable production and worthy of careful preservation. The other paper was from the pen of the versatile and accomplished Dr. V. E. Turner, of Raleigh, who paid his respects to two classes of the profession which he designated "The Merchant Dentist and the Hustler." These papers will not fail to attract attention.

The following is a list of those attending: C. L. Alexander, M. A. Bland, Charlotte; J. A. Blum, W. J. Conrad, H. V. Horton, Winston; R. J. Burwin, Marion; J. H. Durham, J. E. Mathewson, Wilmington; D. E. Everett, Raleigh; J. E. Freeland, Statesville; F. S. Harris, Henderson; H. D. Harper, Kinston; H. C. Herring, Concord; E. L. Hunter, Enfield; J. W. Hunter, R. H. Jones. Salem; T. M. Hunter, Fayetteville; J. B. Little, Newton; H. C. Pitts, High Point; J. F. Ramsey, Asheville; J. M. Riley, Lexington; J. H. London, V. E. Turner, Raleigh; J. S. Spurgeon, Hillsboro; C. J. Watkins, Winston; G. W. Whitsett, Greensboro; E. K. Wright, Wilson; J. E. Wyche, Oxford; J. F. Griffith, Salisbury; J. N. Hester, Reidsville; S. P. Hilliard, Rocky Mount; T. M. Hunter, Fayetteville; William Leach, Durham; R. M. Morrow, Burlington; G. B. Patterson, Fayetteville; C. A. Rominger, Reidsville; H. Smell, Washington; E. J. Tucker, Roxboro.

The visiting dentists are: B. Holly Smith, Baltimore; Henry W. Morgan, Nashville, Tenn.; M. H. Cryer, Philadelphia; R. A. Holliday, Atlanta; J. W. Foreman, Asheville; John S. Thompson, Atlanta. Dr. Smith represents the Baltimore College of Dental Surgery; Dr. Morgan, the Dental Department of Vanderbilt University; Drs. Holliday and Thompson represent the Southern Dental College.

The following gentlemen were elected active members of the Society: L. B. Henderson, Durham; J. M. Ayer, Raleigh; R. W. Reece, Elkin; Robt. Anderson, Calahan; R. L. Ramsey, Salisbury; C. W. Banner, Mt. Airy.

Prof. M. H. Cryer, Philadelphia; Prof. B. Holly Smith, Baltimore, and Prof. Henry W. Morgan, Nashville, Tenn., were elected honorary members and invited to take part in all discussions.

Fifteen operators performed clinical work of almost every character, superintended by the genial, courteous, and hospitable Dr. Hamilton V. Horton, of Winston, who won the esteem of all present.

Besides the address of welcome and President's annual address, the following papers were read: I. N. Carr, Tarboro, J. H. London, Raleigh, on "Dental Education;" C. A. Rominger, Reidsville, on "Dental Pathology and Therapeutics;" Dr. Jones, Salem, on "Dental Hygiene;" H. C. Ewing, Concord, on "The Dentist;" and Sid P. Hilliard, Rocky Mount, on the "Dentist and His Patient."

The Association closed after what the members say was the most interesting session in its history. More business was transacted than at any former session and a larger number of active members added to the roll. The membership now is nearly one hundred.

The following officers were elected for the ensuing year: Frank S. Harris, Henderson, President; H. V. Horton, Winston, First Vice President; J. F. Ramsay, Asheville, Second Vice President; J. E.

Wyche, Oxford, Secretary; J. W. Hunter, Salem, Treasurer; R. H. Jones, Salem, Essayist.

The Association adjourned to hold their next meeting in Raleigh on the second Tuesday in May, 1893. M.

ALABAMA STATE DENTAL ASSOCIATION.

To the Editors *Dental Headlight.*

The twenty-third annual meeting of the Alabama Dental Association was held in McDonald's Opera House, Montgomery, Ala., April 12–16, 1892, with a full attendance of members.

The Association was called to order by President George Eubank, of Birmingham, and opened with prayer by Dr. W. H. Morgan, Nashville. An address of welcome was delivered by Dr. W. G. Bibb, of Montgomery, and responded to by Dr. R. C. Young, of Anniston.

The entire meeting was one of unusual interest, Drs. Morgan, Crawford, and Freeman, of Nashville, Crenshaw and Brown, of Atlanta, contributing much to the pleasure and interest.

The following officers were elected for the ensuing year: C. L. Boyd, Eufaula, President; J. H. Allen, Birmingham, First Vice President; R. A. Rush, Selma, Second Vice President; S. W. Foster, Decatur, Secretary; G. M. Rousseau, Montgomery, Treasurer.

Executive Committee.—T. M. Allen, George Eubank, Birmingham, and J. C. Wilkerson, Selma.

Birmingham was selected for the next place of meeting, the second Tuesday in April, 1893. A full stenographic report of the transactions is in press. S. W. Foster, *Secretary.*

TENNESSEE STATE DENTAL ASSOCIATION.

The Executive Committee has issued the following programme for the twenty-fifth annual meeting of this Association:

PROGRAMME.

Essay by Dr. J. L. Mewborn, Memphis, Tenn. Subject: "Does the Machine Mallet Shatter the Hidden Edges of the Enamel?"

Essay by Dr. B. S. Byrnes, Memphis, Tenn. Subject: "Dental Appliances.''

Essay by Dr. John T. Crews, Humboldt, Tenn. Subject: "Does Climate and Location Modify Dentistry?"

Essay by Dr. W. F. Fowler, Greeneville, Tenn. Subject: "Principles *vs.* Pretensions in the Dental Office."

Essay by Dr. R. R. Freeman, Nashville, Tenn. Subject: "Stumbling-blocks to Young Dentists."

Essay by Dr. G. A. McPeters, Pulaski, Tenn. Subject: "Is Dentistry Becoming Overcrowded?"

Essay by Dr. J. Y. Crawford, Nashville, Tenn. Subject: "Is Honesty the Best Policy in Dentistry?"

Report on "Operative Dentistry," by Dr. Henry W. Morgan, Nashville, Tenn.

Essay by Dr. W. H. Morgan, Nashville, Tenn. Subject: "Methods *vs.* Principles."

Arrangements have been made with the hotels and railroads for reduced fare.

NOTICE.—Be sure and get a certificate from the railroad agent when you buy your ticket coming, showing you paid full fare, so you may get the reduction going back.

On arriving in Chattanooga get from the Committee a badge, so that the railroads and hotels may know you.   J. U. LEE,

W. J. MORRISON,

R. N. KESTERSON.

Committee.

AMERICAN DENTAL ASSOCIATION.

W. W. Walker, New York, N. Y., President; J. D. Patterson, Kansas City, Mo., First Vice President; S. C. G. Watkins, St. Clair, N. J., Second Vice President; Fred J. Levy, Orange, N. J., Corresponding Secretary; George H. Cushing, Chicago, Ill., Recording Secretary; A. H. Fuller, St. Louis, Mo., Treasurer.

The next meeting will be held at Niagara Falls, N. Y., on August 2, 1892.

SOUTHERN DENTAL ASSOCIATION.

Gordon White, Nashville, Tenn., President; E. L. Hunter, Enfield N. C., First Vice President; J. T. Calvert, Spartanburg, S. C., Second Vice President; W. H. Marshall, Oxford, Miss., Third Vice President; D. R. Stubblefield, Nashville, Tenn., Corresponding Secretary; H. C. Herring, Concord, N. C., Recording Secretary; Henry E. Beach, Clarksville, Tenn., Treasurer.

The last Tuesday in July, 1892, Lookout Mountain, Tenn., was selected as the time and place of next meeting.

TENNESSEE DENTAL ASSOCIATION.

D. R. Stubblefield, Nashville, President; S. B. Cook, Chattanooga, First Vice President; W. W. Jones, Murfreesboro, Second Vice President; P. D. Houston, Lewisburg, Recording Secretary; J. L. Mewborn, Memphis, Corresponding Secretary; H. E. Beach, Clarksville, Treasurer.

Place of meeting, Lookout Mountain, Tenn., last Tuesday in July, 1892.

Editorial. . . . .

DENTISTS CLASSED AS MANUFACTURERS.

In some of the large cities the census takers of 1890 attempted to compel dentists to give returns of the number of artificial teeth, amount of rubber, plaster, gold, etc., used annually. They very promptly refused to give the desired information, but there is now a bill before Congress known as H. R. No. 4696, introduced by Mr. Wilcox, amendatory of an act entitled "An Act to Provide for the Taking of the Eleventh Census," which if it becomes a law will so clearly class the dentists that they cannot but comply. The dentist is as much a professional man as the physician, oculist, or practitioner of any other branch of surgery, having been so recognized by nearly every state in the Union by enactment of laws for its regulation, requiring a higher degree of qualification before persons are permitted to practice. The Section of Oral and Dental Surgery of the Nashville Academy of Medicine at a recent meeting adopted resolutions protesting against this great wrong and injustice, and memorialized the Senators and Representatives in Congress from Tennessee, urging that they use their influence to have the bill defeated.

What possible good such information could be, were it possible ever to get accurate statistics, we cannot understand. If the census takers do their duty, and get full statements from the dental dealers, who sell all the supplies used by dentists, there is no need of going to the consumer at all. On the one hand there are only about one hundred, possibly two hundred, dealers; and on the other, nearly twenty thousand dentists. Let the profession appeal at once directly and individually to the Senators and members of Congress to use every honorable means to have this bill so amended that dentistry will be defined as one of the learned professions, and that dentists will not be classed " *as persons engaged in a productive industry.*"

THE MISSISSIPPI DENTAL ASSOCIATION.

This issue contains the transactions of the eighteenth session of the Mississippi Association, held at Columbus May 3–5, 1892.

This is said to have been one of the most interesting and fraternal meetings ever held. The deliberations speak much for the profession of the State. The high standing of its officers and their earnest efforts to keep pace with Association work is evidenced in dividing its membership into sections for the more thorough consideration of papers to be hereafter read before it, and providing for some member, previously appointed, to open all discussions.

The work of State Associations increases in importance yearly, and it should be the pride of every honorable man to hold membership, and give his moral support to such legislation as tends to still further advance the already high standing of our chosen and honored profession. In Mississippi, as in most other states, there are a small number of men who are willing to make the personal sacrifice of time and money, and do the work that has elevated dentistry as only associated effort could have done; while the great body of the profession have been content to remain at home, reap the reward and preach over the chair, and elsewhere, to any willing to listen what is being done, and how rapidly from "a mere trade" dentistry has forced its way to the front ranks of the learned professions.

Men left to themselves become bigoted and narrow-minded, but when they mix with others of their calling, exchange views, theories, and methods of practice, find others equally as competent, and often learn to simplify many harassing, difficult questions.

The value of this work is fully recognized in almost every line of life work, and in America we find associations of almost every character. Why is it then that our associations do not receive the attention and support they should? The answer is not easily given. The Mississippi Association, having divided the work, affords an opportunity for every one to do his part, and we bespeak the largest attendance in the history for the next meeting.

We congratulate the Association and ourselves on the appearance of this copy of its transactions. The report is from the facile pen of "Mrs. M. J. W.," so well known to the profession for her full and accurate report of dental proceedings.

THE SOUTHERN DENTAL ASSOCIATION.

At Chattanooga, beginning July 26, the Annual Convention of this Association will be held. President White has, through his excellent committees, about completed all the arrangements, and feels justified in predicting a very large attendance. His correspondence indicates that much interest is felt, and that the work will be unusually attractive and profitable.

The success of this Association in past years shows that it is doing a good work, and every man within reach of Chattanooga should feel it his duty to add at least his presence to the meeting.

All who have models of irregularities, abnormal anatomical specimens, new devices, improved instruments, or curiosities should bring them. A place for an exhibition of all things of interest to the profession will be provided, and many can contribute in this way to the general success of the meeting.

The Association will be honored with the presence of the members of the Executive Committee of the World's Columbian Dental Meeting, which is to meet at the same time. The character of the membership of this committee, being composed of men of national reputation, and the importance of the work which they are engaged upon will add to the interest of the Association.

THE TENNESSEE DENTAL ASSOCIATION.

The time for the usual meeting of this Association has been postponed by the President and Executive Committee, so that a joint meeting with the Southern can be held in Chattanooga.

Every member of the State Association should exert himself to be present in Chattanooga July 26.

Let every member of the profession who feels a pride in it and in the reputation of his state be on hand, ready to respond to any demand President Stubblefield may make of him, and to do his might toward making the visitors feel welcome and at home.

At the last meeting of the Mississippi State Dental Association (eighteenth annual session) the DENTAL HEADLIGHT was made, by the unanimous vote of the Society, its official organ. Its frontispiece is, therefore, very appropriately graced by the genial and handsome face of its popular retiring President, Dr. D. B. McHenry, of Grenada, Miss., a man too well known to the dentists of his state to need an introduction from our pen. He is a dentist of rare skill and inventive genius, and the profession is indebted to him for several valuable appliances.

The Tennessee State Board of Dental Examiners will hold its annual meeting at Lookout Mountain Inn on July 25, at 10 A.M. Applicants for examination and registration must be on hand at that hour. The members of the Board are: J. Y. Crawford, President; J. L. Mewborn, Secretary; R. B. Lees, S. B. Cook, W. T. Arrington, F. A. Shotwell. The last two were reappointed for three years.

MARRIED.

In Cincinnati, June 20, Dr. L. L. Yonker, of Bowling Green, O., of the graduating class 1889, to Miss Belle Goldamer, also formerly a student of the Department of Dentistry, Vanderbilt University.

Dr. Weldon L. Smith, of the class of 1891, Department of Dentistry, Vanderbilt University, was married to Miss Lizzie Oliver, at the home of her mother, Wellsville, N. Y., April 6. The bridal couple will make Friendship, N. Y., their future home.

Classmates will join us in congratulation and good wishes.

BOOK NOTES.

DENTAL JURISPRUDENCE. By William F. Rehfuss, D.D.S. Published by the Wilmington Dental Manufacturing Company, Philadelphia. Price, cloth $2.50; sheep, $3.50.

A volume the purpose of which, as its author avows in the " Preface," " is to supply to the dental and legal professions a comprehensive treatise, or text-book, covering the subject of dental jurisprudence, showing the relations the dental practitioner sustains to the law."

The work embraces every imaginable dental-legal subject, and quotes many legal cases, giving testimony and decisions.

The following subjects are treated of at length: Dental malpractice, injuries and deaths from anæsthesia, the legal rights of the dental practitioner, and the jurisprudence of patents. The work contains, besides a chapter on the history of dental legislation and the statutes regulating the practice in thirty-nine of the United States, some of which, unfortunately, have been amended, which amendments are not noted, and twenty-six foreign countries, forty-two chapters, a few of which are devoted to the following subjects: " Dental Witnesses—the Distinction between an Expert and Common Witness;" "The Legal Protection Afforded the Dentist by the Degree of D.D.S;" "Malpractice—What Constitutes Malpractice;" "The Degree of Skill Required—Ordinary Skill;" " Responsibilities of the Dentist for an Error of Judgment;" "Negligence, Injuries, and Death Due to Anæsthesia."

It bears the warm indorsement of the Hon. Alex. D. Lander, an eminent attorney at law of Philadelphia, besides the commendation of Profs. James Truman and C. N. Pierce, and Dr. E. C. Kirk, editor of the Cosmos, who carefully read the MSS.

The book is of such value that it will become at once the basis of all legal dental pleadings, and will settle many threatened suits before any action is brought. Dr. Rehfuss is entitled to the honor of giving to dental literature a book that will give it a legal standing which it has not before held, and do much toward elevating it in the eyes of the legal profession.

T̲H̲E̲ DENTAL HEADLIGHT.

| VOL. 13. | NASHVILLE, TENN., OCTOBER, 1892. | No. 4. |

Original Communications. • • • •

DOUBLE COMPOUND COMMINUTED FRACTURE OF INFERIOR MAX-
ILLA, WITH DISLOCATION OF RIGHT CONDYLE.

CASE TREATED BY E. W. KING, M.D., PHYSICIAN TO MENDOCINO COUNTY HOSPITAL;
W. N. MOORE, M.D.; AND W. H. HODGHEAD, D.D.S.

Mrs. H. was thrown from a buggy March 10, 1892, and admitted to the Mendocino County Hospital at Ukiah, Cal.

She was found lying unconscious by the roadside among the rocks and covered with blood.

At the external extremity of the left supraorbital process was a deep, punctured wound; on the left cheek there was a deeper punctured wound; both eyes were closed, the parts around them very much bruised and swollen, and some slight bruises on right breast. The four superior incisors were knocked loose, the left upper cuspid and first bicuspid loosened and split, the other teeth on this side above were missing except the third molar, and that badly decayed and loose. On the left side, below, both bicuspids and the sixth year molar were gone; the others were sound and firmly embraced by the alveolar process. The tongue, gums, and lips were bruised and terribly swollen. On left side, just below the groove for the facial artery, there was an ugly cut communicating with an oblique fracture of the jaw.

This fracture extended from a point on the external alveolar process, opposite the cuspid, to a point on the internal wall opposite the second molar. The gum on the internal wall here was torn away to the extent of an inch.

This end of the segment was displaced backwards, and the exposed point of the alveolar stuck up on a level with the top of the second molar. On this side the inferior dental artery was ruptured.

The fracture on the right side was between the cuspid and bicuspid—a straight break with considerable impaction. The gum on the inner side of the alveolar was cut, communicating with the

10

break. The right condyle was dislocated. The sublingual and submaxillary glands were injured and several fingers hurt.

Such was the condition of Mrs. H. when Dr. Moore first saw her.

We did not attempt reducing the fractures for ten days. During that time the physicians were endeavoring to build her up and I was speculating on the best method by which the segment could be held in place.

External pressure on left side would only have made matters worse. Plaster of Paris was absolutely useless, as was also the inter-dental splint.

We fitted plaster of Paris to the cheeks with malleable wires fastened in it, thinking they would hold the parts in place, but that was a failure.

I next measured the circumference of both lower second molars with tin foil, made a gold crown for each, and fastened them together with a silver bar, to pass in front of all the lower teeth. That might have succeeded, but I could not get it on.

Dr. King administered chloroform in company with Dr. Moore, and Dr. D. A. Hodghead, of San Francisco. While she was anæsthetized I attempted to adjust my crowns, but everything had to give way to a ten minutes' conflict with chloroform narcosis.

After she revived, I made another attempt, with no better results.

For the left side I next took a silver bar, a little more than one-quarter of an inch wide, grooved both ends and curved it, the convex side turned up. My object in curving was because the inclination of either end when in place would then be downward, and that would help to keep it in position. At the front end of the bar I bored two small holes for a wire to pass through.

On the outer corner of the back end I soldered a silver wire to pass along the buccal surface of the second molar and into the space between that tooth and the wisdom tooth. On the back of the inner side I soldered a flat piece of silver (three-eighths of an inch square), at right angles with the bar. We then pressed this appliance into place and passed a strong wire through the holes and around the cuspid, then twisted the wire with flat-nosed pliers. I then wired another silver bar to the labial surfaces of the incisors, passing strong wires from that bar up to and around the right upper cuspid and first bicuspid.

When those wires were twisted they brought the segment into what seemed a natural position. We had no guide above. The right lower cuspid was displaced a little, and will never be again in its natural position.

These appliances were fitted eleven days after she was hurt. Two weeks after that the right submaxillary glands suppurated and the pus was removed through an opening just below the groove for the facial artery on the right side, thus making the fracture on the right side also compound in two places, if you will permit the expression.

Her recovery was good but slow. More than a dozen spiculæ of necrosed bone were removed from time to time, including the left end of the alveolar on the inner side of the separated segment. That piece was half an inch long and nearly as wide.

I hope it may never be my lot to come in contact with a similar case.   W. H. HODGHEAD, D.D.S.

Selections. . . . .

PRESIDENT'S ADDRESS.*

BY GORDON WHITE, D.D.S., NASHVILLE, TENN.

Gentlemen of the Southern Dental Association: I propose to bring before you at this meeting several subjects of interest to our profession. The one I deem of greatest importance has been talked of, in an undertone, by many of us for a long time. We should have no secrets in our professional family. The subject is professional dignity, or rather lack of professional dignity, for the subject is too broad for me to touch on any but abuses known to all.

"Every profession has its scum," says a noted Frenchman. Alas that those whom in the South we term good men should place themselves on a level with that scum by their methods of advertising! True, it is often only a newspaper interview that catches the eye as we glance through the paper, but it is an advertisement none the less. In the secular press of one section we find a column given the dentist who has performed what he considers a very remarkable operation; in that of another section a column and a half is required to properly describe the beauty and perfectness of a cer tain piece of extensive crown and bridge work; while in still another we read not only of the wonderful inventions of our brother, but also of the architecture and furnishings of his office. In one locality we find a college graduate asserting his skill in every known branch of the profession and guaranteeing his work; in another, the familiar poem, "Mary Had a Little Lamb," adapted to the requirements of a dental advertisement. Such advertisements are usually accompanied by broad headlines and not infrequently by a picture of the remarkable individual.

Gentlemen, need I tell you that members of our Association engage in this reprehensible practice? Is it professional? Is it dignified? Does the profession approve it? Does it win public respect? A prominent man who for twenty years has advertised said to me in a recent conversation that he did not remember a single desirable patient who came to him through his advertisements.

*Read at the Southern Dental Association, Lookout Mountain, July 26, 1892.

Why is it that we are so frequently confronted by such advertisements? Are not the schools primarily responsible for this? One reputable (?) college advertises in the newspapers and holds out as an inducement to the uninformed would-be student the fact that the dental graduate is now recognized by the medical profession as occupying the same level as the medical graduate, and further, that their graduates at once step into a lucrative practice, making in ready money so many dollars a day. Another advertises for infirmary patients and holds not even church pews too sacred for the desecration of its handbills. Do such advertisements on the part of the schools give the students a correct idea of the dignity of our profession?

The student while at college should live in an atmosphere of ethics. Does he? It is generally understood that there is one lecture on ethics, delivered usually by the dean at the close of the term, but perhaps not more than one-half of the students hear it.

A worthy professor calls attention to the fact that a student, as a mirror, reflects the idiosyncrasies of his preceptor. What shall we say when a graduate from a college, presumably reputable, with the certificate of the State Examining Board, locates in a town or city and at once calls attention through the medium of flaming handbills to his "New Dental Parlors" and extraordinary low fees? Does he not, as a mirror, reflect the college from which he comes? Are we not agreed that by both precept and example the colleges should sustain and increase our dignity? Are such practices (of both dentists and schools) consistent with our code of ethics? If they are, should not the code be revised? If they are not, should we not feel it our duty to report such violations? Hitherto we have been too timid to report. It is not a personal matter, gentlemen, but we owe it to our profession to aid in every way possible in the suppression of that which will drag us into the mire. The highest court of England quite recently held that a man who joins an honorable and registered society must strictly observe the rules of that society under penalty of forfeiture of his membership and sustained the action of the General Council of Medical Education and Registration in removing the name of a prominent dentist from their membership because of his having advertised his business contrary to the rules of that body. The decisions of that court are a precedent for the courts of other countries. Is not the action of that council a worthy precedent for our Association?

Should not the undignified practice of many reputable men of

placing on their envelopes their business cards be condemned? Do
they not labor under the mistaken idea that it advertises them or
their business? We are professional men, not tradesmen; furthermore, I find upon investigation that in the rarest instances does a
letter pass through more than four or five hands in reaching its destination, and only when it fails to reach its destination is the business card referred to by the busy postman. To be sure, there is
nothing wrong or unprofessional in having our name and address on
our envelope for the safe return of our mail, but is it not shocking
to receive a letter from one of our professional brothers, the envelope
of which is adorned with a cut of his wonderful invention? Strange
as it may seem, I have received such from members of this Association. In a few instances I have received them with a cut of the
writer on both letter head and envelope, but they, I believe, were
not from any of our members.

To my mind the propriety of having price lists is questionable.
There is certainly no reason why a patient may not know the cost
of each operation, but the list, varying as it does from $5 to $50,
practically amounts to no list. To be sure, we have our rates, but
you each know that in like operations the fees are rarely the same.
It seems to me that it would be difficult to explain the difference
satisfactorily to the patient. After all, is the list necessary? We
are not in the mercantile business and do not need the advertisement.
Do not our patients place themselves in our hands because they have
confidence in our integrity and skill? During a practice of thirteen
years I have only once been asked for a list, and during my investigation extending through a number of years I remember only one
man who claimed to adhere to his list. He very frankly said that
he did. If we have a list, should we not adhere to it? If we do not,
are we not practicing fraud and deception? Does it not look unprofessional? The "Cheap John" displays his list on his sign, the dentist of the "upper ten" on his appointment card. Is there any difference save in the fees?

In most, if not in all of our States, laws have been enacted restricting in some particulars the practice of dentistry, and boards of
dental examiners have been appointed. These laws were enacted
for what was conceived to be the protection of the public and our
profession as well. They may not and do not fully accomplish the
desired result, but they are a step toward a higher standard of requirements for the dentist, and the boards in enforcing them should
have the moral support of all dentists. The boards need the sup-

port, for, while it is almost beyond belief that any one would oppose that which even tends toward our elevation, the board of Tennessee has met with opposition.

Our fathers in 1869 organized our Association for advancement in the science and dignity of our profession. Then the spirit of professional interest was stronger than the animal of self-interest, and those loyal, high-minded men did not even dream that one of our membership would ever be so debased as to be valued for personal aggrandizement. It has been said that the professions are made strong by what they include rather than exclude. Let us then include so much love for our grand profession, such high, pure aims in its practice, so much enthusiasm for its advancement, that there shall be no room for any unprofessional act or thought. Let us work to an ideal, and let that ideal be as high as finite conception can reach.

Much has already been said in regard to a home for the Southern Dental Association, and a permanent committee on a "Dental Chautauqua" has been appointed. The idea, as I understand it, is to erect at some desirable summer resort a building in every way adapted to the needs of the Association, and where year after year the meetings may be held. There is much that is desirable in this plan, but is it practicable? It means the outlay of a large sum of money without any return, a dead weight for the Association to carry. Furthermore, men will not go to the same place year after year.

If the Association will have a home, let it by all means be located in some central city and so constructed that a part can be rented, yielding sufficient revenue to pay all expenses. So located, "The Home" will be an object of interest to all dentists passing through that city. Besides, there will always be a number of resident dentists to keep up the interest. My preference is not a Southern but a national home, and, as has been suggested, a national museum, located in Washington or some central city, where from all parts of the country we can send our treasures. Why not unite with the other societies and build a home that will be a credit to our profession and establish a museum that will fittingly preserve our history for this and future ages—in other words, a monument to the dental profession of America?

In 1890, at the meeting of our Association in Atlanta, it was suggested that we be represented at the World's Fair in 1893. The American Association took up the suggestion, and a committee of fifteen has been appointed by the two societies, which committee will meet during our present session. The work of organization is far advanced and the World's Columbian Dental Congress will be held August 17 to 27, 1893, in Chicago, Ill. Let us not forget that it was our suggestion, and that as such it behooves us to give the committee all the support they expect from us. Certainly they have a right to expect our presence, and so far as possible we should attend this congress.

At our last meeting there was a resolution to the effect that the Constitution be so changed as to provide for the election of officers at a much earlier hour, so that the newly elected President might have the opportunity to make his appointments. This idea is excellent. I would suggest that a section on orthodontia be created. Properly, it does not come under the head of any existing section, but is separate and distinct.

It would be well also to thoroughly revise the Constitution. There are some defects not necessary to allude to here, which a committee would readily detect, and of which all the ex-Presidents are aware. The principal one is its vagueness in setting forth the duties of the officers and committees. I do not make this last suggestion to bring about a discussion of the Constitution, for we wish to embody nothing new, but simply to make plain that which we already have.

I have called attention to these things, gentlemen, because of my deep interest in the continued advancement of our profession and the preservation of its dignity.

In the name of those who have shed luster upon that profession, let us be faithful to our sacred trust, transmitting to those who will succeed us an honorable record of duty faithfully performed.

PUS FORMATION, REVIVED.*

BY D. R. STUBBLEFIELD, A.M., M.D., D.D.S., NASHVILLE, TENN.

SUFFICIENT apology for the renewal of an old subject is found in the indisputable fact that the rank and file of the profession fail to a large extent to digest, so to speak, the new matter presented through the journals from the few original investigators that the world possesses. If I shall succeed in rendering the nature and for-

----

* Read at the twenty-third anniversary meeting of the First District Dental Society of the State of New York.

mation of pus more clear to some, I shall deem the attainment of that much a full compensation for the time and patient work incident to the undertaking.

The following substantially direct quotations from a personal communication will open the subject in the middle of things, and fully show why the task was assumed, and why the present method of sequence was observed.

It was written: "It is believed that all bacteriologists claim that pus, wherever found, contains bacteria, and that the bacteria are intimately concerned in causing that pus. They substantiate this position by injecting into a rabbit's leg a virulent, poisonous acid, and thereby develop a suppurating wound. Then, by exclusive argumentation, the same acid, hermetically sealed in a thin glass shell, is put into an incision and healed in. After perfect union the shell is crushed, liberating the acid, and no suppuration is developed. This is due to the exclusion of bacteria. This seems to advocate the theory that bacteria do not float in the blood, and so reach an internal wound.

"Here comes a condition not uncommon in dental practice. A patient falls, strikes a tooth, which, becoming sore, brings about a visit to a dentist. Tooth sound, but from the history and a slight opacity it is surmised that the pulp is dead. A drill hole is made into the chamber, when pus freely exudes. Upon attempting to remove the pulp, however, it is discovered that it is far from dead, and an application of arsenious oxide is necessary before its removal can be effected. Now here we have unmistakably pus running from a locality which has been well protected from the external approach of bacteria. If we are to suppose that they have entered through the circulation, why is it that the point most distant from their entrance way—the foramen—is suppurating, and not that part which is around and about the foramen itself?

"Is it possible that bacteria do not cause suppuration?"

We have in the foregoing extract the gist of what was asked. The author voices the doubtful opinions existing in the minds of so many of our profession, that we feel fully justified in this attempt to remove the foregoing uncertainty. We express the hope that this effort shall be so thoroughly supplemented, not to say corrected and added to, by the discussion, that all may be clear and positive in their understanding of the subject.

We propose as possibly the most natural and effective method, to reply straightway to each question as it occurs, and then to follow immediately with the reasons which have been derived from the best possible sources.

First: "Bacteriologists claim that pus, wherever found, contains bacteria." It may be just as well, at the very outset, to correct the impression prevailing so generally that *bacteria* and the causes of pus, *micrococci*, as we shall see, are the same, and their names, therefore, are interchangeable terms. This is a mistake, although its full effect may be rarely, if ever, felt, because it is so usual, and accepted even among those who ought to strive for exact perspicacity. To be very brief at this time, we may state that *bacteria* or *bacilli* are elongated, rodlike microörganisms, which are the active agents in putrefaction. They exist wherever foul odors are present. On the contrary, *micrococci*, of three varieties, are the causes of pus, and never cause what is commonly called putrescence. Therefore the term permitted above should very properly be changed to *putrefactive-alogists*, unless we shall permit a very loose and indefinite nomenclature to exist. It would be exceedingly difficult to frame a pronounceable, not to say a mellifluous, term to denominate one who occupies himself in the world of microörganisms. We would suggest, however, *micro-organalogist*, as etymologically meeting the demands of the case.

The claim that pus, wherever found, contains the cause of pus is not true. The three forms of micrococci that cause pus demand for food dead—that is, devitalized—but undecomposed organic substances, but they cannot thrive on the products of decomposition. As soon as the supply of dead, undecomposed animal tissue is exhausted, the micrococci starve and perish. Biologists teach that reproductiveness exists in increasing ratio as we go down the scale of organized being. We may not be surprised, therefore, to find that though the micrococci are gone, their *spores* or seed exist in great number, which, while dormant in the pus would awaken through the stimulation of fresh pabulum under favorable conditions.

This brings us to the clause completing the first sentence in the quotation, which reads: "And that the bacteria are intimately concerned in causing that pus." This is true, although it is not a rule without exception. Professor Gerster, in his admirable exposition of the antiseptic and aseptic method in surgery, cites a case illustrating the fact that the introduction of micrococci into the organism does not always produce suppuration. Every manifestation but the crucial one was observed, indicating systemic intoxication from septic infection; but the patient, he says, went off glorying in his triumph of endurance over diagnostic acumen. The rarity, however, of such an escape is so conspicuous that we may be permitted to point the moral from its departure from the rule. Modern surgery declares that inflammation is a physiological process if carried on without septic infection.

The older teaching that all wounds must heal by suppuration, if not by first intention, has been clearly refuted by the well-supported claims of antiseptic surgery. Pus does not always contain micrococci, for they may have starved before the examination; and micrococci, in rare instances, may fail to produce pus, owing to an excessive vitality of the other organisms; but when pus is present we may assume as incontestable that it was caused by the presence of micrococci.

Secondly, our quotation says, "They substantiate this position by injecting into a rabbit's leg a virulent, poisonous acid, and thereby develop a suppurating wound. Then by exclusive argumentation, the same acid, hermetically sealed in a thin glass shell, is put into an incision and healed in. After perfect union the shell is crushed, liberating the acid, and no suppuration is developed. This is due to the exclusion of bacteria." This is only partly true. The injection of a corrosive agent, properly guarded antiseptically, does not result in suppuration, though necessarily followed by destruction of tissue. We think the author used the wrong terms when he spoke of a "virulent, poisonous acid," for we have not found any mention of any experimentation with such an agent. If we interpret him to refer to corrosive agents, like croton oil, corrosive sublimate, etc., we may obtain a most lucid and convincing refutation to the theory that destruction of tissue by chemical agents always produces pus, from Professor Gerster. He says: "The common experience that certain acutely irritating substances, as, for instance, croton oil, oil of cantharides, turpentine, concentrated solutions of corrosive sublimate, and others, brought in contact with living tissues, always should produce suppuration, represented a serious gap in the theory of microbial origin of suppuration. If invariably proved, it would be more than a defect, as it would positively contradict the theory that suppuration is exclusively and *always* the result of the development of microörganisms. The experiments of Councilman, who introduced under the skin of animals small glass globes filled with sundry irritating substances, and then crushed them, all led to suppuration. Scheuerlen and Klemperer, however, in going over Councilman's experiments, showed that his procedure was faulty, inasmuch as pre-

cautions had not been taken to exclude the introduction of microbes
along with the croton oil, etc. They, moreover, positively demonstrated by a large number of successful experiments that wherever
thorough aseptic cautions were observed, suppuration never followed
the introduction of even very considerable quantities of the mentioned substances. Small quantities caused some exudation of plasm,
and then were absorbed outright. Afterward the fragments of the
glass receptacle were found imbedded in a film of connective tissue.
Large quantities of croton oil, for instance, caused a coagulation-
necrosis of a limited mass of tissue, which was found dense, blood-
less, and of a yellow color. These nodes of necrosed tissue were
gradually absorbed, *suppuration never following the experiment.* This
fact is in full accord with other incontestable facts of the same character, as, for instance, the absorption of necrosed ovarian stumps in
the abdominal cavity, if there be no microbial infection present."

This places the foremost belief on the subject clearly and tersely
before you. The complete disproval of the almost classic results
presented by Councilman, and accepted so long as conclusive by the
world, ushers in a new era of belief. The opinion, we believe, in the
profession has been in accordance with the revelations of Councilman, and we rejoice to be able to present the later and better view.
In that learned and carefully prepared monograph on the "Origin
of Pus," by Dr. W. H. Atkinson, who, along with the benefits he
conferred upon our profession, was ever conspicuous in presenting
the cream of new literature and experiments, we find Councilman's
results acknowledged and unquestioned for correctness.

Thirdly, in our quotation we read: "This seems to advocate the
theory that bacteria do not float in the blood, and so reach an internal wound." If we are to understand by *bacteria* the active, proliferating causes of pus—the micrococci—we think we may boldly
advocate the theory that they do not float in the blood, except when
the organism is already the seat of septic infection, diffusing itself
far and wide. Antiseptic surgeons all agree that septic infection
must come from without, although once in it may be the cause of
metastatic abscesses.

We know, also, that the spores, or seed, of the microörganisms
may float in the atmosphere and may be inhaled into the lungs. But
we must speculate purely when we assert that they freely enter the
blood current. If they should, however, we know they are not able
to develop morbid conditions, but are able only to develop in a favorable *nidus* produced by some other cause. Professor Thomas Fille-

brown presented at length "Vitality as a Germicide," and we must all agree with him that the inhibition of vitality upon the adventitious spores that may find their way into the complex avenues of our physiological city renders them harmless adventurers. If, however, a lesion occurs, a change from normal vital resistance is found at any point, the spores begin at once to take on changes from the stimulation of their most welcome pabulum, and infection is established. This conclusion is easily and logically reached if the presence of the spores in the blood is granted. With reference to this point some speculation may arise. If there has been, or is at present, an abscess within the organism, it is not asking too much to grant that the spores may be passed into the circulation by the retrograde metamorphoses going on all the time.

As to whether the spores afloat in the atmosphere and inspired by us can pass the endosmotic wall of the lung cells varied opinions may exist. Our own opinion is that they can go through, and do constantly, with the same impunity that the myriad microscopic life found in the water we drink passes through the wall of the stomach. Still, if it were denied, the presence of abscesses at some time present or remote is so constant that they might be relied upon for spores inside the organism, the measure and extent of the abscesses being always in the ratio of the tissue lowered in vitality, so as to be brought within reach of the spores.

Fourthly, our quotation says: "Here comes a condition not uncommon in dental practice. A patient falls, strikes a tooth, which, becoming sore, brings about a visit to a dentist. Tooth sound, but from the history and a slight opacity it is surmised that the pulp is dead. A drill hole is made into the chamber, when pus freely exudes. Upon attempting to remove the pulp, however, it is discovered that it is far from dead, and an application of arsenious oxide is necessary before its removal can be effected. Now here we have unmistakably pus coming from a locality which has been well protected from the external approach of bacteria. If we are to suppose that they have entered through the circulation, why is it that the point most distant from their entrance way—the foramen—is suppurating, and not that part which is around and about the foramen itself? Is it possible that bacteria do not cause suppuration?"

In the light of what has already been cited from the best sources, this enigma may be easily deciphered. The pulp being a body of a peculiar gelatinous consistency, suffers, like the brain, from the shock or jar of a sudden concussion. In the case recited above the tooth had been subjected to a shock that, while not ·enough to cause a change of molecular nature throughout, was sufficient to cause a molecular change in the extreme portion of its mass. This change rendered it, to that extent, unable to resist the attack made by the morbific agent—the spores of the micrococci—passing, it may be, through the round of the blood.

It may be illustrated by the presence of marauders or freebooters in a seaboard town passing through the streets without harm to any one because of the protection of police or some other municipal force, but the relaxation or absence of such protection is the signal for depredations to commence.

Thus the spores of the micrococci present in the blood current as harmless drift, so to speak, become transformed by favorable opportunity into agents of destruction. The limitation is put upon their action, as it is in the case of all circumscribed inflammatory processes, by the inhibiting power of vitality, which opposes a successful barrier to the diffusion of ptomaines. The line of demarkation is drawn between the sick and the well, and it is maintained with certainty against the disease-producing agent.

Pus is formed by the usual transformations within the limited area; health reigns beyond it.

From the consensus of opinion among the latest and best surgical authorities, it is not possible that bacteria—that is, micrococci—are not the cause of suppuration.

This ends our reply to the interrogatories propounded by one of the most alert minds of the profession; but we shall not deem our task completed until we have more fully met the precise demands of our subject title. To that end, therefore, what is suppuration? and what is pus? To answer these questions intelligibly we must first speak of inflammation, its nature, and phenomena. Literally defined, inflammation is merely a preternatural heat; but the world of surgery has long gone beyond, and we may add below, the mere surface indications that first directed attention to local and systemic phenomena now well understood. The disturbances of nutrition, producing hyperæmia, heat, redness, swelling, and pain, and the extended illustrations necessary to show that several or all of those phenomena must be present to give character and certainty to any inflammation, would consume too much time and space for us here. Our subject being not a research into, but rather a lucid explanation of, the essence of inflammation as understood and explained by the best modern authors, a brief sketch of the leading features of the process will be sufficient for our needs. Whether the cause of inflammation be one of direct injury and irritation of a part, or whether it be due to lesions of the inhibiting nerves or trophic centers remote from the local expression of the morbid process, the pathological changes are practically the same. The activity and violence of these changes will depend in part upon the character and extent of the injury, the presence of certain forms of microörganisms or animal poisons, as well as upon the anatomical character of the part involved, together with the ability of the tissues to resist death and to repair the damages inflicted. It is known that when a vascular living animal tissue is subjected to irritation the vessels within the zone of irritation undergo an instantaneous contraction, followed almost immediately by abnormal dilatation. This contraction is the response of the vasomotor nerves; the dilatation, more complex, is due to paralysis of the inhibitory nerves, to early changes in the vessel walls, or to overstretching of the connective and elastic tissue which supports the vessels. Hyperæmia naturally ensues, and is followed by a more or less complete blood stasis, which is most marked at the center. At this stage leucocytes in unusual proportion appear in the capillaries and venules, to the walls of which they adhere and through which they pass into the intervascular spaces. Following them we find a more or less extensive effusion of red blood disks and liquor sanguinis. This localized structural excitement brings about active cell-proliferation, and this results in the rapid formation of the common embryonic tissue. This embryonic tissue is known to be composed of protoplasmic bodies or cells, spherical, slightly polygonal from pressure, and varying in size from $\frac{1}{3500}$ or $\frac{1}{2000}$ inch, and sometimes even larger. They may be nucleated, but usually appear as slightly cloudy or granular protoplasmic bodies without distinct nucleus or nucleolus.

A diversity of opinion exists as to which of the normal cells are most active in the formation of new tissue; but the safest view seems to be the one that holds that all the cells undergo proliferation when stimulated, and all combine to develop the embryonic tissue as a common product. This conclusion seems rational, since the normal *rôle* of every cell element of the body is one of proliferation and the formation of new elements to replace those that have finished their life history, and it seems reasonable that a more rapid proliferation of the same cells would recur under conditions of increased hyperæmia, and therefore exalted nutrition.

The products of the inflammatory process may be organized into permanent tissue, or failing in this may perish. The peculiar type of the new tissue is probably determined: (1) By the nature of the original cell from which it sprung. Thus the experiments of Goujon showed that the medullo cells and myeloplaxes of bones in young animals, when injected into the muscular tissue, developed into bone even remote from the parent tissue. (2) By the location and function of the new tissue, as is shown in the development of exostoses from a common embryonic tissue near the insertion of tendon into bone.

When the inflammatory process is rapid and severe, the new tissue may perish suddenly, and with it occurs the rapid death or gangrene of the tissue involved. Under milder conditions, however, the supply of nutrition being more gradually diminished, the embryonic cells undergo fatty degeneration and absorption, which is called resolution.

This gives us a *résumé* of the phenomena of inflammation, and presents the subject very briefly. We see from it why inflammation has been very properly called a physiological process when unmodified by septic infection. It has fallen to the lot of the school of antiseptic and consequent aseptic surgery to recognize and explain this process, modified by the introduction of microörganisms and their morbific products, called ptomaines. All authorities agree that septic infection must come from without the organism. The varying intensity of different cases of infection seems to depend in a great measure upon the varying degrees of vitality of different microbial cultures. Recent putrescence is so abundant in all human surroundings that it is an easy matter, almost unavoidable without antiseptic precautions, to introduce into the system considerable masses of active, proliferating microörganisms. The dry spores floating with the air will be easily taken care of by the living, unbroken tissues; for inherent in them we find that mysterious power of nature tending always toward the maintenance of health. Every injury, however, causing a wound, destroys the vitality of those cells that lie in the direct path of the agent producing the lesion. These injured cells, combined with the blood and lymph that is poured out, constitute an area of dead but not decomposed albuminoid substances. This type of organic matter furnishes the richest and most welcome pabulum for the rapid development of myriads of microörganisms. From the moment of their introduction we find the process of inflammation modified. The succeeding steps are marked now by a strength and sympathetic effect upon the entire organism, before unfelt. The existence of microörganisms—a kind of fermentation—is marked by the rapid development of an exceedingly diffusible and poisonous product called ptomaines. By its great diffusibility this agent reaches the adjacent vasomotor nerves, that soon show a toxic impression, and the equilibrium between them and the inhibitory nerves is destroyed. Soon after this the blood is altered, and the vascular walls lose their ability to hold their contents. The surrounding tissues are infiltrated with corpuscles—white and red—and also plasma, as a consequence of this intoxication. Stagnation and infiltration produce a high degree of tension, leading to compression of the larger afferent vessels. The infiltrated portions, devitalized by suppression of normal circulation, readily succumb to the inroads of the millions of microörganisms, and actual necrosis rapidly ensues. The last stage of textural destruction is the final liquefaction of the tissues and infiltrating leucocytes, aided by the exudation of large quantities of lymph serum from the adjacent obstructed blood vessels. This is the formation of an abscess, or a cavity filled with lymph serum, myriads of dead white blood corpuscles, and quantities of shreds of necrosed tissues; in a word, pus. When septic invasion has been of sufficient extent and well advanced, the great tension of the parts will necessarily cause an overflow of the diffusible contents of the focus into the radiating efferent vessels, the veins and lymphatics. The ptomaines, thus entering the general circulation, will at once produce systemic intoxication, manifested by marked rise of bodily heat, rigors, sickness, headache, delirium, and general dejection; in brief, that pervading irregularity of the entire nervous system known as septic fever.

The causes that produce the phenomena above recited are certain microörganisms that may lead a parasitic life in the human organism. Without attempting to consider all that might be dwelt upon, it is sufficient for us now to consider only those that are known to be the active causes of pus formation. The most common source of suppuration, according to Rosenbach, are several varieties of globular fungi called micrococci. These are so few that we may readily retain their names and general characteristics. They are the *Staphylococcus pyogenes aureus*, the *Staphylococcus pyogenes albus*, and the *Streptococcus pyogenes*. The *Staphylococcus pyogenes aureus*, or golden grape coccus, cannot be distinguished from the second variety, the *Staphylococcus pyogenes albus*, or white grape coccus, under the microscope, and must be seen with reflected light to develop the color which distinguishes it. The morbific action is quite indistinguish-

11

able, as both possess the ability to thrive upon animal tissues, causing coagulation, emulsification, and distinct abscess. They occur in colonies or groups, like grapes in the bunch, and hence get the name grape coccus. The third variety, *Streptococcus pyogenes*, or chain coccus, seems to be slightly less active than the other two varieties, except as to its facility in entering vessels and causing progressive gangrene.

These varieties of cocci constitute the living germs that act as the causes of suppuration. Suppuration, as here referred to, must not be confounded with what is commonly called putrescence. Decomposition of tissues, accompanied by the evolution of foul odors, is always due to the fermentative action of diverse forms of elongated, rodlike bodies called bacilli or bacteria. These are seen in association with the cocci pyogenes, where the pus production is complicated by the decomposition, properly speaking, of tissues, evidenced by foul odors. The failure to keep these two terms clearly apart renders many otherwise excellent essays obscure to a certain extent, and we trust we may be permitted, thus incidentally, to make a protest on behalf of the profession at large. As long as morphological microscopists are able to differentiate microörganisms by forms, why not give us the benefit of such separation, uncomplicated by the exchanging of terms until confusion is worse confounded? Let us stick to *coccus*, a globular microörganism, several varieties, and pus-producing; and then, *bacterium* or *bacillum*, a microörganism of elongated, rodlike form, foul-odor producing. If we have form clearly separated, we may hope to more readily associate function— disease-producing power—with the different shapes.

All forms of suppuration owe their origin, as we have previously stated, to infection from without the organism. The epithelial layer of the skin or mucous membrane must be broken to admit the microörganisms. Not all injuries, trivial or otherwise, however, are infected from the beginning. Their frequent spontaneous healing proves them aseptic. Even gunshot wounds have been known to heal without suppuration. An exceedingly interesting case of pistol-shot wound of the tongue and parts posteriorly contiguous is cited by Professor Gerster in proof of this statement.

But when the entrance into the system has been gained, whether superficial or deep, the phenomena recited above as typical of the inflammatory process, modified by microbial infection, appear in more or less pronounced ratio. The intensity and extent of these phenomena depend (1) upon the virulence of the microörganism introduced, (2) on the quantity absorbed, and (3) on the power of resistance of the organism invaded—that is, the *status* of the *vis medicatrix naturæ*.

GROWTH OF THE CEMENTUM.*

BY R. R. ANDREWS, M.D., D.D.S.

In the year 1858, Magitot, a French histologist, claims to have found within the follicle of a developing tooth a special organ for the development of the cementum. In 1861 Robin and Magitot made a presentation of the same facts anew. With the exception of these authors, I am not aware that any other authority has recognized the presence of this special organ, while such men as Kolliker, Waldeyer, Herz, and others have formally denied its existence. In my own investigations I have not been able to trace it with certainty, although there are appearances in a fully formed follicle of a tissue under the calcifying dentine germ, between it and the outer covering of the sacculus, that might admit of the supposition of the existence of such an organ. I have noticed this appearance in sections from embryos from the pig and the calf. At a later stage, where the crown is further developed, there is also to be seen infoldings of the tissue at the base of the germ that may develop into an organ for the growth of the cementum, as stated by Magitot. But in teeth more matured, where the cementum has already commenced its growth, I cannot trace even the outlines of a special organ, although I do not consider my investigations to have been extensive enough to warrant me in denying its existence altogether.

Wedl, whose description of the development of the dentine and the enamel is so minute, has but little to say about the development of the cementum. He tells us that at the margin of the crown the dental sacculus contracts, and upon its inner surface the formation of the cement is effected, increasing gradually as the formation of the root advances. The lower segment of the dental sac becomes therefore the root membrane of the tooth. He believes with Tomes that Nasmyth's membranes belong to the cement. Again, he states that the dentine and cement are connected together by means of a layer composed of an agglomeration of transparent globules of varying degrees of thickness. The spaces intervening between the latter

---

* Read before the Section of Dental and Oral Surgery of the American Medical Association, held in Washington, D. C., May, 1891.

(interglobular spaces) are irregularly notched, and frequently in very close proximity to one another. He considers that the cement commences outside of this layer, but some of my own sections show it as a dividing line. In describing the methods in which hypertrophy of the cement is formed, he speaks of the various sizes of the corpuscles which form it, stating that many of them have a glistening appearance. Smaller corpuscles are sometimes attached to the sides of larger ones, or are blended with them. Large and small ones also occur separately. These so-called corpuscles are, I believe, globules of calcoglobulin, forming, by merging into others, a layer of calcoglobulin that shall form the matrix of the hypertrophied cement.

Tomes tells us that it is difficult to point out any distinguishing structural character between primary bone and that of the cementum. The cells close to the surface of the forming cementum, which were formerly called osteal cells, have now been named osteoblasts. No bone is formed until after the appearance of this osteoblast tissue, and Rollet believes that these osteoblasts are essentially a new growth; they are so distinctly marked off that it almost assumes the character of an epithelium. If we harden a partially formed tooth in chromic acid, and subsequently decalcify and cut it transversely through the root, we meet with the following structures from without inward: On the outside is the outer part of the sacculus, now the periosteum. Internal to this is a layer to which the name cambium has been given, consisting of roundish cells with processes. These lie in a reticulum made up of cells which give out a small number of homogeneous transparent processes. By the inosculation of these processes a network is formed. Between this network and the fully formed cementum lies the osteoblast layer, consisting of much larger cells. As the osteoblasts form a continuous layer and are very numerous, it is obvious that only a small percentage of them ever form lacunæ, or bone cells, or otherwise retain their individuality. As the process of calcification goes on, the outlines of individual cells become lost in the general transparency of the matrix, only a cell here and there remaining as a lacuna. Again, he states that contiguous osteoblasts become fused together by their exteriors, so that their individuality is lost.

Prof. James Tyson, in 1873, writes: "Some difference of opinion existed as to the exact tissue which undergoes conversion into cementum, some alleging that it results from ossification of the tooth sac, while others, among whom are Kolliker and Beale, believe it to originate in a soft stellate tissue, made up of branching and commu-

nicating cells, which are found upon the surface of roots of teeth,
and within the tooth sac. This tissue undergoes calcification, spher-
ules of lime being deposited, which gradually fuse and form a transparent, intercellular substance. In this process not all the cells of
the stellate tissue become lacunæ of the cementum, but some are obliterated by the deposit, and there are therefore fewer lucunæ in the
resulting cementum than in the previous stellate tissue, while the
canaliculi are much more numerous than the prolongations of the
stellate cells, many of the lacunæ have thirty or forty prolongations,
while the stellate cells rarely have more than from ten to twelve.
The cementum is more slowly formed than bone and a more permanent, but probably less perfect, tissue; its matrix is harder and more
transparent, in this respect approaching the dentine."

Klein tells us that the tissue of the tooth sac represents the matrix
from which the cement is formed; its structure and function are that
of the osteogenetic layer of the periosteum, and that the formation of the cement out of that tissue is identical with subperiosteal
bone.

My investigations show that, if we examine the developing tooth
just after the cementum has commenced to form, we shall find that
the matrix of the cementum is made up of masses or layers of that
tissue which we find everywhere on the border land of calcification,
between the organic and inorganic substance. This is a tissue which
Tomes has said was produced solely by the destructive action of
weak acids, but that his conclusions are erroneous is proven by the
fact that this tissue appears in sections where no acid has been used.
It is a tissue formed by the coalescing of minute globular bodies,
calcospherites, into globules and layers of a tissue called calcoglobulin. The minute globules, in forming the matrix of cementum, seem
to come from the osteoblasts, and form the calcoglobulin layer in
somewhat the same manner as I have described and pictured in the
developing dentine. The smaller globules by merging into each
other form larger ones. They have a glistening appearance, like
fat cells, are about the size of the osteoblasts, but are not cells,
though often taken for them. They have no membrane and are
without a nucleus. They are the bodies which Tomes and others
say become fused together by their exteriors. Outside of the layer
formed by the globules, the developing matrix of the cementum, we
see a row of cells, which Rollet stated looked like an epithelium.
They are the osteoblasts or cementoblasts, and the granules that
have been described in their substance are very minute calcospher-

ites, which the cells give out to the forming matrix. Tomes has
called them osteal cells. They are the same in appearance as those
we see around the edges of the developing bone of the jaw. Just
exterior to these cells we find roundish, nucleated cells with innu.
merable processes, reminding one somewhat of a stellate reticulum,
only that the stellate character is not so marked. Just outside these
we find the connective tissue which is really the periosteum. If
there exists at this time a special cement organ, it must be formed
by this slight amount of stellate tissue which is between the periosteum and the layer of osteoblasts that are against the forming cementum.

In preparing my tissue, so as to have it as near life as possible, I
made use of the same methods as were described in my papers on
the development of the dentine and of the enamel, and I find that
the osteoblasts, which are said to be full of a peculiar granular substance, are, when the tissue has been properly prepared, found to
be filled with minute spherical bodies, which have a glistening appearance. Across the developing matrix of the cement are found
numerous fibers, probably connective tissue fibers, that are also found
in developing bone. They were seen and described by Sharpey, and
are named after him, Sharpey's fibers. They become calcified within
the matrix.

As the cementum grows thicker we find that the developing
matrix is infolding in its substance large nucleated bodies which
appear to be connective tissue corpuscles. They are somewhat
larger than the osteoblasts, and are forming the cement cells, or
lacunæ. They have a higher function than the osteoblasts—that
is, they are to give nourishment to the matrix, being connected
with others by means of canals or processes, of which there are
many, some of which run in the direction of the termination of the
dentinal tubes, as though connected with them. When inclosed by
the developing matrix of the cement, minute, glistening, globular
bodies are seen within their outline, or membrane; indeed, it appears
as though these minute globules were deposited on the periphery
of the cell by the cell itself, and here fusing, give the cell its peculiar characteristic shape, which is not as regular as that in bone,
and is oftentimes very much larger. Their processes, probably,
anastomose with the dentinal tubes through the interglobular spaces
of the so-called granular layer, although I have never been able to
trace them. I look upon the granular layer itself as a condition
caused by an arrest of the developmental process, while the first

layers of the dentine were being formed. It has its existence solely from the fact that in the first-forming layer of dentine the globules or calcoglobulin which form its matrix did not fuse together. It is exactly identical to the interglobular spaces found in the crown. The point that I would emphasize in this paper is that the matrix of the cementum is formed from a secretion of the osteoblasts, and this secretion is a multitude of minute globular bodies given out against the dentine.

In a work entitled "General Biology," written by Professors Sedgwick and Wilson, they make a statement that the matrix of a tissue is composed of lifeless matter, which has been manufactured and deposited by the living protoplasm, constituting the bodies of the cells; and again, cells may manufacture a lifeless substance which appears in the form of solid partition walls between the cells, or as a matrix solid or liquid in which the cells lie; and again, the cells are small masses of living matter or protoplasm, which deposit more or less lifeless matter either around (outside) them or within their substance. The lifeless matter which is given out by the osteoblasts to form the matrix of the cementum is in the form of these minute globules, and these fusing together form larger ones, which, merging, form layers of an uncalcified substance that Professor Harting has named calcoglobulin; this is by further calcification to become the calcified matrix.

In speaking of the formation of exostosis, Tomes states that at the junction with the root he finds a substance that is dense. It is torn with difficulty, and under pressure slips about between the two glasses. It is gelatinous, osseous matter, and with it may be seen rounded, amorphous molecules. This is a good description of calcoglobulin. Tomes says that cementum is not developed by a direct metamorphosis of the periosteum, but by the calcification of a new growth. Cells are produced, the individuality of which becomes lost in the process of calcification; the interior of the cells seems stuffed with an opaque and dense substance disposed in large granules, among which the nuclei cannot positively be pointed out. The large granules within the cells are really calcospherites, and careful preparation of the tissue will clearly show them. The merging together of these globules forms a layer, and this calcifying, layer after layer, gives to the cement the peculiar laminated appearance that is so often seen. It is by no means difficult to trace evidences of this globular formation in the cementum of fully formed teeth; indeed, I have several sections of human teeth that show the outlines very clearly; so clearly, in fact, that one might call it the interglobular spaces of the cement, probably caused from some arrest in the full development of the tissues. (*International Dental Journal.*)

CHARGES FOR PROFESSIONAL SERVICE.

THE amount charged for services is one of the most important factors that help to make or mar the fortunes of the modern practitioner of dentistry. This is especially the case with the beginner. Let him be ever so neat, gentlemanly, and thorough in his work, and charge four or five dollars an hour, he will spend most of his time idly, sighing for the patients that never come. While across the way he may see an untidy person constantly busy over an ever-increasing crowd of patients.

The cause of this most unfortunate state of affairs is only too evident. The experienced dentist is not paid merely for excavating a tooth, or for materials placed therein. This fee remunerates for skill, for delicacy of manipulation, for the perfect confidence which he inspires in his patient, and for many other desirable qualities which years of practice alone can give.

If this be true, the failure of the beginner to build up a practice while charging the same price as the older dentist is only what might be expected. Until he has a certain amount of experience outside of the dental college he should keep the fees reduced to such a point that patients will come to him in numbers.

One of the most successful practitioners of Philadelphia, in speaking of his early professional life, said: "I made it a rule to be always busy. If I could not fill my time at two dollars an hour, I charged one dollar. If one dollar could not be paid, rather than be idle, I worked simply for the cost of the materials. Thus, my hand gained skill, my experience enlarged, and the number of my patients increased. True, the low-priced patients, in themselves, did not bring much money, but often, very often, they were the means of sending me their friends, who paid liberally."

The man whose practice consists of a few rich families, charged exorbitantly, occupies a humiliating position, as he is necessarily entirely dependent upon them for his material existence.

A dentist, to be independent, should have a practice so large that the loss of three or four of his most wealthy patients should be a matter of little moment. To bring about this desirable state of affairs he should be sure to give a liberal return for money received.

But it has been said by beginners: "If we start with our prices too low, how shall we be able to raise them?"

The answer to this is clear and simple. If a dentist has eight hundred patients, all of whom pay two dollars an hour for services, he can first raise the price to newcomers, and later, as confidence in his position is established, he can extend the advance throughout the entire practice.

Many established practitioners regard low prices in beginners as a fault, as an attempt to undermine the average scale supposed to be charged by the profession at large. Guided by their protests, many a conscientious beginner fails to collect a practice, where one without regard for dental ethics succeeds.

A professional man is paid quite as much for the manner of performing the work as for the work itself. Leading actors like Booth and Jefferson are not disturbed because large numbers of professionals work for a small sum nightly. This fact nowise jeopardizes their large incomes. So long as their work maintains its high excellence they need not fear competition.

There are hundreds of teeth extracted to-day because the people cannot afford to pay the ordinary charges. This is so much work lost to the profession at large, so much actual money thrown away, so much benefit taken from the community. There are many graduates who would willingly perform this labor for very moderate prices if they could get it, but they are too apt to drive it away by indiscretion.

Almost the first question a young practitioner asks is: "What shall I charge? What do you charge?" These two questions many times have caused him years of sorrow and hopeless longing for the clientele that never comes.

If he had commenced with the idea of doing a great deal of work for a little money, his skill would have increased, his experience enlarged, his reputation spread. Patients would have flocked to him as to the one person whom they would permit to fill their teeth, and then, independent of the whims or idiosyncrasies of the few, he could have charged fair prices with a feeling of security in the rapid development of his reputation and the enlargement of his bank account. (Editorial from *International Dental Journal*.)

ßssociations. • • • • • • • • • •

THE SOUTHERN DENTAL ASSOCIATION.

(Reported by Mrs. J. W. Walker for the DENTAL HEADLIGHT.)

THE twenty-third annual meeting of the Southern Dental Association, which was held at Lookout Mountain the last week in July, will be almost a thing of the past when this issue of the HEADLIGHT reaches our readers. The papers and proceedings, published in full in other journals, have been read by many of our readers. The meeting was one of the most important ever held by the Association, in point of numerical attendance, in the number and value of scientific papers read, and in business transacted.

The opening addresses of welcome by Mr. H. T. Olmstead, on the part of the Mayor and citizens of Chattanooga, and by Dr. D. R. Stubblefield, on behalf of the State Society of Tennessee, and the response by Dr. George J. Friedricks, of the Southern Dental Association, were admirable specimens of oratory, wit, and eloquence.

Dr. Gordon White, Nashville, Tenn., read the annual address as President.*

Drs. Chisholm, Marshall, and Lowrance were appointed a committee to consider the President's address and report what action the Association should take on his recommendations.

The committee to discuss the address of President White reported at a subsequent meeting an indorsement of the address, except as to the advisability of inviting all dental organizations into the movement to build a permanent Chautauqua, and the paper was opened for discussion, in which Drs. B. H. Catching, E. S. Chisholm, R. R. Freeman, H. J. McKellops, W. T. Arrington. J. Y. Crawford. and Francis Peabody participated.

The following new members were elected, who were permitted to assume the floor upon payment of dues: Drs. Thomas B. Hinman, Atlanta; W. J. Morrison, Nashville; N. A. Williams, Valdosta. Ga.; E. N. Wells, Savannah; C. H. McDowell, Griffin; T. A. Pope, Franklin; T. C. West, Natchez, Miss.; S. W. Foster, Decatur, Ala.; E.

* See page 154.

G. Grant, Columbia; W. E. Walker, Bay St. Louis, Miss.; U. D. Billmeyer, Chattanooga; R. D. Griffis, Paris, Tex.; W. T. Arrington, Jr., Memphis; J. N. Jones, Jacksonville, Fla.; A. W. Palmer, Knoxville, Tenn.; J. N. Simpson, Rock Hill, N. C.; A. R. Melindy, Knoxville, Tenn.; George W. Evans, New York; E. C. Kirk, Philadelphia; J. N. Crouse, Chicago, honorary member. Additional new members elected later on.

The courtesies of the floor were extended our visiting dental surgeons, and especially to the members of the Tennessee Association.

Dr. W. W. Walker, New York, addressed the Association in the interest of the World's Columbian Dental Congress, and made an eloquent appeal for funds with which to carry out the objects proposed. At a later session the sum of $200 was voted to this object.

The report of the Committee on Dental Education, Dr. B. Holly Smith, Baltimore, Chairman, dealt largely on the relations between the dental colleges and the State Boards of Examiners, and was discussed at great length by members of various State Boards, including Drs. Chisholm, of the Alabama Board; Browne, of the Georgia Board; Wright, of South Carolina; Barton, of Texas; Turner, of North Carolina; Jones, of Florida; and Smith, of Maryland.

In the absence of any report from the Committee on Prosthetic Dentistry, the subject was discussed by Drs. R. R. Freeman, A. P. Johnston, W. C. Browne, E. S. Chisholm, W. H. Marshall, B. H. Catching, L. P. Dotterer, George Evans, of New York; II. E. Beach, S. B. Cook, and others.

Many points of practical importance were brought out in this discussion, which were illustrated in the clinics to which Wednesday and Thursday mornings were devoted.

Dr. J. N. Crouse, Chicago, addressed the Association in the interest of the Dental Protective Association, and resolutions were framed indorsing the objects of the Association, and a committee of three in each Southern state appointed to further the objects of the Association and increase the membership in the South.

Dr. B. Holly Smith addressed the Association on what is known as the Wilcox bill and the recent attempt on the part of the Census Bureau to class dentists as mechanics and manufacturers in the collection of statistics. Suitable resolutions were adopted, and a committee appointed to raise funds toward defraying the expenses of the recent "fight" on this question, before Congress.

Dr. W. J. Barton, Chairman of the Committee on Pathology and Therapeutics, read an able report on this subject, which was followed by *a talk* from Dr. John S. Marshall, Chicago, on the subject of electricity as a therapeutic agent. He has found the application of the electric current very valuable in the treatment of hypertrophical conditions of the pulp and peridental membrane, the reduction of endoplasms. The action of such remedies as aconite and iodine is rendered more prompt by application through the sponges of the electrodes, abortive action being hastened. The current also affords an unfailing test in ascertaining the vitality of a tooth.

This was followed by papers from Dr. S. W. Foster, Decatur, Ala., on the care of children's teeth, and one from Dr. L. P. Sharp, Knoxville, Tenn., entitled "A Plea for Nitrous Oxide in Cardiac Weakness."

Dr. John C. Story, Dallas, Tex., read a paper on the subject of the proposed Dental Chautauqua, or a permanent location for the Southern Dental Association.

Dr. B. H. Catching, who had been announced for a paper, spoke briefly on the adverse effects of carbolic acid and other agents sealed in under a permanent filling over a nearly exposed pulp.

The reading of these papers was followed by a lengthy and interesting discussion of the various topics presented, by Drs. Freeman, Noel, Barton, Beach, Jones, Chisholm, Richards, Catching, Beadles, and Browne.

Dr. B. Holly Smith, of the Committee on Dental Education, read a paper received from Dr. Louis Ottofy, Chicago, entitled "Post Graduate Study," being an outline of a proposed Chautauqua system of home reading and study for the busy practitioner.

A large number of papers were presented by the Committee on Operative Dentistry: "Some of the Causes of Failures in Gold Fillings," by Dr. J. H. Allen, Birmingham, Ala.; "Philosophy in Dental Operations," by Dr. E. S. Chisholm, Tuscaloosa, Ala.; "The Application of Vapor in the Treatment of Diseased Tissues," by Dr. Peabody, Louisville, Ky.; a paper without title from Dr. Theo F. Chuplin, Philadelphia, which might be called "Practical Points of Interest;" a paper from Dr. T. H. Parramore, Hampton, Va., on the "Sponge Graft as a Pulp Capping," and one from Dr. William H. Cooke, Texas, on "Combination Fillings of Amalgam and Oxyphosphate."

Dr. L. G. Noel, Nashville, Tenn., exhibited and explained a great variety of apparatus for the administration of anæsthetics.

The Clinic Committee and the Committee on Dental Appliances made very interesting reports. At the last session the reports of the remaining committees and a number of papers which were unavoidably crowded out for lack of time were read by title and ordered published in the transactions.

Dr. E. C. Kirk, editor of the *Cosmos*, was elected a member of the Association, and addressed the body on the value of systematic work in the matter of papers and discussions, and outlined a plan for securing better results and more scientific work.

The election of officers was held in the afternoon of the second day, with the following result: President, B. H. Smith, of Baltimore; First Vice President, R. K. Luckie, of Holly Springs, Miss.; Second Vice President, S. B. Cook, of Chattanooga; Third Vice President, L. G. Dotterer, of Charleston, S. C.; Corresponding Secretary, D. R. Stubblefield, of Nashville; Recording Secretary, S. W. Foster, of Decatur, Ala.; Treasurer, H. E. Beach, of Clarksville, Tenn.; Members of the Executive Committee, Gordon White and W. R. Clifton.

After the customary resolutions of thanks to the hotel and railroads for reduced rates, to the city press for excellent reports in the daily papers, to the ladies for their presence and assistance, etc., the Association adjourned to meet in Chicago during the Dental Congress in August, 1893—the day being left subject to the action of the Executive Committee.

NATIONAL ASSOCIATION OF DENTAL FACULTIES.

THE ninth annual meeting of the National Association of Dental Faculties was held at the Cataract House, Niagara Falls, commencing Monday, August 1, 1892.

Twenty-six colleges were represented as follows: Baltimore College of Dental Surgery, R. B. Winder; Boston Dental College, J. A. Follett; Chicago College of Dental Surgery, Truman W. Brophy; Harvard University, Dental Department, Thomas Fillebrown; Kansas City Dental College, J. D. Patterson; Missouri Dental College, Dental Department of Washington University, W. H. Eames; New York College of Dentistry, Frank Abbott; Ohio College of Dental Surgery, H. A. Smith; Pennsylvania College of Dental Surgery, C. N. Pierce; Philadelphia Dental College, J. E. Garretson; University of Iowa, Dental Department, A. O. Hunt; University of Michigan, Dental Department. J. Taft; University of Pennsylvania, Dental Department, James Truman; Vanderbilt University, Dental Department, W. H. Morgan; Northwestern College of Dental Surgery,

B. J. Roberts; Louisville College of Dentistry, Francis Peabody; Indiana Dental College, J. E. Cravens; Northwestern University Dental School, E. D. Swain; Dental Department of Southern Medical College, William Crenshaw; Dental Department of University of Tennessee, J. P. Gray; School of Dentistry of Meharry Medical Department of Central Tennessee College, G. W. Hubbard; University of Maryland, Dental Department, John C. Uhler; Columbian University, Dental Department, H. C. Thompson; Royal College of Dental Surgeons of Ontario, J. Branston Willmott; American College of Dental Surgery, John S. Marshall; University of Denver, Dental Department, George J. Hartung.

The *ad interim* committee reported that it had investigated a charge preferred against the University of Maryland, Dental Department, by the College of Dentistry of the University of California, of graduating a person in less time than the rules demanded; that it found that no rule of the Association had been violated, and had so reported to the parties in interest; that it had dismissed an effort for the reinstatement of the American College of Dental Surgery, Chicago, as not within the jurisdiction of the committee. with the advice to reorganize the college before attempting to influence the Association to change its action, which reorganization has since been accomplished.

The committee also stated that its value in settling such matters has been made so clearly apparent that it recommended that it should be made a standing committee. to be elected by the Association, instead of being appointed by the President.

The report was received and placed on file, and the recommendation with regard to the *status* of the committee was adopted.

The following resolutions, laid over from last year, were adopted:

1 *Resolved*, That in case of charges against any college no final action shall be taken until all parties concerned shall have at least thirty days' notice.

2. That at all future meetings of the National Association of Dental Faculties the delegates shall consist of members of Faculties, and demonstrators will not be received.

The following resolutions, also over from last year, were laid on the table:

1 *Resolved*, That after June, 1893, the yearly course of study shall be not less than seven months, two months of which may be attendance upon clinical instruction in the infirmary of the school, now known as intermediate or infirmary courses.

2. That after the session of 1892–93, four years in the study of dentistry be required before graduation.

The following resolutions lie over under the rules:

Offered by Dr. Winder:

Resolved, That hereafter graduates of pharmacy be placed on the same footing as graduates of medicine, and be entitled to enter the second-year or junior class, subject to the examination requirements of each college.

Offered by the Executive Committee:

Any college failing to have a representative present for two successive sessions without satisfactory explanation shall be dropped from the roll of membership of this Association.

The Chair, having been asked for a ruling upon the admission of graduates of pharmacy to the junior class, decided that under the rules they could only be admitted to the first-year or freshman class.

The Executive Committee offered a report recommending the restoration of the American College of Dental Surgery to full membership, which, after an explanation by Dr. Marshall of the reorganization of the college, was unanimously adopted.

The Executive Committee reported on the application of the Western Dental College, of Kansas City, recommending that it lie over for one year. The report was adopted.

The report of the Executive Committee recommending the rejection of the application of the Tennessee Medical College, Dental Department, of Knoxville, Tenn., for irregularities in conferring the degree of D.D.S. and in the reception of students, was adopted.

The application of Howard University, Dental Department, Washington, D. C., was laid over for one year.

The following applications for membership, also reported by the Executive Committee, lie over under the rules: United States Dental College, Chicago; Homœopathic Hospital College, Dental Department, Cleveland; Detroit College of Medicine, Department of Dental Surgery.

The report of the Executive Committee recommending that the Baltimore College of Dental Surgery be censured by the Association for conferring the degree of Doctor of Dental Surgery upon Charles F. Forsham, M.A., LL.D., of Bradford, England, *in absentia* and honorarily, in violation of the rules of the Association, was adopted.

Dr. Truman offered an amendment to the rule regarding the conferring of the degree of Doctor of Dental Surgery honorarily, absolutely prohibiting the exercise of that privilege to the members of the Association, but the amendment was lost, after discussion, it being the general sense that the present rule is a sufficient safeguard against the unworthy bestowal of the honor.

Dr. Cravens offered the following amendment to the Constitution, which goes over under the rules:

Amend Article VII., so that it shall read as follows:

ARTICLE VII. Any reputable dental college, located in any state of the United States, may be represented in this body upon submitting to the Executive Committee satisfactory credentials, signing the Constitution, conforming to the rules and regulations of this body, and paying such assessments as may be made.

The Association adopted a protest against the classification of dentists as manufacturers, as provided in House Bill No. 7,696, known as the Wilcox bill, and against the collection of statistics from dentists under its provisions, on the grounds that dentists are not manufacturers in any sense, not being engaged in the manufacture, fabrication, or sale of any product having a merchandisable value; that all the laws heretofore passed in the various states and territories and the District of Columbia distinctly recognize dentists as professional men; and that the attempt to collect statistics would be an injustice not only to them but to their patients, and that such statistics if collected would be valueless to the government because showing the products of a class of men not engaged in manufactures.

The following, offered by Dr. Winder, was also adopted:

Resolved, That the National Association of Dental Faculties recommends that their alumni write and demand of the Census Bureau of the United States the return of all statistical reports, as, under the recent agreement between the dental profession and said Bureau, lawyers, physicians, and dentists are exempted from making statistical reports for the census of 1890; and that a copy of this resolution be forwarded to the chief of the Census Bureau.

A communication from the Post Graduate Dental Association of the United States, suggesting the establishment by the colleges of short courses of training and teaching especially designed and arranged for practitioners, was received and referred to the Executive Committee.

The manuscript of a Compend of Materia Medica and Pharmacy for Dental Students, by Dr. E. L. Clifford, of Chicago, was referred to the committee on text-books, with power to act.

Dr. Marshall offered the following resolution, which was adopted:

Resolved, That the Secretary be instructed to notify the National Association of Dental Examiners that the National Association of Dental Faculties considers it out of its province to legislate upon the relative values of the L.D.S. and D.D.S. degrees.

The following were elected officers for the ensuing year: J. D. Patterson, Kansas City, President; H. A. Smith, Cincinnati, Vice

President; J. E. Cravens, Indianapolis, Secretary; H. A. Smith, Cincinnati, Treasurer; F. Abbott, of New York; J. Taft, of Cincinnati, and A. O. Hunt, of Iowa City, Executive Committee; James Truman, of Philadelphia; Frank Abbott, of New York, and Thomas Fillebrown, of Boston, ad interim committee.

The President appointed as the Committee on Schools Drs. J. A. Follett, Boston; S. H. Guilford, Philadelphia; E. D. Swain, Chicago; C. N. Pierce, Philadelphia; T. W. Brophy, Chicago.

Adjourned to meet at the call of the Executive Committee.

NATIONAL ASSOCIATION OF DENTAL EXAMINERS.

THE eleventh annual meeting of the National Association of Dental Examiners was held at Niagara Falls, commencing Monday, August 1, 1892.

The sessions were presided over by the Vice President, Dr. Magill, the elected President, Dr. L. D. Shepard, of Boston, explaining his resignation from the State Board of Massachusetts, which necessarily carried with it his resignation of the presidency of the Association. The resignation was accepted with regret, and Dr. Shepard was unanimously accorded the privileges of the floor.

The following State Boards were represented at the sessions: Colorado, George J. Hartung; Georgia, D. D. Atkinson; Iowa, J. T. Abbott, J. B. Monfort; Indiana, S. T. Kirk; Maryland, T. S. Waters; Minnesota, L. W. Lyon; Massachusetts, E. V. McLeod; New Jersey, Fred A. Levy; Ohio, Grant Mollyneaux, Grant Mitchell; Pennsylvania, W. E. Magill, Louis Jack, J. A. Libbey; Tennessee, J. Y. Crawford; Wisconsin, Edgar Palmer; Kansas, A. H. Thompson.

The following Boards were admitted to membership: Virginia, J. Hall Moore; North Carolina, V. E. Turner; Oklahoma, D. A. Peoples; South Dakota, C. W. Sturtevant; District of Columbia, Williams Donnally.

At the instance of the Committee on Colleges, the following communication was sent to the National Association of Dental Faculties:

NIAGARA FALLS, August 1, 1892.

To the National Association of Dental Faculties.

Gentlemen: Whereas, a very considerable abuse has arisen by the improper use by students of the various certificates of the schools, such as the "standing" and "passing" certificates, to support students and graduates under age in their attempt to illegally engage in practice; we therefore ask

12

your Association to request the various colleges to have their "standing" and "passing" certificates of such uniformity of terms in each case that they can be used for no other purpose, and that they be printed in few words and small type, and be signed only by the Dean.

Respectfully,   NATIONAL ASSOCIATION OF DENTAL EXAMINERS,
                                           Fred A. Levy, Secretary.

A Committee of Conference was appointed, consisting of Drs. Truman, Marshall, and Swain, on the part of the Faculties' Association, and Donnally, Palmer, and Monfort, on the part of the Examiners' Association, which, after consultation, agreed upon a favorable report.

Dr. Lyon offered the resignation of the Minnesota Board, which was laid upon the table, as it had evidently been offered as the result of a misunderstanding, and the Board was requested to withdraw it.

The following resolution, offered by Dr. Crawford, was adopted:

Resolved, That when a member of any State Board becomes a teacher of a dental school his resignation from his Board should follow:

A resolution protesting against the classification of dentists as manufacturers and the collection of census statistics from them under the provisions of House Bill No. 7,696, commonly known as the Wilcox bill, was adopted. The resolution was similar in terms to those adopted by other dental societies.

The Committee on Colleges reported that they had received reports showing that the actual number of students in attendance at the last sessions in the schools recognized by the Examiners' Association was 2,881; of graduates, 1,357. In the schools not recognized by the Association the students were 236; graduates, 96.

The report also considered desirable advances to be made in educational methods, and offered the following memorial, which the Secretary was directed to transmit to the National Association of Dental Faculties:

The National Association of Dental Examiners would respectfully memorialize the National Association of Dental Faculties to authorize two advances in the system of dental education.

These are: First, that your Association require the universal enforcement of a higher grade of preliminary education of candidates for matriculation. This proposition lies at the foundation of dental education, in which is involved the quality of the graduates of the future, upon which depend the advancement, the standing, and the dignity of the dental profession.

The second proposition is that complete preparation be made in each school for laboratory technique in the studies of histology, pathology, and in each of the departments of dental surgery and dental prosthesis, and that this method of teaching be made a requirement of the schools.

The committee also reported the following amended list of colleges which they recommend as reputable:

Baltimore College of Dental Surgery, Baltimore, Md.

Boston Dental College, Boston, Mass.

Chicago College of Dental Surgery, Chicago, Ill.

College of Dentistry, Department of Medicine, University of Minnesota, Minneapolis, Minn.

Dental Department, Columbian University, Washington, D. C.

Dental Department, National University, Washington, D. C.

Northwestern University Dental School, formerly Dental Department of Northwestern University (University Dental College).

Dental Department of Southern Medical College, Atlanta, Ga.

Dental Department of University of Tennessee, Nashville, Tenn.

Harvard University, Dental Department, Cambridge, Mass.

Indiana Dental College, Indianapolis, Ind.

Kansas City Dental College, Kansas City, Mo.

Louisville College of Dentistry, Louisville, Ky.

Missouri Dental College, St. Louis, Mo.

New York College of Dentistry, New York City.

Northwestern College of Dental Surgery, Chicago, Ill.

Ohio College of Dental Surgery, Cincinnati, O.

Pennsylvania College of Dental Surgery, Philadelphia, Pa.

Philadelphia Dental College, Philadelphia, Pa.

School of Dentistry of Meharry Medical Department of Central Tennessee College, Nashville, Tenn.

University of California, Dental Department, San Francisco, Cal.

University of Iowa, Dental Department, Iowa City, Ia.

University of Maryland, Dental Department, Baltimore, Md.

University of Michigan, Dental Department, Ann Arbor, Mich.

University of Pennsylvania, Dental Department, Philadelphia, Pa.

Vanderbilt University, Dental Department, Nashville, Tenn.

Western Dental College, Kansas City, Mo.

Minnesota Hospital College, Dental Department, Minneapolis, Minn. (defunct).

St. Paul Medical College, Dental Department, St. Paul, Minn. (defunct).

American College of Dental Surgery, Chicago, Ill.

The report was adopted.

The following officers were elected for the ensuing year: W. E. Magill, Erie, Pa., President; J. Y. Crawford, Nashville, Tenn., Vice President; Fred A. Levy, Orange, N. J., Secretary and Treasurer.

Adjourned.

STATE BOARD OF DENTAL EXAMINERS.

THE State Board of Dental Examiners held its second annual meeting at Lookout Inn, Monday, July 25, all the Board being present. The Board is composed of J. Y. Crawford, of Nashville, President; Dr. J. L. Mewborn, of Memphis, Secretary; and Drs. R. B. Lees, of Nashville; S. B. Cooke, of Chattanooga; W. T. Arrington of Memphis, and F. A. Shotwell, of Rogersville.

The reports of the officers showed that 407 dentists had registered during the year. This is the largest percentage ever registered in the United States during the first year of a registration law. Only ten or fifteen dentists in Tennessee have failed to comply with the law, and they will be vigorously prosecuted by the Board of Examiners.

During the morning D. M. Barrow and G. M. Dayton, of Chattanooga, and J. N. Fann, of Georgia, presented diplomas and was duly registered as dentists. A number of other candidates were examined. The Board will not recognize any diplomas except from reputable colleges requiring the full three years' course.

The utmost harmony prevailed in the Board, and business was rapidly transacted. Drs. W. T. Arrington and F. A. Shotwell, whose terms expired, were reappointed by Gov. Buchanan members of the Board.

On Wednesday morning the Board elected Dr. S. B. Cook, of Chattanooga, President; and Dr. J. L. Mewborn, of Memphis, Secretary for the ensuing year.

Editorial. . . . .

JAMES C. ROSS, D.D.S.

This venerable and eminent member of the dental profession died in Nashville October 4, after a brief illness of pneumonia, in the seventy-eighth year of his age.

Dr. Ross was born in Hamilton County, O., July 16, 1815, and during his minority was apprenticed to a cabinetmaker, but upon entering his twenty-second year he abandoned this trade and began the study of dentistry in the town of Lawrenceburg, Ind., under the tutelage of his brother and Dr. John Harris, a brother of the distinguished Prof. Chapin A. Harris. He received his degree from the Ohio Dental College, and afterward received an "ad eundem degree" from the New Orleans Dental College. He practiced in Frankfort, and later on in Lexington, Ky. In 1847 he moved to Nashville, Tenn., where he has since resided.

He was at one time President of the Tennessee Dental Association, also of the Nashville Dental Society. He participated in the organization of the Department of Dentistry of Vanderbilt University, was chosen President of the Faculty, and for seven years faithfully and ably discharged the duties of Professor of Operative Dentistry and Dental Hygiene. Failing health forced him to resign, but he has since sustained honorable connection with the institution as Emeritus Professor. Dr. Ross was a warm-hearted, honest man, in the highest sense a Christian gentleman, for many years steward and treasurer of Elm Street Church, in this city. He was a conscientious, hard-working dentist, an exponent in all his acts of his highest conception of what was ethical and professional. He enjoyed an extensive practice, and always held the confidence of his patrons and professional brethren. For nearly sixty years he has stood shoulder to shoulder with representative men in every effort to promote the best interests of dentistry, and there will be many who will learn of his death with feelings of deepest sympathy for those immediately affected by his loss. The world has been made richer by such a pure and guiltless life, and in his departure he leaves dentistry advanced to a position which could not have been anticipated when he entered the profession.

Dr. Ross was the head of a large and interesting family, the father of eleven children, and it was in his domestic relations that he exhibited the noblest traits of character. His wife and three children survive to mourn the loss of a model husband and loving father. To them we tender our sincerest condolence and heartfelt sympathy. His exemplary life is the richest legacy he could possibly have bequeathed to them and to the world.

WORK FOR THE COMING YEAR.

PRESIDENT W. W. WALKER, in his annual address to the American Association, at Niagara Falls, suggested a plan of work for local and state societies through which it is intended to obtain the weight of opinion on a selected number of questions of interest to the profession each year, to be formulated by a committee of the Association. It met with the hearty indorsement of the committee to which the address was submitted and a standing committee on State and Local Organizations was created.

The President appointed on this committee Drs. Edward C. Kirk, Philadelphia, Chairman; J. N. Crouse, Chicago; and Louis Jack, Philadelphia, who reported the following set of questions, which they recommended for discussion among the state and local societies during the coming year, in accordance with the plan before described:

1. Should examining boards have power to grant certificates of qualification to undergraduates.

2. Should immediate root-filling be practiced while purulent conditions exist at the apex?

3. What are the best materials to enter into the composition of temporary fillings, to be retained for a minimum of three years? .

4. What are the best methods for obtunding sensibility of the dentine by . either local or general means? Should arsenic ever be used?

5. What are the best forms of partial lower dentures and the methods for constructing the same?

6. Corrective dentistry: its present status; what are the simplest and most universally applicable forms of apparatus and most efficient retaining fixtures? .

7. To what extent and under what conditions is the collar crown a cause of pericemental inflammation?

8. In cases of congested pulp, should the arsenical application be made without preliminary treatment?

9. What are the advantages and disadvantages of the use of the matrix (a) with gold, (b) with plastics?

10. The ætiology of pus formation?

That a consensus of opinion may be had, let all State Associations and every local society properly consider these questions and send to the Chairman of the committee a full report. In advance of these meetings much work must be done by every individual member of these organizations and a well-matured opinion formed before any attempt is made to answer them. There is enough here, if properly used, to occupy the attention of any Association an entire week, and unless each question is thoroughly investigated in the quiet hours at the office or at home before it is sprung in societies, will involve the loss of much valuable time, and in the end the plan will prove a failure.

JOINT MEETING OF THE SOUTHERN AND STATE DENTAL ASSOCIATIONS.

THE joint meeting of the Southern and State Dental Associations at Lookout Inn in July was a wonderfully successful affair. The presence within our borders of so many distinguished men had the effect to bring out members of the profession of our state in large numbers. The papers and discussions were quite up to the wide-awake, outreaching spirit that characterizes such a gathering of representative men of to-day. Members of the Executive Committee of the World's Columbian Dental Congress and the State Board of Dental Examiners attended regularly, and took an active part in the discussions. The social features were a dance on Wednesday night, in the dining room of the Inn, tendered the visitors by the citizens of Chattanooga, and the banquet on Thursday night, also at the Inn, by the Tennessee Dental Association and dental dealers to the Southern and Congress Committee, were both enjoyable and creditable affairs.

THE DENTAL PROTECTIVE ASSOCIATION OF THE UNITED STATES.

THE objects of this Association are so well known that it is not necessary to repeat them here. Organized to protect its members against corporations, such as the International Tooth Crown Co., which seek to exact licenses, fees, and annuities, it has already saved many thousands of dollars to the profession.

We must, therefore, insist that members of the dental profession in Tennessee and other Southern states should send their names to Dr. J. N. Crouse, of Chicago, for blanks.

A recent circular reveals the fact that there are but twenty-two dentists in Tennessee who have connected themselves with the Association, and nearly half are residents of Nashville.

The expense is so small that no one should hesitate. The power of such an organization for good is incalculable.

Send in your names. Indifference and delay may cause the loss of many dollars and much annoyance and trouble.

DR. S. B. COOK, OF CHATTANOOGA.

THROUGH the personal efforts of this genial and obliging gentleman, whom the dentists of Tennessee will ever honor, the Lookout Inn was thrown open in July for the entertainment of the Southern and State Dental Associations. The broad-gauge railroad from Chattanooga to the mountain top and the Inn were in Chancery, and it was necessary to obtain consent of fifteen attorneys, representing as many claimants. The task was not only accomplished, but he secured a hotel proprietor and had everything thoroughly cleaned up and put in perfect running order before the guests began to assemble.

There was a hotel full of dentists and their charming wives. The elegance of everything, the unsurpassed grandeur of the scenery, bracing atmosphere, delightful cool nights, freedom from noises and counter attractions combined to make this the most desirable place of meeting which the Associations had ever visited. The members of the State Association were so pleased that it was voted to meet there again in 1893.

That Dr. Cook should have without any help perfected the arrangements is but another evidence of his popularity. In testimony of the appreciation of his labors he was elected President of the State Association, President of the Board of Dental Examiners, and First Vice President of the Southern. Did one man ever before have so many distinguished honors thrust upon him? He owes them to his modesty, unselfish spirit, and recognized devotion to the profession.

DOUBLY HONORED.

DR. J. Y. CRAWFORD, ex-President of the Southern Dental Association, at the last meeting of the National Board of Dental Examiners was elected one of the Vice Presidents; also at the recent meeting of the American Dental Association he was chosen its First Vice President. Thus honors are showered thick and fast upon this distinguished Southerner. No dentist in the South has worked harder for the cause of organized dentistry than J. Y. Crawford, M.D., D.S., of Nashville.

OUR genial friend, Dr. Gordon White, ex-President of the Southern Dental Association, has recently returned from a trip abroad. After the conclusion of his duties as presiding officer of the Southern Dental Association, he prolonged his vacation by a tour of Europe. The Doctor speaks with great enthusiasm of the benefit to be derived, both physical and mental, by the voyage across the great Atlantic, and recommends to all his tired and weary brethren the invigorating atmosphere of "old ocean," to say nothing of its grandeur and soul-inspiring qualities. He had the pleasure of visiting the London Dental Hospital and Museum. With the latter he was greatly interested, especially with the elaborate and valuable contribution of "Tomes." Otherwise he learned nothing new at the hospital clinic. The operations were inferior in character, some of them obsolete, as compared with American dental surgery; all were disappointing. The dental chairs and surgical appliances were both antiquated and inconvenient. A case of alveolar abscess, he learned, was treated by lancing the face over the seat of the disease. The Doctor purchased an admirable dental syringe of "Ash," for which he paid only $2.35, the same article selling in this country for $3.75, thus affording a practical illustration of the onerous tariff system, of which the poor dentist is compelled to bear his share.

Dr. White was much impressed with the great metropolis, its stupendous proportions and vast resources. He visited the art galleries, museums, cathedrals, etc. He remarked the scrupulous cleanliness everywhere maintained; the admirable system of sanitation adopted, rendering it the healthiest city in the world.

THE editor of the HEADLIGHT wishes to acknowledge the receipt of an invitation to the marriage of Dr. Edward C. Kirk to Miss Helen Theodora, daughter of Mrs. Samuel Clements, of West Philadelphia. The marriage took place Thursday, October 6, in the Church of the Saviour at 12 o'clock noon. While Dr. Kirk receives our warmest and best wishes, Miss Clements wins a most estimable and worthy man. As the cultured editor of the *Dental Cosmos* and a successful practitioner of dentistry he ranks with the foremost in the profession.

WORLD'S COLUMBIAN DENTAL CONGRESS.

W. W. Walker, Chairman, 67 West Ninth Street, New York City; J. S. Marshall, Treasurer, Argyle Building, Chicago, Ill.; A. O. Hunt Secretary, Iowa City, Ia.

Executive Committee.—L. D. Carpenter, Atlanta, Ga.; J. Y. Crawford, Nashville, Tenn.; W. J. Barton, Paris, Tex.; J. Taft, Cincinnati, O.; C. S. Stockton, Newark, N. J.; L. D. Shepard, Boston, Mass.; W. W. Walker, New York City; A. O. Hunt, Iowa City, Ia.; H. B. Noble, Washington, D. C.; George W. McElhaney, Columbus, Ga.; J. C. Storey, Dallas, Tex.; M. W. Foster, Baltimore, Md.; A. W. Harlan, Chicago, Ill.; J. S. Marshall, Chicago, Ill.; H. J. McKellops, St. Louis, Mo.

The meeting will be held in Chicago in 1893.

OFFICERS OF THE AMERICAN DENTAL ASSOCIATION.

J. D. Patterson, Kansas City, President; J. Y. Crawford, Nashville, Tenn., First Vice President; S. C. G. Watkins, Mont Clair, N. J., Second Vice President; Fred A. Levy, Orange, N. J., Corresponding Secretary; George H. Cushing, Chicago, Recording Secretary; A. H. Fuller, St. Louis, Treasurer; W. W. Walker and S. G. Perry, New York, and D. N. McQuillen, Philadelphia, Members of the Executive Committee. Next meeting to be held in Chicago.

OFFICERS OF THE SOUTHERN DENTAL ASSOCIATION.

B. Holly Smith, Baltimore, President; R. K. Luckie, Holly Springs, Miss., First Vice President; S. B. Cook, Chattanooga, Second Vice President; L. P. Dotterer, Charleston, S. C., Third Vice President; D. R. Stubblefield, Nashville, Tenn., Corresponding Secretary; S. W. Foster, Decatur, Ala., Recording Secretary; H. E. Beach, Clarksville, Tenn., Treasurer; W. R. Clifton, Waco, Tex., and Gordon White, Nashville, Tenn., Members of the Executive Committee. Next meeting to be held in Chicago.

OFFICERS OF THE TENNESSEE DENTAL ASSOCIATION.

S. B. Cook, Chattanooga, President; W. W. Jones, Murfreesboro, First Vice President; W. J. Morrison, Nashville, Second Vice President; P. D. Houston, Lewisburg, Recording Secretary; D. R. Stubblefield, Nashville, Corresponding Secretary; H. E. Beach, Clarksville, Treasurer. Next meeting to be held in Chattanooga first Tuesday in July, 1893.

STATE BOARD OF EXAMINERS.

S. B. Cook, President, Chattanooga; J. L. Mewborn, Secretary, Memphis; J. Y. Crawford, Nashville; W. T. Arrington, Memphis; A. F. Shotwell, Rogersville; R. B. Lees, Nashville. Next meeting July, 1893, at Lookout Mountain.

STATE NOTES.

DR. J. M. MILLEN, of Covington, recently visited Chicago.

Dr. H. E. Beach, of Clarksville, passed through Nashville after a pleasant and beneficial visit to East Brook Springs.

Dr. Gordon White has returned from a trip across the waters. After the adjournment of the Southern he went directly to New York and thence to England. Cholera prevented a continental visit. He enjoyed the sail and was much improved.

Dr. Ed Cabiness was in Nashville September 25th, journeying home from Red Boiling Springs, where he had gone some weeks before for much-needed rest.

Dr. Lyman C. Bryan, of Basil, Switzerland, has just concluded a business visit to his old home, Nashville, where as a boy he enjoyed the reputation of being an upright, honorable, and all round good fellow. His old schoolmates made his stay a pleasant one.

Dr. S. B. Cook, of Chattanooga, President of the Board of Examiners, made an official visit to Clarksville, early in September, to blister some violators of the law regulating the practice of dentistry. He was agreeably surprised to find that between the time of his last letters and his arrival the offenders had, through the advice of friends, accepted terms of compromise before offered, and have registered with the Secretary of the Board. Instead of having three suits to bring, he found three legally licensed practitioners to welcome him with smiles and fraternal greetings. May his administration throughout be marked by the success and conservatism that so successfully registered over four hundred dentists the first year!

SOUTHERN NOTES.

DR. E. C. KIRK, of Philadelphia, is the editor of the *Dental Cosmos*, which has the largest circulation of any dental paper in the world.

Among the many learned gentlemen in attendance upon the convention of dentists were several whose reputations are national.

Dr. A. W. Harlan, of Chicago, is one of the most finished writers known to dental literature.

Dr. J. N. Cruse, President of the Dental Protective Association, of Chicago, was a respected attendant upon the meetings.

Dr. A. O. Hunt, of Iowa, and Dr. Jonathan Taft, Dean of the Dental Department of Ann Arbor College, were conspicuous figures in the gathering.

Dr. John C. Harly, of Dallas, Tex., is a diamond in the rough. He is a noted dentist and was the originator of the idea of a Columbian Dental Congress.

Dr. V. E. Turner, of Raleigh, N. C., is a graceful speaker and a man well known among all dentists. He is the Chauncey Depew of the dental fraternity.

Mrs. J. M. Walker, of Bay St. Louis, Miss., reported the convention for a number of the dental magazines. She has officially reported the convention for a number of years.

Mrs. S. B. Cook came in for a very great share of praise from all the delegates for the cordial and elegant manner in which she received the visitors and made them feel at home.

Dr. H. J. McKellops, of St. Louis, is one of the best known men in the country, and his prominence in dental circles makes him an authority on all subjects. He was a close friend to the late John McCullough.

Dr. Mackall, of Cecil County, Md., is the oldest graduate in dentistry now living, having taken his degree in the first class that graduated from the Baltimore Dental College, which is the oldest institution of the kind in the world. Dr. W. H. H. Thackston, of Farmville, Va., is the second oldest living graduate.

Among the ladies attending the meeting were Misses Mewborn, of Memphis, daughter of Dr. J. L. Mewborn, member of the State Board; Mrs. M. A. Sullivan, of Texas; Mrs. L. D. Carpenter, of Atlanta, Ga.; Mrs. A. O. Hunt, of Iowa City; Mrs. E. Fredricks, of New Orleans; Mrs. R. R. Freeman and Miss Freeman, of Nashville; and others.

Necrology.   . . . .

DR. JAMES C. ROSS.

Died, at Nashville, Tenn., Oct. 4, 1892, of pneumonia, James C. Ross, D.D.S., in the seventy-eighth year of his age.

TRIBUTE OF RESPECT.

A meeting of the Faculty of the Dental Department of Vanderbilt University was called at the office of Dr. T. A. Atchison Tuesday afternoon, October 4, 1892, to take action regarding the death of Dr. J. C. Ross. Dr. Ross had been selected as one of the teachers in that department at its organization, and severed his connection voluntarily when failing strength warned him to retire. After his resignation he was created Emeritus Professor, which position he held at the time of his death.

The following preamble and resolutions were adopted unanimously by the Faculty:

Death is at all times a sad event. To reach the end of life, to roll up the scroll of time, and to go out into the mysterious beyond must forever bring a shock upon those who are left behind. But when the term of years is full, and more also; when the flagging of energies, spent with usefulness, tells that the eventide of life is fully come, it is not strange that death may be looked upon as almost a benediction. Our associate, Dr. James C. Ross, had lived a singularly blameless, useful, conscientious life, and his reward is as certain as the promises of Christ are strong and enduring.

Whereas for a number of years our association with Dr. James C. Ross was entirely amicable and satisfactory; and whereas Providence has seen fit to call him to his rest after a long and useful life; therefore be it

Resolved, That we, his associates in the Dental Department of Vanderbilt University, desire to express our deep regret by these resolutions, and do hereby tender our sympathy to his bereaved family in their affliction.

Resolved, That we attend his funeral in a body.

Resolved, That a copy of these resolutions be spread on our minutes on a page set apart as a memorial page to his memory.

[Signed]   W. H. MORGAN,   D. R. STUBBLEFIELD,
                  T. A. ATCHISON,   A. MORRISON,
                  R. R. FREEMAN,   HENRY W. MORGAN,
                      O. H. MENEES.

Dr. W. A. Small.

Died, at Oakland, Tenn., of typhoid fever, on September 12, 1892, Dr. W. A. Small, in the sixtieth year of his age.

Dr. Small was born in North Carolina November 18, 1833. He began the study of dentistry at eighteen years of age, and after completing his studies practiced a few years; but his health becoming poor, he resorted to farming in 1860. In 1861 he joined the army, and remained until the close of the war. He practiced continuously in this state since 1879, and was a modest gentleman who strove to adorn his profession and to be a living exemplification of Christian character.   J. L. M.

The following testimonials and resolutions were presented, read, and adopted at the recent meeting of the American Dental Association :

In Memory of Dr. John Allen.

In the dispensation of an all-wise and overruling Providence, Dr. John Allen, of New York, on the 8th day of March, 1892, at the age of eighty-two, passed from this to a higher and better life, having attained a fullness and ripeness of age beyond that of the common lot of men.

Dr. Allen stood as a representative man in the profession of his choice. In the line to which he gave special attention he was the chief, and was so recognized not only in this, but in the countries of the world wherever prosthetic dentistry is known and practiced. He it was who brought to its present high state of perfection that variety of substitutes known as continuous gum dentures. Though his chief attention and labor were devoted to this special work, he was interested and took part in the various lines of thought and effort that were employed for the development, growth, and establishment of dental science and art. He was ever ready to defend, and sought to elevate the profession to a higher plane of usefulness.

Dr. Allen was one of the organizers of the Ohio College of Dental Surgery, a professor and an efficient teacher in that institution. In the subject of dental education he always manifested a warm interest. A writer of more than ordinary ability, he has added many valuable contributions to the literature of the profession. He was an active member of this Association from almost the time of its organization, and did much to promote its welfare. He was also a member of, and an active worker in, a number of other dental societies. He was a man of purest character and highest integrity; one not only respected but loved by all who knew him; in manner, most affable; in bearing, dignified; in spirit, gentle and sympathetic.

The loss of such a one is always an occasion of sadness and sorrow, but we have the consolation of the knowledge that his career was rounded, full, and complete, and his death closed a life filled with good works for his fellow-men.

In view of the above,

1 *Resolved,* That we will ever cherish the memory of our departed brother and seek to establish and perpetuate the high principles that were so fully illustrated in his noble life.

2. That the traits so preëminently characterizing the life of him we now commemorate are worthy not only of our high regard, but most earnest emulation.

3. That this testimonial be placed on a memorial page of the transactions of this body and a copy, properly engrossed, be sent to the family of the deceased; also that a copy be sent to the dental journals of this and other countries for publication.

In Memory of C. A. Kingsbury, M.D., D.D.S.

Within the last year Dr. Charles A. Kingsbury was called from this to a higher life, in the seventy-second year of his age.

Dr. Kingsbury many years ago became identified with this Association, and retained his membership to the time of his death, and though he was not always present at its meetings, so highly was he esteemed by the membership of the body that it was a pleasure to all to have his name upon the roll of members.

Dr. Kingsbury entered the practice of the profession in 1839, in Philadelphia, and continued actively engaged in its pursuit during his life. He studied dentistry in Trenton, N. J. He was intimately acquainted with the leading men of the profession almost the whole of his professional career, and imbibed, in a large measure, the interest and enthusiasm of those men for dental science and art. Indeed, that association, in a degree, shaped his professional life. He was familiar with all things that entered into the development and progress of dentistry for about fifty years. He was a man of liberal learning and broad culture; one whose sociability was a predominant characteristic. In his early life he was a teacher, and after many years practice of his profession he was for a time a successful teacher in one of the dental colleges in the city of his home. He was highly esteemed by all who knew him. He was a man of sterling characteristics, genial, kind, and sympathetic in his association with his fellows. In his death not only this association, but the entire profession, loses another of the pioneers who was ever devoted to its interests, ever contributing of his resources to its upbuilding.

1 *Resolved,* That we will ever cherish the memory of our departed brother as one whom we delight to honor and to emulate in his leading characteristics.

2. That this statement and resolution be placed upon the memorial page of the proceedings of this body; that a copy, in proper form, be transmitted by the Secretary to the family of the deceased; and that it be sent to the journals for publication.

THE DENTAL HEADLIGHT.

VOL. 14.   NASHVILLE, TENN., JANUARY, 1893.   No. 1.

Original Communications. · · · ·

IDEAS ABOUT BRIDGE WORK.

BY L. G. NOEL, M.D., D.D.S., NASHVILLE, TENN.

A QUICK and accurate method of adjusting a cap that is to be used as an abutment for a bridge occurred to me some time ago, and I will describe it to the readers of the HEADLIGHT. It may be of use to some who, like myself, have tried many other plans with a feeling of dissatisfaction.

For clearness we will suppose that you have a case where a patient has lost the second bicuspid, first molar, and second molar. Here you have the first bicuspid and wisdom tooth for abutments. We will suppose that both of these teeth have good antagonists. With suitable corundum disks and stones we cut off the bulging parts of these teeth until they are reduced to nearly straight lines.

We do not cut the grinding or antagonizing surfaces, for reasons that we shall presently show.

Measuring from grinding surface to gum line at deepest points, we cut strips of crown gold plate wide enough to make a band in the usual way; perhaps allowing extra width so that we may drive on, and, after adjusting with files and stones to the gum line, we may have enough to strike off with a stump corundum wheel to a level with the masticating surface.

Having fitted a band, we are ready to get a correct adaptation of gold to the masticating surface, so that our cap may be so closed as to protect the cement from washing out. We take a small piece of pure gold plate—same thickness as that used for the band—and first annealing, to soften, we burnish it into the sulci, as perfectly as we can, bending down over our band and marking, so that we can put the two together accurately after removing from the tooth.

We now remove and catch the cap to the band with some little bits of solder, *placed upon the outside*. In doing this it is not necessary to try to close all the openings. We then replace it upon the tooth and have the patient bite it to an accurate adjustment.

We now remove it and blow a quantity of solder clippings—18 carat preferred—into the cusps upon the inside of the crown, filling it over level to a thickness to withstand wear in mastication. When returned to the tooth, the crown will be too long. This is corrected by grinding away enough of the masticating surface to let the border of the band under the free edge of the gum, as originally fitted.

When both crowns have been finished in this way, we are ready to take the impression and complete the bridge as usual.

It gives a perfect adjustment in less time than any method I have seen.

Pure gold plate may be used for the bands also, or we may use the platinum-faced gold plate.

SUCCESS OF A DENTIST AFTER GRADUATION:*
ON WHAT DOES IT DEPEND?

BY ROBERT R. FREEMAN, M.D., D.D.S., PROFESSOR OF MECHANICAL AND CORRECTIVE DENTISTRY, VANDERBILT UNIVERSITY.

I WOULD answer: His love for his profession, stimulated by love of his fellow-men.

There is a success that comes only at the end, after the battle is fought, the race is over, and the victory won, when is proclaimed the welcome plaudit: "Well done!" But man wants the gratification and satisfaction of seeing day by day results attained, hopes realized which maketh the heart glad, that success which will afford us present happiness in the comfortable assurance that we are moving onward, and that with us it is well.

Twenty-five years have made a great change in the status of our profession. The young man of to-day who has been faithful in his preparation and is a graduate in dentistry starts out upon his life work upon a plan which the dentist of but a few years ago had scarcely hoped for, and only dreamed of. While dentistry by a few scattered landmarks gathered from history and tradition claims ancient origin, it is hoped that the great Columbian Congress will demonstrate the fact that she stands to-day a peer, and to the front, among the world's most noble benefactors.

Proud are we of the South, and especially in Tennessee, that

---

* Read before the Dental Section of the Nashville Academy of Medicine, December 23, 1892.

sons of our own soil have played no insignificant part in these great achievements.

There are still those entering our profession, quacks and humbugs, who, having failed in other fields, seek dentistry as a refuge, believing it will afford them easy means of accomplishing their selfish ends. Some such do seem to blossom as the green bay tree, but their end, as saith the preacher, will be vanity and vexation of spirit. Their very memory soon fadeth from the minds of men. Not so with Allen, or an Atkinson, or even Ross, of noblest memory. We may measure our success by comparison. He best succeeds who best loves. We may not all hope to be first, but there are high places for every one.

> In good works grow not weary;
> Duty done, the crown of life is won.

The best benefit which humanity receives is not always the product of the greatest man with the broadest brain. Such may conceive a principle, give it out, and pass on to other investigations, while the humbler, taking up the thought, makes its application and distributing its benefits sends its blessings forth.

> If you cannot shine yourself,
> Set some other life aglow;
> Blessings they may shower upon you
> As you journey here below.

The man who would succeed must be ever diligent and watchful. This is no age for laggards.

The young man who does not catch the cadence of the march, and keep step with the onward progress, soon is altogether hustled out; and there is no easy place for him, not even in the rear.

Rise above simple sordid money getting; seek something higher. Think not of selfish interests; be a blessing to mankind.

Habit is a great factor in the make-up of the life of every man. It may mar or it may make him. It will prosper or will break him. It stands upon no neutral ground. He must its slave or master be.

Who has not seen some start with bright prospects, with almost every assurance of success, and then seen them lost among the rabble in the rear, through some pernicious habit. Dentistry demands of its votaries a pure, clean life, and he who would be successful must walk upon this lofty plane. Its benefits are extended both to the humble and the great, so that princes and rulers may be our companions if we so desire; while the poorest will not hesitate, if we be worthy, to count us his brother and his friend.

In the South especially many of our young members have thought it necessary, in order to succeed, to itinerate around from place to place. This, I believe, is a great mistake. If you would have people interested in you, make yourself a useful fixture in their midst. Interest yourself in their welfare, person, Church, and state. Have a pleasant office, and let it be as respectable as your means can possibly make. In fixtures and appliances what you want are such as with which you will be enabled to do your work. If your means will afford, purchase and try the new inventions; but in no case fail to avail yourself of the well-tested and desirable.

There are many things like the traditional razor, made only to sell, and a man can waste much money in useless instruments. It is a very false economy to deprive yourself of needed appliances, although you may have to secure them at great personal sacrifice.

The successful dentist avails himself of the counsel of his brethren. There is no truer saying that as " iron sharpeneth iron; so a man sharpeneth the countenance of his friend."

While our dental journals are the exponents of these counsels, it is attendance on the meetings of our Dental Associations, and the contact there enjoyed, which has developed and implanted the great American standard of dentistry to which the whole world bows in respectful obeisance to-day.

> For emblazoned in its folds,
> The world must needs confess,
> Is a sentiment which goes:
> Our motto is " Success."

FRACTURE OF THE SUPERIOR MAXILLA.

BY R. M. WALKER, D.D.S., CHICAGO, ILL.

As this is of much more rare occurrence than are fractures of the inferior, it may prove of some interest to your readers. The patient, a gentleman about twenty-five years of age, was kicked by a horse. Fortunately, he was not shod. Upon recovering from the shock sufficiently to know what ailed him, he said he felt something wrong in his mouth, and his first thought was that he had a mouthful of gravel, but on closer examination found, as he said, the whole side of his face knocked in.

I was called, in company with a physician, and upon reaching the house found his face considerably swollen, but not cut or bruised to any great extent. On examining the mouth I found the fracture to be on the right side from the canine back. The entire side had been broken loose up nearly to the floor of the antrum, and was turned over so the grinding surface of the teeth nearly touched the roof of the mouth, and was only held by the gum and tissues.

I had no trouble in pushing the broken part back to its place, and having him close his mouth found he could do so fairly well. As it was late at night, and I had been called in a hurry, and not informed as to the nature of the case, I concluded to put a bandage around his head to hold the jaws firmly together till morning. In the morning I took a plaster impression, and as he could only open his mouth about half an inch, it was no easy job; but I succeeded in getting a good one on the first attempt, and removed it without disturbing the broken parts. After getting one of the lower, I returned to the office and made a rubber plate covering the roof of the mouth and crowns of the teeth. On each side I vulcanized a wire, but so as to come out of the corners of his mouth, and then turned back toward the ears.

When this was adjusted, I fastened a strap from a skullcap to each wire and drew them up tight. This kept the plate perfectly solid, and he was able to eat a good square meal as soon as it was completed. After the first week he was able to remove and replace the plate himself, and after the second week I removed the wires. He then wore the plate for another week, and on removing it, I found everything perfectly solid. the occlusion perfect, and no sign of what had happened. About nine months have passed, and so far the teeth show no signs of giving trouble, though they are undoubtedly dead.

EROSION OF TOOTH STRUCTURE.*

BY MARVIN M'FERRIN, D.D.S., NASHVILLE, TENN.'

WE have for our subject the consideration of a condition which often confronts us, and which is of very great importance for us to study, especially in the sense of prevention, as we know little of the causation, and hope by careful research and study some day to find a manner of combating the conditions which give rise to erosion.

---

* Read before the Dental Section, Nashville Academy of Medicine.

Erosion of tooth structure is a process of destruction without such evident cause as we see in case of caries.

The cavities of erosion are generally somewhat regular in form, often saucer-shaped, the enamel being more widely destroyed than the dentine, and extending over a limited space. It commonly attacks the necks of the teeth, more usually those of the upper jaw; and Dr. Black says that four-fifths of the cases he has seen are as far front as the first bicuspid, and one-half of them on the labial surfaces of the incisors.

Many cases resemble mechanical wearing of the tooth by the toothbrush, though we often find it attacking positions of the tooth where we know the brush does not reach.

A case of wasting of the front teeth by which a separation of three-eighths of an inch was brought about between the incisors in the course of two years is related in Harris's "Dental Surgery."

Erosion is rarely solitary. Though we sometimes find it confined to one tooth, in many of these cases we might find it extended to others at a later date. Erosion and abrasion are often found combined, though it appears incidentally so, for they seem not to have anything common in their ætiology. I have at present a patient with a lower molar very much worn by abrasion, with an additional destruction by erosion. There is no evidence of decay about the tooth, but certain portions of the grinding surface are destroyed to a greater depth than the occlusion reaches. The tooth is quite sensitive, and I expect to build it up with gold.

I have a patient suffering from a cavity of erosion, encircling the mesial and palatal surfaces of an upper lateral incisor, about a line above the edge of the gum. It is a decided opening, making a dig in the direction of the nerve, and is quite sensitive, as cavities of erosion usually are. This is not the first trouble I have had with erosion in the case of this patient. In this case I have used the nitrate of silver treatment, which seems to be relieving the sensitiveness, so that I hope to stop the progress of the cavity without suffering the loss of the nerve.

We sometimes find erosion cutting its way through a tooth, just as if it has a grudge against the tooth and wanted to get it beheaded. Dr. Black has found secondary dentine in cases of near approach of cavities of erosion to the pulp. He has also noted spontaneous cessation of erosion after it had made considerable progress.

Mr. Tomes relates a case of rare form of erosion, wherein the destruction occurred, not in isolated spots, but the whole of the exposed portion of the tooth being attacked. As the morbid action progressed the enamel was slowly removed, making the teeth shorter and thinner, assuming a yellowish and translucent appearance so as to mark the position of the pulp cavity by the difference in color

Syphilitic markings of the teeth are accomplished by arrestation of development. Though certain markings are spoken of as dental erosion due to syphilitic diathesis, it is not correctly considered as belonging to this affection.

In such cases the tooth may appear worn, corroded or honeycombed on portions of its surface. There may be pits from pin point to pin head size, or larger, often white or yellow, darkening with age, more frequently affecting the centrals. Due to syphilitic affection, we often find markings of the cutting edge of the tooth; the first molars often having their first two-thirds well formed, with the grinding surface so badly deformed as to entirely obliterate the cusps.

Of these syphilitic markings, however, none save the Hutchison notch are usually considered pathognomonic of syphilis, as they (the others) may be dependent on other disturbances.

The manner of destruction by erosion seems to be an acid dissolution of the tooth structure. Our best authorities seem to favor the belief that it is through the agency of acids. Dr. Black claims to have obtained results the same as erosion with acids on teeth out of the mouth.

We would like to have the testimony of those present as to whether they have often found erosion accompanying rheumatic diathesis.

It is a satisfaction to know that cavities of erosion are as amenable to treatment by filling as are cavities of caries. We believe nitrate of silver to be invaluable in the treatment.

The great desideratum is the prevention, though it does not appear that we shall ever be better able to prevent erosion than to prevent caries.

Selections. . . . .

INCIDENTS OF DENTITION.

Two weeks ago I was called to a child that was said to be suffering from diarrhea and in great danger. During the long journey which I had to make there, from Paris to Bordeaux, I happened to come across, among my journals, the last issue of the *Bulletin Medical*, which contained the discussion at the Academy of Medicine on the report of Dr. Ollivier on the accidents of dentition. It seemed that the learned teachers were far from holding one opinion. While M. Magitot contended there were no diseases of dentition, and that "they ought to be erased from the text-books on medical nosology," M. Pamerd demonstrated, by means of numerous observations, that the diseases of dentition are a well-established fact, and that every child that cuts its teeth is sick.

I was not then interested in the differences as to the ætiological side of the discussion, but all seemed to agree that incision of the gums brought about very generally the disappearance of the accidents.

As I could diagnose it since my arrival, my little patient presented the symptoms attributed to dentition, which had been recalled in the course of the learned discussion.

This child, which was of a strong constitution, left the breast on the fourth month, and continued taking its milk from the bottle. This change of nourishment did not seem to affect the child much, except occasioning diarrhea for a few days, but then it soon regained its good humor. The first teeth, the lower incisors, appeared toward the seventh month. We were in the ninth month then.

Since ten days the trouble commenced, so much so as to become soon alarming; gastro-intestinal trouble, repeated diarrheas, vomiting, insomnia, rise of temperature. The treatment was mostly internal and general. The milk, which the child could no longer digest, was replaced by grogs (three teaspoonfuls of whisky, in a large glass of liquid, half boiled water and half Vichy), limewater.

lactic acid, etc. When I saw the child, it was very much run down, and the attending physician did not hide from me the seriousness of its condition.

Influenced by the reading on my way there, I called the attention of the physican to the condition of the gums. The mucous membrane was red, warm, stiff, and had on the interior border of the superior maxillary two mammillated surfaces, indicating that two superior central incisors were on the way of making their appearance. Very often the child would bring its hands there. I proposed lancing the gums, which I did in the presence of the physician. The horizontal incision was made deep enough to reach the crowns of each upper incisor on the labial surface. The incidental bleeding was quickly and easily arrested by borated wadding. A favorable change set in at once, the ordinary milk diet was resumed, insomnia disappeared, and after ten days both teeth appeared outside the gum, and the child was cured.

What we notice in this case is as follows:

1. The contemporaneity of the troubles and the eruption of the two upper central incisors. This agrees with the remark of Constantin Paul, who says that the upper incisors provoke toubles oftener than the lower.

2. Though the child changed to artificial nursing between the fourth and ninth months, no bad consequences were noticed then.

3. There was local irritation, which resulted in a gingival inflammation limited to the erupting region.

4. The immediate disappearance of the trouble after the lancing of the gum.

It is desirable that our colleagues make similar inquiries in their practice, and no doubt a collection of sincere observations emanating from practicing dentists will greatly assist the study of this debated question. (Prof. Ch. Godon, of Paris, in *Revue Internationale d'Odontologie*.)

THE DENTAL ASPECT OF IT.

BY A. BETHUNE PATTERSON, M.D.

In the June issue of the *Dental Cosmos* there is an editorial on the use of the lancet during first dentition. The writer labors to show the importance of more general information upon the conditions which call for the use of the lancet in teething, and deplores the ignorance manifested by medical men in anatomy, physiology,

histology, and pathology; and also in the ætiology of disease, especially as it bears upon pathological conditions found in the jaws of
teething children, which conditions, to the editor's mind, are responsible for many, if not all, of the diseases of infantile life.

He would force practitioners to accept his views by placing
them before the profession as facts. He says in reference to the
therapeutic value of the lancet: "This is simply a question of fact,
regardless of any theories involved."

With this flood of light from the dental side, it is presumed that
dentists and medical practitioners will supply themselves at once
with gum lancets, the panacea of all the ills of infantile life, or remain "culpable" and "ignorant" obstructionists to modern therapeutics. The editor continues: "Gum lancing is a therapeutic
measure of unquestionable value, relieving the pulp irritation set
up by backward pressure of the developing root. . . . But
why, in the light of all that has been written on the subject from a
dental standpoint, should any be ignorant? . . . Doubly culpable is a writer of acknowledged authority who condemns the operation of gum lancing as a therapeutic measure."

The author referred to is F. Forchheimer, M.D., Professor of
Physiology and Clinical Diseases of Children in the Medical College
of Ohio, and a member of the Association of American Physicians
and also of the American Pediatric Society, etc. This author condemns gum lancing, and the editor of the *Dental Cosmos* "records a
protest against the fallacies here promulgated, in the hope of preventing the serious consequences which must necessarily follow
such antiquated teaching, an adherence to which is largely responsible for the high rate of infantile mortality." Exactly how our
learned critic arrived at these remarkable conclusions and the
"facts" is of great interest to practitioners of scientific medicine.
Sections of the tissue in question, properly investigated under the
microscope, reveal the only solution to the problem. And by this
means only can a full explanation be given of that much-dreaded
condition characterized by "unyielding and overlying gum tissue."
Hypertrophy and atrophy, and in fact all other morbid conditions
of the cellular tissue, nerves, and blood vessels, can be understood only
by aid of the microscope. It is to be inferred that the editor has
not entered very far into the study of pathological dentition, for he
says: "There may be produced a hyperæmia sufficient possibly to
produce the protrusion of a part of the mass from the incomplete
aperture of the root, giving abundant cause for extreme constitu-

tional disturbance." This does not show that the editor is very confident in the position he has assumed. Again, we are informed that "in the adult, irritation of a dental nerve may give rise to otalgia, otorrhœa, deafness, amaurosis, hemicrania, neuralgia, hysteria, chorea, epilepsy, tetanus, and it is not only possible, but highly probable, that a like irritation may be the occasion of grave and even fatal disorders in the infant." What proof have we of such statements? Such conclusions would have claimed precedence some twenty years ago. But the great hurricane of progressive medicine has swept away such wild and speculative theories. I hold that it is not in good taste, neither is it fair, to brand practitioners and "authors of high authority" with deplorable ignorance, and especially is this out of taste in an editor who guards his opinions and expressions with such terms as "may be," "possibly," "highly probable," etc. Such opinions the editor expects medical men to accept as a lucid explanation of the pathological condition which in his mind is responsible for the present high rate of infantile mortality.

It is upon a clinical and scientific standpoint, keeping before me the laws of modern pathology, that I venture to take issue with the learned editor. Research in the ætiology of disease has opened the way to a clear understanding of at least some of the pathological conditions. The investigations have led us into the field of bacteriology. By the study of microörganisms we have accounted for many phenomena which before could only be speculated upon.

In considering gum lancing, the question of inflammation cannot be ignored, neither the so-called "reflex neurosis"—a bulwark behind which a vast amount of ignorance takes refuge.

To enter into a discussion of a number of diseases that are daily sending the little ones to an untimely grave is not the object of this paper. A brief reference to only a few of the fatal diseases of children will prove that pressure upon formative pulp cannot be the "occasion of grave and even fatal disorders in the infant."

Mothers and doctors have been bothering themselves for ages about nature's way of getting out teeth. Nearly all of the morbid conditions of this period have been traced to this physiological process. Swollen and inflamed gums have attracted the attention of the illiterate as well as the educated. So commonly is this pathological condition accompanied by grave enteric and nervous disturbances that they have been readily associated as cause and effect, and many have been the arguments knit together to satisfactorily demonstrate the plausibility of the theory.

Medical men are agreed as to the ætiology of inflammation. This pathological condition has its origin in microörganisms. It would be a waste of time, too, in the light of what has been developed from a medical standpoint, to offer proof of this statement. Swollen and inflamed gums are not alluded to in the editorial as characterizing "pathological dentition," and as these signs are the only evidence of a morbid condition it is necessary that they be carefully considered. There are two ways by which bacteria find their way into the tissue in question: through the circulation and through the buccal cavity, which cavity is the habitat of several different species of bacteria. While some of them appear to be harmless, others are inflammatory. The latter find their way into the gum tissue by accidents which are traumatic in character. The disposition of an infant to bite on any hard substance that it can get into its mouth is due, in my opinion, to the itching sensation which results from external pressure compressing the nerves supplying the intervening and absorbing structures. This is an invariable and constant condition. Mothers and nurses, eager to supply every apparent comfort and whim, have confidence in rubber rings, knife and toothbrush handles, friction and pressure upon the overlying gum, as a means of assisting the exit of the teeth. Thus supplied with the would-be auxiliaries, the infant, unmindful of the consequences, comes down with too much force, with the result of a rupture of the epithelium and underlying tissues. Fissures are thus formed through which pyogenic germs find their way into the underlying structure, the cells of which have already been crippled by the traumatism. At once the operation of development commences which is signaled by the common evidences of inflammation—viz., redness and swelling, etc. Another way by which inflammation is frequently set up is as follows: In the cutting of a molar, one of the cusps makes its appearance before any of the others, leaving an overlying flap which forms a pocket for the reception of particles of food, accompanied by inflammatory germs which here find conditions favorable to development. The result is inflammation. This condition calls for a lancet, the indications being those of an abscess. The resulting hemorrhage would wash out the bacteria and foreign matter, and relief would follow.

The editor, strong in his faith in " the light that has been developed on the dental side of this question," is uncompromising in his belief that "the present high rate of infant mortality is due to the irritation of the dental nerve set up by backward pressure upon the formative pulp." He is confident that his learned colleague (Dr. J. W. White) has proven conclusively that "in the adult the irritation of a dental nerve may give rise to otalgia, otorrhœa, deafness, amaurosis, hemicrania, neuralgia, hysteria, chorea, epilepsy, tetanus," etc. Dr. J. W. White concludes: "It is surely not only possible, but highly probable, that a like irritation may be the occasion of grave and even fatal disorders in the infant."

Reasoning from this standpoint, it would be logical to conclude that the gratification of the infant to bite on hard substances would increase the backward pressure, thus increasing the "distressing symptoms," and the diseases incident to first dentition, especially fits, diarrhœa, cholera morbus, etc., would be intensified.

In the study of the cause of disease at the present day we find bacteriology and pathology moving along hand in hand dispensing light, the rays of which are daily penetrating the hovel as well as the palace, carrying comfort and good cheer. The observations of pure cultures of microörganisms in the tissues of animals, as well as in the tissue of human beings, have demonstrated conclusively the part they play in the cause of disease. So no doubt remains in the minds of medical men as to the cause of tetanus, cholera morbus, and the various forms of diarrhea. Bacteria of a well-known shape, having color which is reflected in the characteristic green stools of teething infants, find their way into the alimentary tract, and are responsible for the enteric troubles which help to swell the high rate of infant mortality.

Further comment upon modern etiology would be akin to supererogation, and, as a parting word to our dental editor, we must say we will regard with reverence the "light" that has been developed on the dental side of the issue as a beacon, warning us of dangerous ground upon which the fondest hopes have been wrecked, and to say we are now headed for that revolving light in midocean of progressive medicine, a field which, when diligently cultivated, yields a harvest rich in good to suffering humanity.

DERMATOL, A SUBSTITUTE FOR IODOFORM.

BY DRS. R. HEINZ AND A. LIEBRECHT.

DERMATOL prepared according to our statements is a basic bismuth gallate in a suitable form. The remedy is a saffron-yellow, exceedingly fine powder unaffected by light and air and not hygroscopic. Very like iodoform in external characters, it has the advantage over it of absolute odorlessness. Insoluble in ordinary solvents—and therefore its antibacterial properties are manifested only when it is mixed with the culture medium—when applied to the human body its eminent drying action is effective in the same direction, preventing bacterial development by deterioration of the nutrient soil. Dermatol is absolutely nonirritant and, on account of its insolubility and extraordinary constancy, in contrast to other bismuth compounds perfectly nonpoisonous. It can be applied to the same extent as iodoform.

Up to the present it has been employed in surgical, gynecological and dermatological practice, and it has proved, above all, an excellent vulnerary.

It does not produce on the tissues treated an irritating or injurious influence, as do many antiseptics, which prolongs the process of healing; on the contrary, it tends to reduce the symptoms of irritation, diminishes in a striking manner the wound secretion, and furthers the formation of granulations. The properties of dermatol are recommended in the treatment of affections associated with much secretion, such as burns, eczema, ulceration of the feet, and ulcers in general.

It is believed to be applicable to opthalmology as well as diseases of the ear and nose.

Accurate limitations of the field of application, as also detailed description of practical methods of application, naturally cannot yet be given. At any rate the hitherto favorable experience, and the absence of any unpleasant side action, ought to stimulate the further trial of the compound in various departments of medicine.

Moreover, dermatol appears to be suitable for internal treatment. In view of its nonpoisonousness large doses could be given without hesitation, and as a matter of fact 30 grains pro die have been very well borne. Dermatol appears to be valuable not only instead of bismuth subnitrate in gastric affections, but also in diseases of the intestines; above all it should be tried in catarrhal and ulcerative processes associated with profuse diarrhea.

The communications made here apply only to the preparations put forward in practically suitable form and in a state of absolute purity by the "Farbwerke vorm. Meister Lucius & Bruning Hoechst on the Maine." We ask those who repeat our experiments to use only this preparation, the more as injurious admixtures (lead, arsenic, free gallic acid, and so forth), which are certainly avoided in the "Dermatol Hoechst," might manifestly have an important effect upon the results.

THE ADVANTAGES OF NITRATE OF SILVER IN DENTAL PRACTICE.*

BY A. M. HOLMES, D.D.S.

WHEN I received the request from the President of the Sixth District Dental Society to read a paper at this meeting, I hesitated, and was at a loss to know what to contribute that had not been written and talked over and over in society meetings; still, duty required that I should respond, and I decided to give something of my personal experience in the use of nitrate of silver in the treatment of diseases of the teeth, with the hope that, although it had but recently been up for discussion in dental societies, by reason of the able papers of Dr. Stebbins, the subject had not become threadbare, and that you would find something of interest in its consideration. The character and scope of the discussions that I have read on the use of this remedy for the treatment of diseased teeth have been such as to impress me with the belief that its benefits are not generally understood and appreciated.

Nitrate of silver is conceded to rank as one of the most efficient and reliable remedies in medicine and surgery; and when its merits are fully known, it is believed that it will be found equally efficient in the treatment of a large class of diseases of teeth. Take, for instance, decay in temporary teeth. We all know from individual experience how trying it often is to fill the teeth of small children, in the ordinary way of making such operations; how they resist all efforts to excavate and fill sensitive cavities. By the use of nitrate of silver these operations are more easily made.

In approximal cavities in the posterior teeth, where the child is not too nervous and timid, cut away the walls to a V shape, prepare a piece of gutta-percha of the proper size to fill the space, soften it by heat, and cover the parts that are to come in contact with the diseased surfaces with powdered crystals of nitrate of silver, and carry it to the place in the tooth or teeth prepared for its reception, packing it firmly, and leaving it there to be worn away by use in

* Read at the Twenty-fourth Annual Union Dental Convention of the Sixth, Seventh, and Eighth District Dental Societies of the State of New York, held at Binghamton, October, 1892.

mastication. When that takes place, the surfaces of the teeth treated will be found black and hard, with no sensitiveness to the touch
or to change of temperature, and they will remain so indefinitely.
In case the child is so timid and fearful as to prevent this course,
dry the cavity, take out such softened dentine as the patient will
permit, carry the crystals on softened gutta-percha into the cavity,
and pack it, leaving it to the time when it is desired to replace it
with a more thorough operation. On removal of this filling, the
dentine will usually be found hard, without sensitiveness, and need-
ing but little excavation for the final filling.

I have treated diseased pulps with nitrate of silver crystals very
frequently since early in my practice, especially in temporary teeth,
where devitalizing pulps with arsenious acid is unsafe, applying
the crystals direct to the exposed pulp, usually with relief to the
patient.

Nitrate of silver is a resolute remedy. It cauterizes the surfaces
of the soft tissues to which it is applied, but does not penetrate
them as does carbolic acid, nor does it involve the entire pulp in an
inflammatory condition, tending to destroy the whole mass, as does
arsenious acid.

In cases of extreme sensitiveness about the necks of the teeth at
the margins of the gums, where the tendency is to softening of the
tissues of the tooth, a condition very annoying to the patient and
troublesome to the dentist, nitrate of silver has proved more successful with me than any other remedy, in checking the progress of the
disease and relieving the patient. The salt may be applied directly
to the sensitive part without pain to the patient. A good method
that I have practiced is to cover the parts after the nitrate is applied with a phosphate filling material of a creamlike consistency.
That hardens and prevents the washing away of the remedy, and
the surrounding parts from coming into contact with the salt.

Erosion, or wasting of the teeth, is checked by nitrate of silver
more perfectly than by any other remedy that I have ever used.
The salt is applied to the affected parts, and covered with a phosphate filling to protect and retain it in place until it is firmly established in the dentine. In cases where the progress of the disease
has gone so far as to require restoration by filling, this preliminary
treatment is very beneficial in preventing a further waste of the
tooth substance, and consequent failure of the operation.

In cases of superficial decay in soft teeth, where dark surfaces
are not objectionable, nitrate of silver is very beneficial. By re-

moving the softened portion of the tooth, polishing the surface, and rubbing the salt into the dentine, using a warm burnisher, and varnishing the parts to protect them and to hold the remedy until it is taken into the organic matter of the tooth, there will succeed a dense, hard surface, free from sensitiveness in mastication or change of temperature. In filling cavities in the class of teeth having an excess of organic matter, with which there is so much trouble from chemical or electro-chemical action between the walls of the cavity and the filling, an application of nitrate of silver will effectually prevent these unfavorable results. The remedy is taken up by the dentine, penetrating the surface sufficiently to prevent any such action between filling and tooth.

This treatment will at times result in a darkish hue to the walls of the cavity about the filling. This I explain to patients, that they may know that it results from the treatment, and that it is a proper and favorable condition for permanency of the operation. In crowns and bridges, where dentine is uncovered, it is beneficial to use this remedy on the teeth and roots used to sustain the bridge or crown, as a protection against thermal change and decay. The use of nitrate of silver may be varied by applying the rubber dam, using a strong solution of the salt, and evaporating the moisture by the use of a hot air syringe. When used in this way, a solution of soda can be applied to the parts to neutralize any acid remaining. In the class of cavities extending so far beneath the soft tissues as to render the use of the rubber dam or matrix impracticable, and a leakage from the surrounding tissues is liable to enter the cavity while introducing the filling and injure the permanency of the operation, cauterizing these tissues thoroughly with nitrate of silver will effectually prevent such a result.

After treatment of diseased pockets, and removal of the deposits from the roots of teeth, nitrate of silver has proved more successful in restoring a healthy condition to the parts than any other remedy that I have used in the treatment of pyorrhœa. The finely pulverized crystals may be applied by a small spatula of wood or plantinum, slightly dampening the end of the instrument and applying it to the salt. The crystals will adhere sufficiently to be easily placed in the space between the gums and the roots of the teeth. After the remedy has been left for a few moments in contact with the parts, it may be washed away with water, by the use of a syringe.

In cases of the extirpation of pulps, where the canal is sensitive at or near the|apex|of the root, nitrate of silver crystals carried to

2

the sensitive part and left there for a few hours usually relieve the trouble, and the canal can be filled without pain or danger of un-favorable results.

These are some of the many cases in which nitrate of silver crystals are advantageous in dental practice. I will not detain you longer, for it was not the purpose of this paper to cover the entire field of this remedy. It is a powerful agent. It acts promptly, with great uniformity, and leaves its track in darkened surfaces when applied to the teeth. This should be considered, and its employment governed accordingly. (*Dental Practitioner and Advertiser.*)

FAILURE IN CROWN WORK.

BY H. B. MEADE, D.D.S., BUFFALO, N. Y.

MANY articles have been written concerning the failures made in banding roots for crown work, and many dentists have abandoned the use of collars or bands entirely, because a few cases have come under their observation in which a band was poorly adjusted, there being a V-shaped space between the band and root, the parts being in a highly inflamed condition. If bands are to be fitted in this manner, it would be better to discontinue the use of them at once; but with the exercise of intelligent care, no such space is necessary.

We must bear in mind that we are dealing with one of the most delicate tissues of the human anatomy when we extend a band so far up the root that it interferes with the peridental membrane. If we take into consideration the anatomical form of any of the anterior roots, after the crown has been excised and ground down to the gum line, and then imagine the fitting of the band to noth-ing more nor less than a cone, commencing at its base, we will at once see how the V-shaped space is produced, as the band must necessarily be as large as the end of the root, that it may be driven on.

To overcome the V-shaped space, and have the parts remain in a healthly condition, the band must not be driven on a root that has had no other preparation than merely the crown excised and ground down to the gum-margin. The remaining portion of the enamel that lies under the free margin of the gum must be re-moved, care being taken so to shape the root that its sides will be parallel, and the band must not extend farther than the free mar-gin of the gum. A band adjusted in this manner (using pure gold) will cause no inflammation. The end of the root should be ground concave from labial to palatal surfaces, giving it the same curvature as the gum line. This will make the fitting of the porcelain crown to the band an easy matter. (*Dental Practitioner and Advertiser.*)

FORCES WHICH MAKE FOR PROGRESS.

THE growth of civilization and the elevation of peoples in a moral and intellectual sense is an ever instructive problem. Nations have moved in the circle of development from savagism to a higher cultivation, and returned again to nearly the first estate, apparently but little better for the civilizing gymnastics. The circle has been the favorite figure of many of the philosophers of the ages. It presents many discouraging features to the optimist who is unwilling to see anything in the future but an unending advancement to higher and still higher perfections. The scientific thought very naturally drifts into this conception, as any loss from a point gained seems an impossible idea, and yet the world cannot fail to remember its "lost arts" and past periods of active scientific development.

The forces that underlie progress are subtle in their character, but within certain limits are irresistible. The gentle zephyr becomes a cyclone. The slight shock of the Leyden jar indicates a dynamic force to influence a universe.

The lines of force in the physical run parallel with the moral and intellectual, while the circles are ever changing and the orbit of thought presenting new phases, the impetus given being ever onward to a more orderly sequence and larger development.

The contemplation of the last decade of the century, thought-engendered by the new year, carries us back, very naturally, to the forces which have developed dentistry through its ten periods, each containing experiences of vital interest to the profession.

The dentistry of 1800 to 1810 was remarkable only for the fact that it exerted no perceptible influence for good or ill. The places where men labored as dental surgeons were practically shops, but they were building better than they knew, for the energy which was to travel through the century was even then silently at work.

Moving along decade by decade, the power widens and the environments improve, one period witnessing one advance and another still greater growth, one a period of college development, another of law, one great professional ability, and another enlarged theoretical acquirement. The force of intellectual progress has steadily advanced, few perceiving it, and to some it may be unintelligible, but it is ever making for a larger growth in the direction of a fuller professional life.

The world needs to be oftentimes reminded that the forces that make for progress are not impelled in one direction, but radiate everywhere, seemingly at times in diverse and antagonistic lines, yet the manifestations are ever tending to increasing mental power and professional strength. It is not alone in the classic shades that the world looks for its masters, but rather in the great, active world, where frictional energy is being constantly developed, evolving new ideas, new relations, new activities.

He is most wise who can read the indications of accumulated power. Sooner or later this increase will result in the bursting of bonds, and revolutions will arise in the political, moral, religious, and even in the scientific world. When Wendell Phillips said "Revolutions are not made, they come," he spoke a great truth. They are the accumulated force of long periods of apparent inertia.

The conservative or not deep-thinking element in dentistry must certainly see that the close of the last decade of the century means the centralizing of all these periods of active thought, and the breaking away from the old, and with this change revolution of ideas will come. Selfish indulgence in ease, commercial alliances, want of professional enthusiasm, all must give place to a higher conception of duty than now exists.

The future of dentistry is yet an unsolved problem, but in whatever direction it may be led it will advance only in the broadest and best sense, by a close adherence to those laws which have developed individuals and nations, and in these are embodied the forces which increase moral and scientific power. (Editorial in *International.*)

PERFECT OCCLUDING GOLD CROWNS.

BY W. H. WHITSLAR, M.D., D.D.S., CLEVELAND, OHIO.

OFTEN cases present that irregularity of occlusion which renders the ordinary use of the various dies in the market useless. Among the various expedients in such cases, the following may suffice: Fit the band to root and trim on the occluding edges, with flat file, to line where side walls begin to curve upon occluding surface. Return band to root and fill to overflowing with plaster Paris well mixed so as to set quickly. Have patient close teeth together naturally. When the plaster is hard, remove band and trim so as to make presentable surface, retaining the imprint of the occluding cusps. Use this model for making an impression in Mellotte moldene, from which make a fusible metal die. Perfectly adjust to the band the resulting swaged cap and solder. Finish. Result, a perfect occluding crown.

Another Way.—Fill band with moldene or plaster and proceed to get "bite" as before, then insert this model into fusible metal almost cold and drive gold plate into this mold with lead. (*Ohio Dental Journal.*)

DIED WHILE HAVING A TOOTH EXTRACTED.

A NEGRO woman recently died in Macon, Ga., while having a tooth extracted. The woman had been suffering severely from an aching tooth, and to get relief her husband called in a physician. It had been the custom of this doctor to administer chloral to lessen the pain in extracting teeth, and in this case he gave the woman fourteen grains of chloral. While waiting for the chloral to take effect he visited another patient, and on his return found the woman in a stupefied condition. With the assistance of the husband he then proceeded to extract one of the teeth. While extracting it the woman struggled and screamed. The husband pointed out to the doctor another offending tooth, and insisted on its extraction. The woman resisted and struggled, but the husband held her while the doctor extracted it. The woman became quiet, and it was soon discovered that she was dead. (*Dental Luminary.*)

Extracts.

DEATH IN A DENTIST'S CHAIR.

Mrs. Philip Sebest, a comely married woman, met death in a dentist's chair in Elmira, N. Y., recently. The cause of her death is not fully accounted for by the physicians, as the autopsy showed her heart to be in a normal condition, and a combination of causes was the verdict of the doctors. Shortly before 5 o'clock Mrs. Sebest went to the office of Dr. G. H. Preston, on West Water Street, to have some teeth extracted. He administered nitrous oxide gas, and placed his nippers upon the tooth to be extracted. Mrs. Sebest said to the dentist that she felt the pain, and wanted him to administer more gas, which he refused to do. When the tooth was drawn, she uttered a scream, and spat some blood in the cuspidor, gasped, and fell back in the chair, dead. Drs. C. W. M. and M. M. Brown were summoned, and also Coroner Westlake, who made an examination. It is claimed that Dr. Preston had before, on several occasions, administered gas to Mrs. Sebest without injury, and it is thought that the excessive heat, pain, and excitement brought on a stroke of apoplexy. Mrs. Sebest's sister was with her at the time the operation took place, and certified to the fact of the dentist's refusal to administer gas a second time. (*Items of Interest.*)

HOW TO DRINK MILK.

Some complain, says a contemporary, that they cannot drink milk without being "distressed by it." The most common reason why milk is not well borne is due to the fact that people drink it too quickly. If a glass of it is swallowed hastily, it enters the stomach and then forms in one solid, curdled mass, difficult of digestion. If, on the other hand, the same quantity is sipped, and three minutes at least are occupied in drinking it, then on reaching the stomach it is so divided that when coagulated, as it must be by the gastric juice, while digestion is going on, instead of being in one hard, condensed mass, upon the outside of which only the digestive fluids can act, it is more in the form of a sponge, and in and out of the entire bulk the gastric juice can play freely and perform its functions. (Exchange.) ——

ARTIFICIAL INDIA RUBBER.

Dr. W. A. Tilden discovered some months ago that isoprene, which can be prepared from turpentine, under certain circumstances changes into what appears to be genuine India rubber. Bouchardat had also found that the same change could be brought about by heat. The material so produced resembles pure Para rubber in every way, and, whether it is genuine rubber or not, it may be equally good for all practical purposes. It is said to be capable of vulcanization. (*Scientific American.*)

LEGAL RESTRICTIONS RESPECTING THE ADMINISTRATION OF ANÆSTHETICS IN FRANCE.

In a late issue of the *Revue Odontologique* it is stated that the proposed new law now pending in the French Parliament for regulating the practice of medicine makes it illegal for dentists who are not provided with French diplomas in medicine or dental surgery to practice anæsthesia without the assistance of a doctor or an officer of the public health.

Even the existing law has been construed by the tribunals, in a case recently tried, to prohibit the use of cocaine as an anæsthetic agent by dentists who are not doctors of medicine. A dentist and his assistant were held for injury by negligence and for illegal practice in administering cocaine in a case which resulted in nervous disorder and palpitation of the heart, and they were punished by fines and costs of suit. (Exchange.)

HOW TO PROCURE AN IMPRESSION OF THE MOUTH WHEN THE PATIENT IS INCLINED TO NAUSEA AND VOMITING.

Get your druggist to make you some lozenges with one-quarter grain of cocaine to each lozenge. Before taking the impression allow the patient to dissolve one of these lozenges in the mouth and swallow the spittle. If one is not sufficient, give the patient another lozenge, allowing time for the lozenge to dissolve slowly, and you will find that you can take an impression with plaster of Paris without inconvenience to the patient or yourself. (C. V. Snelgrove, L.D.S., Toronto.)

Or better: Permit the patient to inhale the fumes of spirits of camphor from a napkin or handkerchief until all sense of taste and largely feeling is destroyed; then insert the plaster, and while that is in the mouth continue the inhalation. (ED.)

TRIONAL AND TETRONAL.

M. A. RAMONI, after experimenting on fifty-one insane men in the Roman Lunatic Asylum, comes to the following conclusions: 1. They are superior to sulphonal and chloral. 2. The patient awakes more easily; there are no unpleasant after effects, such as nausea, vomiting, loss of appetite, etc. 3. Action of the drug is rapid (thirty to sixty minutes). 4. Trional is superior to tetronal, the sleep produced by the former being sounder and more lasting. 5. The sleep (after either of the drugs) lasts on the average six to eight hours, and is not disturbed by dreams. Dose adopted, 30 grs. 3 times a day. (*American Therapeutist.*)

HINTS WORTH REMEMBERING.

To remove the stains of tincture of iodine, from either the hands or napkins, apply strong ammonia. The spots will immediately come out clear.

The stains of nitrate of silver, on either the hands or napkins, can be easily removed. First cover the spots with tincture of iodine, wait a few moments, then apply strong ammonia, and rub well.

In using argenti nitras in treating children's teeth one may accidentally get a few grains of the powder on his hands and not discover it till the hands are washed, when the black stains will be well set. By proceeding as above they will disappear at once. (George A. Maxfield, D.D.S., in *International.*)

WHY WOMEN ENDURE PAIN SO STOICALLY.

PROF. LAMBROSO, an Italian physicist of distinction, declares that women are less sensitive to pain than men, and actually feel less of it in given operations. Experiments on one hundred women led to the conclusion that they were not more than one-half as sensitive to pain at the tip of the forefinger as the average man. The above is confirmatory of a well-known fact to surgeons and dentists. It is a wise provision of nature that the sex to whom pain is a birthright should enjoy protection from its shock and immunity from its sharpest pangs. (*Med. and Surg. Record.*)

CIRCULATION OF PERNICIOUS LITERATURE.

UPON this subject Mrs. L. S. Rounds, President of the Illinois W. C. T. U., said in her annual address: "Never in the history of our country was pernicious literature so systematically and widely circulated as now. The mails are burdened with tons of reading matter that barely escapes the legal definition of impurity, and hence avoids confiscation, but which in point of real truth is intensely dangerous to the minds and morals of the young. Mr. Thomas K. Cree, Field Secretary of the International Committee of the Y. M. C. A., says: 'In Great Britain Zola's grossly impure French novels are not allowed to be sold, and the publishers are fined and imprisoned. In this country we carry all his books in the mails by the ton at one cent a pound, while Bibles, miscellaneous, and school books cost eight cents a pound!'

" It will surprise you to know that five thousand tons (ten million pounds) of paper-covered books are carried in the mails annually from New York City alone, at one cent a pound. The amount of second-class mail is a matter of record, but the exact amount of books so mailed cannot be known, yet it is the opinion of those best able to know that five thousand tons is a fair estimate for New York, and half as much for other cities.

"All over the land are scattered far and wide engravings, photographs, pamphlets, leaflets, and microscopic charms, all teaching with Satanic skill every stage of impurity, while over this seething mass of printed pollution there hangs to-day a criminal apathy on the part of good people which will in the near future give place to a mournful wail for the children debauched and hopelessly ruined."

HOT WATER FOR HEMORRHAGE.

DR. JULIUS SCHEFF, JR., of Vienna, according to the current number of *Ash's Quarterly Circular*, recommends strongly the use of hot water for arresting hemorrhage after tooth extraction. "We are accustomed," he writes, "to stop hemorrhage by the method that has been used for generations—viz., by the direct application of cold water to the wound. Practitioners started with the idea that heat caused expansion of and induced increased bleeding from the vessels; but, on the other hand, cold caused contraction. In an ordinary case of extraction, hemorrhage from the arteria dentalis, or from the gums and periosteum, soon ceases; but it frequently happens, even when the patient does not suffer from hematophilia, that there is difficulty in arresting the flow of blood." Dr. Scheff mentions three cases occurring in his practice in each of which there was a history of profuse hemorrhage after extractions. "I allowed one patient," he says, "to take a great quantity of cold water, and yet there appeared not the slightest diminution in the bleeding. I then took a glass syringe and continuously applied hot water, in drops, to the wound, from which the blood previously trickled without cessation. After a few seconds the bleeding diminished, a coagulum was formed, and the bleeding finally ceased. With the second patient I used hot water at once, and the flow of blood was arrested. In the third case the wound had been bleeding for a long time. I plugged the alveolus with iodoform gauze, and on removing the plug the wound bled afresh. I then employed hot water; the hemorrhage ceased and did not recur." Dr. Scheff applies the hot water by means of a syringe, injecting it by drops into the socket of the tooth. The arrest of hemorrhage in surgical operations by the application of heat is a recognized resource, and it would therefore seem that this principle might with advantage be applied in cases of tooth extraction, especially as the mouth is able to bear a very high temperature without inconvenience. In fact, water so hot that it causes pain when the finger is inserted in it will in many cases be tolerated in the mouth. (*Lancet.*)

ANOTHER CASE OF INSANITY CAUSED BY NITROUS OXIDE GAS.

Miss Lizzie Lots. the beautiful daughter of Mr. Otto Lots, a prominent citizen of Covington, Ky., is now at College Hill Sanitarium receiving treatment for her mind. The family claim that Miss Lots became insane through taking nitrous oxide gas for the extraction of two teeth while visiting at Indianapolis.

Her sister noticed her condition of mind, and immediately sent her home. She seemed to be all right for a few days, but occasionally claimed that she had gas in her head. She would tell her father that he was to be the next President of the United States, and that he would reappoint Mr. Hardeman as postmaster of Covington. Again, she would talk on all sorts of financial schemes of great magnitude. Her condition finally became so alarming that her folks were compelled to remove her to the sanitarium for treatment. She now speaks of the sudden spells she had at home. Her father is confident that her malady was caused by the taking of gas and pulling of her teeth. He declares that the girl was never sick in her life before, and that her mind became impaired immediately after she had her teeth extracted.

Mr. Lots is borne out in his statements by his other daughter, who lives at Indianapolis. She is a favorite child, and her venerable father is almost prostrated over her condition. He has given her all the medical attention that he can, but grave fears are entertained as to her ultimate recovery. She is well known among the people of Covington, and had a successful business. Her father is enraged over the treatment she received at the hands of the dentist in Indianapolis, and he has employed Senator Goeble, of Covington, to bring an action for damages against him.

Miss Lots is a striking beauty and highly accomplished in music. She is about twenty-two years of age, and a brunette of purest type. She is very popular among all her acquaintances, who will be surprised to learn of her distressing malady. Her parents reside at 1003 Greenup Street, Covington. (*Dental and Surgical Microcosm*.)

SMUDGED GOLD.

THE common method of annealing gold mats, pellets, or cylinders, by holding them over or in the flame of an alcohol lamp or Bunsen gas burner, is a practice which, while ordinarily successful, is liable to occasion defects in the fillings.

The resulting imperfections are not often observable in flush-finished fillings, although some of these subsequently scale at marginal points on their surfaces; but elaborate building or contour work not infrequently meets with most disappointing disaster, due to the smudging of the gold by the incomplete combustion of the flame fuel. Yet the real cause of the calamity is unnoticed, and fault found with the gold, or the possible presence of a leak in the dam or other source of moisture suspected, whereas the first thought following the surprising failure should be: "The flame is at fault." Clearly one of the most important preliminaries to a gold operation should be a careful scrutiny of the annealing flame, to be sure beyond a peradventure that there is not a trace of smoke; that the combustion is perfect. The wick of the alcohol lamp is usually too tight in its tube, and not loose enough in its assemblage of fibers to permit a free flow of the fluid fuel. Of course the appearance of a single glow point at a fiber end of the wick is a certain sign of smoke, and should at once be remedied. When a lower grade than

95 per cent. alcohol is used, the residual fluid after a few hours' burning becomes so watery as to lessen combustion and cause the charring of the wick-end. The sight of a blackened wick-end leaves no doubt as to the probable character of the annealing and the operative work done by the use of that lamp.

The illuminating gas of divers cities differs in quality, and even in the same city varies from time to time in its heat and light giving properties; therefore the ordinary Bunsen burner is liable to vary in its degree of combustion; but the habit of closely observing the flame and keeping it regulated to the blue point of complete combustion will tend to the avoidance of the risk of smudging, the main thing being to be sure that the burner is a good one. It is well to keep at hand a piece of white porcelain—for instance, a small butter plate—and by occasionally holding it for a minute or two over the flame gain an assurance of the entire absence of smoke. When the gas is of a poor quality, the impurities and the gaseous products of their combustion contaminate the gold to a degree incompatible with a perfect welding or cohesion.

The mica method of annealing is preferable, as avoiding all risk of a smudge; but many practitioners are confirmed in the habit of flame annealing, and will probably continue to employ the means to which they have become accustomed, and which it is believed may be satisfactorily modified in the particulars herein mentioned. (W. Storer How, in *Dental Cosmos.*)

HARRIS MEMORIAL COMMITTEE.

At the meeting of the Virginia State Society a committee was appointed to carry out suggestions contained in the paper of Dr. Cockerille published in this number. This is a laudable undertaking, and should meet the approval of every lover of our profession. Chapin A. Harris deserves a monument at the hands of the dentists. What an unselfish life! his only ambition to make of dentistry a profession, giving freely of his knowledge to whoever wanted to learn. It is to be hoped that the different State Societies will take an interest in this movement and appoint similar committees.

Now is the time to discuss it, and let it be brought before the World's Columbian Dental Congress, which will undoubtedly be the grandest gathering of dentists ever held, made possible by Harris and his compatriots. Contrast the present with the past, when, after having all overtures to establish a chair of Dentistry in one of the medical schools of Baltimore spurned, the Baltimore College of Dental Surgery was established, the success of which may be called the beginning of dentistry as a profession. His name should be dear to every dentist, and it should be a labor of love to aid in erecting a suitable monument to this great man in dentistry, having done so much and received so little—name not even mentioned by the leading biographers. Sad commentary. Yet his works will live long to do him honor. (Editorial in *Southern Dental Journal.*)

SUDDEN DEATH.

Mrs. Kate Ledbetter, daughter of Peter McCue, who lives near Oakford, died in Dr. Solenberger's dental office Wednesday evening, while under the influence of an anæsthetic, which, at her request, had been administered by Dr. Whitley, preparatory to having some teeth pulled. Coroner McAtee held an inquest, at which it appeared in evidence that the usual precautions had been taken, before and during the administration of the anæsthetic, which Dr. Whitley stated was hydrobromic ethyl, used for short operations. The coroner's jury was composed of Z. A. Thompson, foreman; A. E. Estill, E. H. Bigelow, Mat Hainsfurther, E. R. Oeltjen, and H. C. Levering. They returned the following verdict:

" We, the undersigned jurors, sworn to inquire of the death of Mrs. Kate Ledbetter, on oath do find that she came to her death by paralysis of the heart. We further exonerate all parties from blame."

The body was taken to the residence of J. D. Roberts, where it was prepared for burial by Undertaker Conant, and taken to Oakford on Thursday morning's train, the funeral occurring yesterday.

The deceased was the wife of Frank Ledbetter, a Wyoming ranchman, and was here on a visit. Her husband was coming through to Chicago with a shipment of stock, and was to stop here on his way home. Mrs. Ledbetter was about thirty years of age. (From *Democrat*, Petersburg, Ill., in *Medical and Surgical Microcosm.*)

PYORRHŒA ALVEOLARIS.

Anything that will add to the resources of the dentist in arresting the flow of pus from the pockets around roots of teeth must be considered advantageous to the recipient and user as well.

For a period of ten months we have been using the following so

lution in the manner indicated: After the roots have been cleansed
of all deposits (when present), the edges of the alveolar process
have been scraped with small spoon excavators, breaking down the
necrotic process as far as possible. Following this process the pockets have been syringed with H_2O_2 until the *débris* has been removed.

Now, take twelve minims of oil of cassia and add to sixteen
ounces of distilled water. Agitate this from time to time for a few
days at a temperature of 70° F., or upward. Very soon the oil
will be dissolved in the water.

To each ounce of the above add five minims of the officinal dilute
sulphuric acid. Agitate this until thoroughly dissolved.

This solution is to be injected into the pockets carefully and
slowly, having previously dried them as well as possible with paper
cones.

The solution is astringent and stimulating, and according to the
latest experiments it is a bactericide of positive value. Should the
teeth feel sensitive, the mouth may be rinsed with lime water or
soda water or any other alkaline fluid, as weak ammonia water or
soap water.

We have continued this treatment at intervals of four days for
from four to five weeks with most excellent results.

In all cases where the teeth are very loose they must be made
firm by wiring with pure gold wire or banding them with narrow gold or platinum bands cemented to the teeth.

When the acidity is too pronounced, the treatment is alternated
with a 2 per cent. solution of zinc iodide·in water. When there is
much inflammation in the beginning of the treatment, washing the
pockets with boroglycerine water one to ten for four or five days
consecutively will be of advantage. When great pain is felt on account of the depth of the pockets, inject one minim of vinum opii
into each pocket, when the pain will quickly subside. Holding hot
water in the mouth from three to five minutes will also relieve
pain. (Editorial in *Dental Review*.)

———

SYNTHETIC chemistry seems to have no limits. The latest product which has been successfully made from coal products is camphor. It promises to be cheap, and the specimens submitted respond to the most crucial tests. (*St. Louis Medical and Surgical
Reporter.*)

CARBOLIC ACID.

[The following letter from Sir Joseph Lister, giving to carbolic acid as an antiseptic preference over the bichloride of mercury, will be read with interest and much satisfaction by many dentists who have persistently advocated it and used it exclusively for the treatment of devitalized teeth.—Ed.]

My Dear Sir: Your letter has been forwarded to me to this place. I have no hesitation in answering your question to the effect that the presence of the minute quantity of free chlorine cannot possibly interfere with the antiseptic action of the bichloride. If it had any effect at all, it would be to enhance the antiseptic efficacy. It might possibly make the solution act slightly more upon the steel of the instruments. I may remark that, as the result of recent investigations, I have for some months past abandoned the use of the bichloride in favor of our old friend, carbolic acid. It has been shown that a 1 to 40 solution of carbolic acid is really superior in actual germicidal power for such organisms as cause inconvenience in surgery, as compared with any solution of bichloride that could be used for surgical purposes. . . .

Believe me sincerely yours, Joseph Lister.

P. S.—For purifying instruments and sponges, and the skin of the part to operated upon, a 1 to 20 solution of carbolic acid is, of course, used.

Glenelg, N. B., September 24, 1892.

Although Sir Joseph Lister has abandoned the use of bichloride of mercury in favor of carbolic acid, the former is still largely used, though we may expect many to follow the example of the great surgeon in giving it up. There appears to be some uncertainty as to the effect of heat upon aqueous solutions of mercuric chloride. In "Martindale" there is a statement, concluding with a note of interrogation, that "heat reduces the salt to calomel." Mr. Rushton Parker, one of the honorary surgeons to the Royal Infirmary, was anxious to be assured on this point, and as the results of many experiments, performed quantitatively, Mr. Johnson could not detect the slightest reduction of the chloride in such solutions as 1 in 500, 1 in 1,000, 1 in 2,000, etc., even after submitting to prolonged boiling. (From an extract of paper read by J. R. Johnson at a meeting of the Liverpool Pharmaceutical Students' Society. From the *Chemist and Druggist.*)

———

Prof. Da Costa says that gelsemium is especially useful in neuralgias of the dental organ. Begin treatment by giving 5 drops of the tincture three times a day, and increase to 10 drops three times a day until the patient sees double, and then stop the administration of the drug. (*Coll. and Clin. Rec.*)

NOSE AND THROAT.

In a recent lecture before the Chemists' Assistants' Association, London, by William Hill, M.D., London, the throat was described in detail, and the pharynx and the larynx pointed out as the two most important parts. The nose has a very important connection with the throat and its disorders. It contains a series of bones called the turbinated bones, which expose a large surface of warm blood, and cause the air inhaled to be warmed ready for the lungs; moreover, the cilia of the nose cause the secretions to move, and reject the solid particles it has collected. The nose is the proper organ for breathing, not the mouth. The larynx, which is the air passage, is bounded at the upper extremity by the vocal cords, and has, therefore, the double function of breathing and of phonation. The epiglottis, by altering its form, causes the food to pass down the pharynx, and keeps it from the larynx. In speaking of proper breathing, the author pointed out that diaphragmatic breathing was the proper method, and not clavicular. It was reported that Rubini had broken his clavicle during singing, by persisting in this method of breathing. Throat diseases are often caused by germs, by inhalation of sewer gas, etc. Fortunately there are other organisms in the throat always ready to attack these germs. The throat was well provided with tonsils, both faucial and lingual. The tonsils produce phagocytes, or leucocytes, amœboid corpuscles which actually swallow up the germs. Why, then, should tonsils be cut out? Because, when they become enlarged and horny, they lose this function, and by removing the horny surface, the newly exposed portion can go on producing the corpuscles. The decay of teeth is largely due to germs. This shows the importance of keeping the teeth in order. Obstruction in the nose is the cause of many throat disorders. Care must be exercised in the use of both alcohol and tobacco. Many people can use these luxuries with impunity in moderation; others cannot. People liable to throat disorders should be very chary of eating piquant or hot dishes. Irritating remedies, too, such as cayenne and (except in special cases) tannin lozenges or nitrate of silver, should be avoided. Hot tea, too, is bad.

THE HUMAN HEART.

THE workings of the human heart have been computed by a celebrated physiologist, and he has demonstrated that it is equal to the lifting of 120 tons in twenty-four hours. Presuming that the blood is thrown out of the heart at each pulsation in the proportion of sixty-nine strokes per minute, and at the assumed force of nine feet, the mileage of the blood through the body might be taken at 207 yards per minute, 7 miles per hour, 168 miles per day, 61,320 miles per year, or 5,150,880 miles in a lifetime of 84 years. In the same period of time the heart must beat 2,869,776,000 times. (*The American Therapeutist.*)

WHEN we see a young dentist peering round here and there, sparing no pains to find "the lay of the land" in every direction, and insinuating himself into the good graces of every one he meets, and becoming a favorite with old dentists, concealing nothing and learning of all, we make up our mind that this young man is going to succeed. But when we see one coming home from college in a dandy suit, to be admired and waited on, and sitting down in the parlor in egotism and self-consciousness, to wait for rich patients, we do not count on him. (*Items of Interest.*)

WAS Carlyle thinking of teeth when he characterized the present age as one of sham? If so, he knew what he was talking about. In a compensation case, a dentist in Ludgate Hill deposed that he had 20,000 patients on his books, and that he had supplied over 100,000 sets of teeth during the time he had been in practice. (*Pall Mall Gazette.*)

DR. W. W. KINKEAD, of Nashville, Tenn., in the *Columbus Medical Journal*, gives the following example of a reliable treatment for an acute cold, and for the incipient stage of inflammations of the air passages, as tonsillitis, bronchitis, etc.

R. Atropinæ sulphatis......................................gr. $\frac{1}{60}$-$\frac{1}{100}$.

Morphinæ sulphatis......................................gr. ss.

Aquæ destillatæ......................................f$\frac{2}{3}$ ij.

M. Sig.: One teaspoonful every half hour.

"WHAT is the matter with the baby?" asked a lady of a little girl whose baby brother she had understood to be ailing. "O nuthin' much," was the answer. "He's only hatchin' teeth."

Editorial. . . . .

ASSOCIATION OF SOUTHERN MEDICAL COLLEGES.

THE journal of the American Medical Association, in its issue of December 10, devotes a little more than a column to the "Association of Southern Medical Colleges," recently formed in Louisville, in which it attempts to show that there is no need for its existence, and making, as we conceive, many statements at variance with the facts.

The language used is far from courteous or polite, when the character of the promotors of the new association is considered. Every Southern school of any consequence was represented at the Louisville meeting, and the intention of the Convention was to elevate the standard of these institutions and in the interest of advanced medical education.

That these colleges are not recognized as "high-grade schools" is not true, as some of them at least, and most of them could if they so desired, hold membership in the Association of American Medical Colleges, organized in this city in May, 1889. Indeed, there is very little difference in the Constitution and By-laws of the two Associations, and the graduates of the Southern colleges stand before Boards of State Medical Examiners with as much credit and honor as those from the boasted "high-grade schools."

The article invites attention to the requirements for entrance examination, and adds: "It will readily be inferred that the people of the South are not desirous of a very high grade of intellect in their medical men. . . . The statistics of the various State Boards will classify them where they belong, and their *clientele* will consist of the scavenger element."

Such language is inexcusable and outrageous. The South wants all the intellect it can command, and that of the broadest and most humanitarian kind, for the material from which to make her medical men, and we claim that it has always had it. Such brilliant men as Marion Sims, John Weyth, William Polk, and Page are notable among the graduates from these "low-grade schools" and representatives of this scavenger element" from which the Southern schools draw their pupils. If the South can furnish men to teach in these high-grade institutions, write text-books for the world, and bring to the profession so many startling and wonderful revolutions in surgery and practice, it will never lack for material to make men who will be an honor to medicine and a blessing to suffering mankind.

INSALIVATION.

THE importance of thorough insalivation of food is either too little understood or sadly neglected. In our intercourse with our patients we find scarcely any who know anything about the necessity of mixing the food with saliva. The failure to teach it is largely the cause of the rapid manner our American people have fallen into of devouring their food.

In the first place writers of physiology are at fault in that they have not taught that it is one of the steps in digestion. We hold that it is and a very vital one, so recognized. In substantiation of this statement we quote from a recent little publication by Dr. Isaac B. Davenport, of Paris: "We have not learned the full signification of all that part of digestion which ought to take place in the mouth, and which in its beginning is coincident with mastication, insalivation, aeration, the sense of taste, and beginning of deglutition. Man's life may be sustained for a time by transfusion of blood, by rectal enemas, or by direct introduction of food into the stomach, but something needed is left out of the process of digestion by all these modes. It is not enough that the food reaches the stomach simply well reduced. Of course thorough reduction of food is essential, and digestion is active or sluggish according to whether the reduction is perfect or imperfect. Insalivation is directly related to mastication. Dalton shows that on the side engaged in the act of mastication the corresponding parotid gland secretes three times as fast as that of the opposite side, and besides facilitating the reduction of food, it is more and more evident in physiological studies that the thorough mixing of saliva with the food in the mouth is essential, and that the saliva is a true digestive of certain food elements. Until we can exactly calculate the importance of a normal mixing of saliva with the food in the process of digestion and know all the remote effects traceable to it we must assume that that process is essential and that the best interests of the body require its perfect performance. We may also suppose (and the supposition is proved true by clinical and personal experience) that the well-being of the individual requires that mastication be perfectly performed, not simply for the mere reduction of food, but that insalivation may also be completely accomplished, and to that end both sides of the dental arches ought to be equally competent to perform their functions."

Finally we would add that the saliva does not admit of substitution. No other fluid can take its place without resulting sooner or latter in disordered digestion. All who practice sipping water, coffee, tea, or other beverages during the process of mastication, instead of relying upon the normal secretion of the salivary glands, which nature furnishes in such abundance, are doomed to suffer with dyspepsia. ˙The laws of nature, when broken, are irreparable. The saliva, also in accordance with well-established chemico-physiological law, is the normal stimulant to gastric secretion, its alkaline reaction exciting the flow into the cavity of the stomach of the acid gastric juice. _____

THE WORLD'S COLUMBIAN DENTAL CONGRESS.

The officers of this Congress were selected by the Executive Committee at its meeting in October, and we give a full list of them in this issue. Elsewhere we present a biography of the President, Dr. L. D. Shepard, of Boston. The selection—as far as experience, ability, culture, strength of character, voice, and *personnel* go to make a finished presiding officer—cannot but prove a wise one. With eight representative men as Vice Presidents, the meetings of the Congress will run harmoniously and successfully.

The burden of the preparations has been great and nobly borne by this Executive Committee, and the elaborate arrangements and multitude of detail have been, in the main, prepared without any great friction or discord.

We bespeak in advance the coöperation and brotherly assistance of those who attend and take part in this meeting. Leave ambition and self-gratification or personal aggrandizement at home when you go to Chicago next August. Go to add to the good feeling and to support and not find fault with those who have labored so to make this an occasion of dentistry, for American dentistry, and American dentists.

It is estimated that from three to five thousand dentists will attend this meeting. Not all can realize their expectations, but let it not be you who is first to complain or find fault. Professional men are often too sensitive and feel slighted when Associations are not conducted as they are accustomed to seeing them or as they desire, or they are not treated with just the amount of consideration that they think is due them. This grows out of the fact that with pupils or patients their wills are the law and their judgments are final. It begets a disposition to rule; and when crossed or differed with, they at once think ruination is but a short distance ahead. Leave self behind, and go to add to the meeting your presence, enjoy yourself and make some one else do so.

THE DENTAL PROTECTIVE ASSOCIATION.

At the risk of being charged with riding a hobby we want once more to call on the profession of Tennessee and the South to support this organization. If you are not already a member, send Dr. Crouse, of Chicago, your application for membership, accompanied by $10, and fall in line. It is your duty to yourself and family, and you will never cease to rejoice over it.

There are no salaried officers to grow rich off the Association. The work is done as a work of love. There are no annual dues as some think. The money is all spent in defending suits for damages, and if you are sued secures to you defense by the Association's council without a dollar's additional cost until all the funds of the Association are exhausted. The attorney of the Association is a man of experience and ability, fitted for this work, therefore superior to those who could be had ordinarily.

The report of the fourth annual meeting will be found in this issue and is a source of gratification and pleasure to members of the Association.

"THE DENTAL HEADLIGHT FOR 1893."

A happy New Year's greeting is extended to all friends and patrons of the DENTAL HEADLIGHT. The year 1893 is pregnant with bright promise for the future of dentistry. We congratulate our *confrères* and the profession generally upon its auspicious opening. The standard of higher dental education has been proudly reared by nearly all the dental schools of America, and we still aspire to even greater achievements. If we mistake not, the World's Columbian Congress of Dentists will inaugurate a new and grander era in the history of dentistry. Dentistry has already proven a very David amongst the scientific Goliaths, and its motto is: "Onward and Upward." We ask your hearty coöperation in our endeavor to make the DENTAL HEADLIGHT the leading organ of dentistry in the South, and the faithful exponent of the best sentiments of the profession throughout the world.

The editor of the HEADLIGHT passed a most agreeable day in Chattanooga during the holidays mingling with the dentists of that wide-awake and progressive city.

Dr. Louis Ottofy has severed his connection with the *Dental Review*, to take full control of the *Weekly Dental Tribune*, a paper which he begins with the new year.

We wish to acknowledge with thanks receipt of the "Pearsons Dental Appointment Book" for the vest pocket, which is issued annually by R. I. Pearsons & Co., Kansas City, Mo.

Dr. S. B. Cook, President of the State Board of Dental Examiners, and Dr. M. D. Billmyer, of Chattanooga, recently spent the day in Nashville attending the Section of Oral and Dental Surgery of the Nashville Academy of Medicine.

WORLD'S FAIR NOTES.

"THE eighth and greatest wonder of the world" is what the World's Fair buildings and grounds, even in their present incomplete condition, are pronounced by Maj. Woods, Executive Commissioner of the Connecticut World's Fair Board.

In lighting the World's Fair, 92,622 incandescent lamps, of 16-candle power each will be used, according to present estimates. The contract for furnishing and maintaining these lights has been let to George Westinghouse, Jr., for $339,000. This is more than $1,300,000 less than the Edison-Thompson-Houston electrical combine, or trust, first asked for the work. This immense saving was effected by rejecting the bids and readvertising. Mr. Westinghouse is required to file a bond for $1,000,000 by June 10 to guarantee the faithful execution of his contract. In addition to the incandescent lamps, about 5,000 arc lights of 2,000-candle power each will be used. The contract for these was let some time ago at $20 per lamp.

BOOK NOTES.

567 USEFUL HINTS FOR THE BUSY DENTIST. By William H. Steele, D.D.S. Published by the Wilmington Dental Manufacturing Company, Philadelphia. Price $2.50.

Another collection of valuable extracts, many of which are original, which answers many perplexing questions that arise daily in the early practice of every dentist. One of the benefits of the work is that it affords an opportunity of obtaining the opinion of most writers upon dental subjects without the delay or trouble of writing to them. With ample index and an admirable arrangement there is no difficulty in getting the views of any number of men on any given topic. Such works are invaluable to dental students and busy dentists, for whom Dr. Steele has intended his book.

THE ANGLE SYSTEM OF REGULATION AND ROTATION OF THE TEETH. Third edition revised and enlarged by Edward H. Angle, Former Professor of Histology and Orthodontia and Comparative Anatomy of the Teeth, Department of Dentistry, University of Minnesota. Published by The Wilmington Dental Manufacturing Company, Philadelphia. Price 75 cents in paper.

A work of fifty pages full of thought and good suggestion well illustrated. That portion of Chapter I. under the head of "Fundamental Principles" is worth the price of the little book. One point is assured: a regulating appliance made by the Angle System once adjusted is assurance that the patient will come back and it is not at all likely that it will be removed until the services of a dentist are obtained. Those not familiar with the ingenious appliance of Dr. Angle and his methods should at once procure a copy of the book.

METHODS OF FILLING TEETH. By Rodrigues Ottolengui, M.D.S. With illustrations. Published by the S. S. White Dental Manufacturing Company, Philadelphia, and Claudius Ash & Sons (limited), London. Price $2.

This work is written in an attractive, easy style, and cannot fail to interest the reader. It is an exposition of the author's own practice of dealing with all classes of cavities. In a hasty examination much more is found to approve than to criticise, and we feel that the work on the whole can be recommended. The author does not set up any claim of originality, and certainly none can be vouchsafed, for most of the methods were in vogue before he was a dentist. His preparation of cavities in the anterior teeth, involving the loss of a portion of the labial plate of enamel near the gingival margins, making the cavity circular instead of cutting to an angle, is faulty and violates mechanical laws which are most important to recognize. The latter treatment overcomes the tendency of mastication to force the plug from position. Then, too, his failure to recognize the value of noncohesive gold, leads to the mistake of too deep retaining pits and grooves in the cervical region cutting off portions of enamel and dentine from their nutrient supply. The author in his "Preface" states that it is his purpose to describe only such methods as he himself has tested, "believing that the student will be more benefited by adopting a single mode of practice than by essaying the various methods of many men." As long

as there are "many men" there will be "various methods," and it
is only by contrasting them that the seeker for truth can get what
he is after, and no one writer can hope to meet the demands of the
great variety of conflicting opinions and †perplexing cases in a
"system" of operating.

QUESTIONS AND ANSWERS FOR DENTAL STUDENTS. By Ferdinand J. S. Gorgas,
A.M., M.D., D.D.S. In three parts, for Freshmen, Juniors, and Seniors.
Published by Snowdon & Cowman, Baltimore. Price $2.50 each.

These works consist of questions and answers on anatomy, physiology, dental histology, materia medica and therapeutics, chemistry, oral surgery, dental pathology, operative dentistry, prosthetic
dentistry, metallurgy, plastic and metal work, porcelain-crown,
cap-crown bridge work, and atmospheric pressure, deformities of
the palate. etc., and is altogether the most complete and comprehensive set of quiz compendiums we have seen. They have been
examined by some of the prominent teachers of the profession, and
pronounced good. They unhesitatingly recommend them to students. As books of review and reference they cannot fail to serve
a valuable purpose to any one. Our medical brothers should possess a complete set.

FLINT'S HUMAN PHYSIOLOGY. D. Appleton & Co., New York.

This, the fourth edition of this standard text-book, is just received. The text has been entirely rewritten and brought fully up
to the present advanced state of the science. Old illustrations have
been discarded, and many new electrotypes and lithographic plates
are introduced. Much contained in the old text-book, especially
historical detail and theoretical disquisition, has been omitted.
Altogether, the present edition is a vast. improvement upon its
predecessors, and we earnestly commend its perusal by both students and the profession.

TRANSACTIONS OF THE TWENTY-EIGHTH ANNUAL MEETING OF THE ILLINOIS
STATE DENTAL SOCIETY, held at Springfield, May, 1892. The Dental Review Company. H. D. Justi & Son, Chicago.

LUTHER DIMMICK SHEPARD, A.M., D.D.S., D.M.D.

LUTHER DIMMICK SHEPARD was born September 11, 1837, at
Windham, Me., where his father, the late Rev. John W. Shepard,
was pastor of the Congregational Church. He received his literary,
classical. and professional education at the Nashua (N. H.) High
School, Phillips Academy, Andover, Mass., Amherst College, and the

Baltimore College of Dental Surgery. As a boy he was handy with tools; and having a friend in a dental office, while he was in the Nashua High School, he became interested in dentistry, and in leisure hours and in vacation took a course of three months' instruction with Dr. P. Brooks, paying therefor fifteen dollars, and received from him a certificate in 1856 as a qualified dentist. He itinerated somewhat, but spent most of his vacations while in college in practice in Bristol, N. H. He also had patients in Amherst among the students and Faculty, so that he became known to the village dentist, the late Dr. Chester Stratton, of Amherst, with whom in 1862 he entered into partnership, which continued with most pleasant relationship till January 1, 1865, when he removed to Salem, Mass., and became a partner with the late Dr. W. L. Bowdoin. In 1867 he moved to Boston and became associated with the eminent dentists, the late Dr. Joshua Tucker and Dr. George T. Moffatt. This connection lasted nine years.

Dr. Shepard early became convinced that his three months' pupilage was not sufficient professional education, and so he supplemented it by a course and graduation at Baltimore. He was from the start a subscriber to dental journals, and in 1863 attended his first dental meeting, that of the American Dental Association in Saratoga. Several of those present had just come from Philadelphia, where they had attended the third annual meeting of the American Dental Association. From them he became convinced that the plan of a delegate association presented advantages, and on his return to Amherst he assisted the late Dr. F. Searle, of Springfield, Mass., in getting up the Connecticut Valley Dental Society, which was organized in Springfield November 10, 1863. Dr. Searle was elected President and Dr. Shepard Secretary. He served as Secretary three years and was then elected President, and has attended most of the meetings ever since.

In 1864, at Niagara Falls, Dr. Shepard joined the American Dental Association as a representative of this society. He has attended most of the meetings of the American Association since that date, has held several officers, was President in 1880, and has been fifteen years on the Executive Committee. In 1878 he declined to join the seven who signed the majority report of the Committee on Sections, and in connection with Dr. J. N. Crouse, presented a minority report which was adopted entire by unanimous vote and is the plan upon which the Association has worked ever since. After fourteen years' trial, and in view of the recent efforts to change the plan, it may be interesting to quote the opening sentences of the minority report: "The undersigned members of the Committee on Sections did not sign the report with the majority for various reasons. They were and are in doubt as to the wisdom of dividing the Association into sections; but as the opinion of the members in favor seems to predominate, they would recommend that the experiment be tried."

On removing to Salem in 1865, Dr. Shepard joined the Massachusetts Dental Society and the Merrimac Valley Dental Association (now the New England Dental Society), and has been President of each. In 1879 a joint convention of the eleven New England Dental Societies was held in Boston, of which he was elected President by a nearly unanimous vote. He was the orator at the annual meeting of the Massachusetts Dental Society in 1870, and again in 1892. He joined the American Academy of Dental Science in 1869, was Secretary several years, Treasurer a number of years, and Vice President, resigning from the Society in 1882.

The scheme and plan of the Dental School of Harvard University was worked out by a joint committee from the Massachusetts Dental Society, consisting of Drs. E. C. Keep and E. C. Rolfe and Dr. Shepard, and from the Harvard Medical School, consisting of Drs. H. J. Bigelow and H. I. Bowditch and Calvin Ellis. Of this joint committee, Dr. Shepard is the only survivor. On the organization of the school in 1868 Dr. Shepard became a member of the Faculty as Adjunct Professor of Operative Dentistry, and on the resignation of Professor Moffatt in 1879 was elected Professor of that chair, serving until 1882, when he resigned. At the Commencement of the university in 1879 Dr. Shepard received the honorary dental degree. Common as it is now, it is worthy of note that this school was the first to be organized as a part of a university. It should also be known that during the thirteen years that he was professor the professors received no remuneration, but freely gave their time and labors, preferring that all the revenues from students should go to the support of the new and unendowed school. It can also be shown that this school repeatedly raised its standard, to its financial detriment, and so during these years was a powerful factor in the progress which has been so marked throughout the country. At a meeting of *alumni* and friends of the school held in 1889, to celebrate the twentieth anniversary of the graduation the first class, Dr. Shepard gave the historical and anniversary address.

On the passing of the dental law in Massachusetts in 1887 Dr. Shepard was appointed to the Board of Registration for three years. He was elected President of the Board on its organization, and continued President until 1892, when he felt compelled from press of engagements to resign.

In 1887 Dr. Shepard, as the representative of the Massschusetts Board, joined the National Association of Dental Examiners, and in 1891 presided as President *pro tem.* at the annual meeting, and was elected President for the ensuing year.

His two addresses on "Dental Colleges" and "Dental Laws," delivered before the First District Dental Society of New York, at New York City, and before the New York State Dental Society at Albany, in 1889 and 1891, published in the *Dental Cosmos*, have been widely read by the profession.

Dr. Shepard was a member of the International Medical Congress at London in 1881, at Washington in 1887, and at Berlin in 1890, and at the latter was one of the four honorary Presidents from America. It was while he was absent in Europe, in 1890, that he was appointed by the American Dental Association as one of the members of the General Executive Committee for the World's Columbian Dental Congress and also elected Chairman of the General Finance Committee. He has attended every meeting of the Executive Committee since 1890.

For the information of those who think that Dr. Shepard has not added anything to the literature of the profession we present a list of papers and essays read by him before one society only, the Connecticut Valley Dental Society, from 1864 to 1884: "The Importance of the Dental Collegiate Course;" "Popular Education," published in the *Dental Cosmos;* "Are You a Reading Man?" published in the *Dental Cosmos;* "Dental Colleges;" retiring President's address; "The Use of the Mallet in Consolidating Gold;" "Separating Teeth;" "Controlling the Flow of Saliva;" "Examination of the Teeth;" "Mouth Mirrors, Their Value and Use," published in the transactions of the American Dental Association, 1872; "How Shall Children's Teeth between the Ages of Twelve and Eighteen Years Be Treated?" "Nutrition;" "The New Departure, review of its claims with especial consideration of the question to what extent the conscientious dentist shall adopt its principles in his practice," published in the *Dental Cosmos*, 1878; "Report of Case of Replantation with Specimens and Models;" historical address on twenty-first anniversary of the Society. (*Dental Tribune.*)

Associations. . . . .

PERMANENT OFFICERS OF THE WORLD'S COLUMBIAN DENTAL CONGRESS.

PRESIDENT.—L. D. Shepard, Boston, Mass.

VICE PRESIDENTS.—W. W. H. Thackston, Farmville, Va.; A. L. Northrop, New York City; W. H. Morgan, Nashville, Tenn.; W. W. Allport, Chicago, Ill.; W. O. Culp, Davenport, Ia.; C. S. Stockton, Newark, N. J.; Edwin T. Darby, Philadelphia, Pa.; H. J. McKellops, St. Louis, Mo.; J. Taft, Cincinnati, O.; J. H. Hatch, San Francisco, Cal.; J. B. Patrick, Charleston, S. C.; J. C. Storey, Dallas, Tex.

SECRETARY GENERAL.—A. W. Harlan, Chicago, Ill.

ASSISTANT SECRETARIES.—George J. Friedrichs, New Orleans, La.; Louis Ottofy, Chicago, Ill.

TREASURER.—John S. Marshall, Chicago, Ill.

EXECUTIVE COMMITTEE.—W. W. Walker, Chairman, 67 West Ninth Street, New York City; A. O. Hunt, Secretary, Iowa City, Ia.; L. D. Carpenter, Atlanta, Ga.; J. Y. Crawford, Nashville. Tenn.; W. J. Barton, Paris, Tex.; J. Taft, Cincinnati, O.; C. S. Stockton, Newark, N. J.; L. D. Shepard, Boston, Mass.; W. W. Walker, New York City; A. O. Hunt, Iowa City, Ia.; H. B. Noble, Washington, D. C.; George W. McElhaney, Columbus, Ga.; J. C. Storey, Dallas, Tex.; M. W. Foster, Baltimore, Md.; A. W. Harlan, Chicago, Ill.; J. S. Marshall, Chicago, Ill.; H. J. McKellops, St. Louis, Mo.

The meeting will be held in Chicago August 17–29, 1893.

_____

OFFICERS OF THE AMERICAN DENTAL ASSOCIATION.

J. D. Patterson, Kansas City, President; J. Y. Crawford, Nashville, Tenn., First Vice President; S. C. G. Watkins, Mont Clair, N. J., Second Vice President; Fred A. Levy, Orange, N. J., Corresponding Secretary; George H. Cushing, Chicago, Recording Secretary;

A. H. Fuller, St. Louis, Treasurer; W. W. Walker and S. G. Perry, New York, and D. X. McQuillen, Philadelphia, Members of the Executive Committee. Next meeting to be held in Chicago.

OFFICERS OF THE SOUTHERN DENTAL ASSOCIATION.

B. Holly Smith, Baltimore, President; R. K. Luckie, Holly Springs, Miss., First Vice President; S. B. Cook, Chattanooga, Second Vice President; L. P. Dotterer, Charleston, S. C., Third Vice President; D. R. Stubblefield, Nashville, Tenn., Corresponding Secretary; S. W. Foster, Decatur, Ala., Recording Secretary; H. E. Beach, Clarksville, Tenn., Treasurer; W. R. Clifton, Waco, Tex., and Gordon White, Nashville, Tenn., Members of the Executive Committee. Next meeting to be held in Chicago.

OFFICERS OF THE TENNESSEE DENTAL ASSOCIATION.

S. B. Cook, Chattanooga, President; W. W. Jones, Murfreesboro, First Vice President; W. J. Morrison, Nashville, Second Vice President; P. D. Houston, Lewisburg, Recording Secretary; D. R. Stubblefield, Nashville, Corresponding Secretary; H. E. Beach, Clarksville, Treasurer. Next meeting to be held in Chattanooga first Tuesday in July, 1893.

GEORGIA STATE ASSOCIATION.

S. M. Roach, President; X. A. Williams, Valdosta, First Vice President; —— Hinman, Atlanta, Second Vice President; S. H. McKee, Talbotton, Recording Secretary; L. D. Carpenter, Atlanta, Corresponding Secretary; A. H. Lowrance, Athens, Treasurer; Executive Committee, Drs. Wells, Sims, Rosser, McDonald, and Hinman. Permanent place of meeting in the future, Atlanta. Next meeting, Tuesday, May 9, 1893.

TENNESSEE STATE BOARD OF EXAMINERS.

S. B. Cook, President, Chattanooga; J. L. Mewborn, Secretary, Memphis; J. Y. Crawford, Nashville; W. T. Arrington, Memphis; A. F. Shotwell, Rogersville; R. B. Lees, Nashville. Next meeting July, 1893, at Lookout Mountain.

SECTION ON ORAL AND DENTAL SURGERY OF THE PAN-AMERICAN MEDICAL CONGRESS.

José Joaquin Aguirre, Santiago, Chili; R. R. Andrews, Boston; E. A. Baldwin, Chicago; George Beers, Montreal, Canada; S. B. Brown, Fort Wayne; Emegdio Carillo, City of Mexico, Mexico; William Carr, New York; H. B. Catching, Atlanta; George J. Fredericks, New Orleans; Ricardo Gordon, Matanzas, Cuba; J. H. Hatch, San Francisco; A. O. Hunt, Iowa City; Louis Jack, Philadelphia; H. J. McKellops, St. Louis; Francis Peabody, Louisville; J. C. Storey, Dallas; J. Taft, Cincinnati; J. B. Willmot, Toronto, Canada.

EXECUTIVE PRESIDENT.—M. H. Fletcher, M.D., D.D.S., 65 W. Seventh Street, Cincinnati, O.

SECRETARIES.—John S. Marshall (English-speaking), Chicago, Ill.; Ramón Campuzano (Spanish-speaking), Philadelphia, Pa.; N. Etchepareborda [Tacuari 355], Buenos Ayres, Argentine Republic; Dr. Wilson, Lapaz, Bolivia; Benicio de Sá, Rio de Janeiro, U. S. of Brazil; Luke Teskey, Toronto, Canada; Guillermo Vargas Parédes [Carrera 7, Núm. 638], Bogota, Republic of Columbia; J. Louis Estrada, Guatemala City, Guatemala; George Herbert, Wailuku Maui, Hawaii; Rafael Rico [Escuela de Med.], City of Mexico, Mexico; A. Lacayo, Granada, Nicaragua: Andres G. Weber [Corrales 1], Havana, Cuba; Angel Guerra, Montevideo, Uruguay.

Meeting to be held in Washington, D. C., September 5, 6, 7, and 8, 1893.

THE FOURTH ANNUAL MEETING OF THE DENTAL PROTECTIVE ASSOCIATION.

THE fourth annual meeting of the Dental Protective Association of the United States was held at the Grand Pacific Hotel, Chicago, last Monday afternoon. A committee consisting of Drs. Louis Ottofy, of Chicago, T. S. Hacker, of Indianapolis, and F. O. Hetrick, of Ottawa, Kans., was appointed to examine the condition of the books of the Association. It was found that the expenses of the Association are kept at the minimum, that the cases now pending in various courts of the country have made favorable progress, and that the membership is constantly increasing. The issuing of a membership certificate was left discretionary with the Board of Directors, as was also the publication of the report of the annual meeting and of a financial statement. The members of the Board of Directors, consisting of Drs. J. N. Crouse, E. D. Swain, and T. W. Brophy, were unanimously reëlected. (*The Dental Tribune.*)

WORLD'S COLUMBIAN DENTAL CONGRESS.

ORDER OF BUSINESS.

Thursday, August 17.—10 A.M.: Meeting of the General Executive Committee. 11 A.M.: Opening of the Congress. Reading of the resolutions creating the Congress by the Secretary General. Address of welcome by John Temple Graves, of Georgia. Responses. Responses from foreign countries. Address of the President. Adjournment. 1:30 P.M.: Papers to be read in the Sections. 5 P.M.: Adjournment.

Friday, August 18.—9 A.M.: Clinics. 10 A.M.: Meeting of the General Executive Committee. 12 M.: Address before the whole Congress. 1 P.M.: Adjournment. 2:30 P.M.: Papers to be read in Sections. 5 P.M.: Adjournment. 8 P.M.: Bacteriological exhibit.

Saturday, August 19.—9 A.M.: Clinics. 10 A.M.: Meeting of the General Executive Committee. 12 M.: Address before the whole Congress. 1 P.M.: Adjournment. 2:30 P.M.: Garden party. 8 P.M.: Conversazione.

Monday, August 21.—9 A.M.: Clinics. 10 A.M.: Meeting of the General Executive Committee. 12 M.: General address before the whole Congress. 1 P.M.: Adjournment. 2:30 P.M.: Papers before the Sections. 8 P.M.: Biology. Lantern exhibition.

Tuesday, August 22.—9 A.M.: Clinics. 10 A.M.: Meeting of the General Executive Committee. 12 M.: General address before the whole Congress. 1 P.M.: Adjournment. 2:30 P.M.: Papers before the Sections. 8 P.M.: Bacteriological and biological exhibit. 8 P.M: Conversazione.

Wednesday, August 23.—9 A.M.: Clinics. 10 A.M.: Meeting of the General Executive Committee. 12 M.: Address before the whole Congress. 1 P.M.: Adjournment. 2:30 P.M.: Papers to be read before the Sections. 8 P.M.: Public address under the direction of the World's Congress Auxiliary.

Thursday, August 24.—9 A.M.: Clinics at hospitals. Clinics at the Art Institute. 10 A.M.: Meeting of the General Executive Committee. 12 M.: General address before the whole Congress. 2:30 P.M.: Papers to read in the Sections. 8 P.M.: Dinner to the whole Congress. (Subscriptions by members from the United States only.)

Friday, August 25.—10 P.M.: Visit in a body to the Medical and Dental Exhibits at the World's Fair Grounds. 12 M.: Closing addresses to the Congress. Luncheon by the members in the restaurant. (Name to be supplied.)

NOTICE.

NASHVILLE, January 1, 1893.

I want the address of every dentist in the State of Tennessee who proposes to attend the Columbian Dental Congress. Do not fail to sit down at once and send me a postal card to this effect.

HENRY W. MORGAN,
Chairman of State Committee.

211 North High Street.

IN MEMORIAM.

OCTOBER 4, 1892, at a called meeting of the dentists of Nashville for the purpose of taking action in reference to the death of Dr. James C. Ross, on motion of Dr. Crawford, Dr. Jones was called to the chair, and Dr. Hickman requested to act as Secretary.

After short talks by Drs. Freeman, Crawford, White, Lees, Jones, and others, eulogistic of the life and character of Dr. Ross, the following committee was appointed to prepare suitable resolutions expressing the high esteem in which the deceased was held by the members of his profession.

1. *Resolved*, That in the death of Dr. James C. Ross our profession and the community has suffered an irreparable loss, in all the qualities that go to make the pure, true, noble, upright Christian gentleman, who was ever ready to aid in the promotion of the best interests of his fellow-men.

2. *Further*, that we tender our sincere condolence and sympathy to his bereaved family, and commend them to the Giver of all good.

3. *Further*, that this resolution be given to the DENTAL HEADLIGHT for publication.

W. W. P. JONES,   R. R. FREEMAN,
J. Y. CRAWFORD,   *Committee.*

DIED, January 4, 1893, at the home of his parents, near Woodville, Tex., Dr. A. Lincoln Pedigo, of the graduating class of 1891–92, Department of Dentistry, Vanderbilt University. Dr. Pedigo made many friends during his stay at college, and it is with sincere regret that we chronicle his untimely death. He gave bright promise of success in his chosen profession. We tender his family our heartfelt sympathy in their sad bereavement.

C. L. BOYD, D.D.S.,

EUFAULA, ALA.

PRESIDENT OF THE ALABAMA STATE DENTAL ASSOCIATION.

Original Communications. . . . .

FILLING TEETH WITH GOLD.*

BY W. K. SLATER, D.D.S., NASHVILLE, TENN.

THERE has been no subject that has been so thoroughly and ably handled as the one under consideration. Since the time when dentists first organized themselves into societies and began to read papers for their advancement, the present subject has been a favorite one; and again, there is no one operation in dentistry that is so universally performed, and so thoroughly understood, both practically and theoretically, as the operation of filling teeth with gold. It would be a hard task indeed for one to present a paper adhering closely to the subject, that would be at once interesting, instructive, and new.

The ancients were familiar with gold, its working qualities, and its value. The Egyptians were skilled enough in working it to draw it into threads which they wove into cloth that equaled any done in modern times. The Etruscans wrought it into jewelry that, try how he may, the best goldsmith of any time since then has never been able to equal. As you know, these and many others of their time, and in fact almost every people, who have succeeded them have known more or less of its value and have been able to work it accordingly. But it was left for modern American dentists to bring it to its present state, bordering on perfection, in that best of all spheres to which it has yet been adapted—namely, the filling, and if rightly introduced, necessarily the preservation of the teeth, to the comfort, health, and consequent longevity of man.

Statistics show that the average life of man is to-day longer by some years than it was a few decades ago; and who will say that the world is in no small degree indebted to the dentist for this great blessing, on account of his ability to preserve the teeth that they may perform their functions physiologically.

The forms in which gold is offered to us are innumerable, but they

are either cohesive or noncohesive. The relative merits of these are
for the first, readiness with which we can, by proper care, restore to
their original shape and former usefulness teeth, portions of which
have been lost from the various causes, and for the second case of
adaptation in all simple cavities.

It was as noncohesive the first gold was offered to the profession,
and was for a long time the only kind; but finally cohesive gold was
ushered in with such flattering commendations that the claims of the
old and always reliable kind were threatened with annihilation; but
glad are we to say that as the prosthetic art is returning to gold in
another form, so also is noncohesive gold being more and more used.

We will describe three classes of cavities, one of which is best
filled with noncohesive, another with cohesive, and the third with
the two in combination. Let us suppose a grinding surface, or any
simple cavity with four strong walls. Such fillings are best made of
noncohesive gold, in the form of ribbons, cylinders made by rolling
the ribbons around the point of an instrument, or in mats made
from suitable-sized pieces torn from the sheet with the foil carriers
and rolled in the fingers. This last seems to be the most rational
method, as the filling can be almost introduced while one is preparing his gold for either of the other ways. If we use cylinders
that are ready prepared, this last is not such an affliction.

These mats should be stood on end in the cavity and partly consolidated against the distal wall with a foot plugger, allowing the
end to protrude from the cavity, and forcing them in until no more
can be made to enter. The protruding ends may now be driven
down and consolidated, looking carefully for any soft place that
might occur. If one is found, insert a sharp-pointed instrument
and force in a tight-rolled mat; pack as before, burnish well and
finish. In cases where extensive contours are necessary all cohesive gold should be used, adding bit by bit until the tooth is the
same shape it was originally.

The best place to illustrate the use of the two golds in combination is in compound fillings between molars and bicuspids. The old
way was to make permanent separations; now we cut through the
grinding surface, extending the cavity back into a fissure sufficiently far to obtain secure anchorage; then beginning at the cervical
wall, fill solidly with noncohesive gold from one to two-thirds of
the cavity, finishing off with cohesive gold and leaving the tooth as
nearly as is in our power in the same shape as when intrusted to
the patient's care by nature.

DENTAL THERAPEUTICS.

BY W. J. MORRISON, D.D.S., NASHVILLE, TENN.

Mr. President and Gentlemen of the Dental Section of the Nashville Academy of Medicine: While the subject, dental therapeutics, which has been assigned to me for this occasion is so vast that with one of ability it could be made one of the best and most instructive papers that has ever been presented before this Section, I feel that I am compelling you to lose valuable time in listening to a paper which makes me blush at the bristling quotation marks, and brings me to a realization of my insignificance in the knowledge and proper application of that which is so important for the successful practitioner of the dental science.

Therapeutics is that branch of medical science which comprises the doctrine of management of disease. Much more attention has been given by the profession to this subject than to pathology, thus placing ourselves in the relative position of the cipher before the one, for to treat disease by name and not by nature, and to use medicines in the same way, is practicing fraud upon the confiding sufferer. By far the larger number of teeth upon which we are called to operate are in such a pathological condition as to demand a careful investigation of disease, and to require a change of condition by such therapeutical treatment as a necessary means of saving them. Ignorance, even in the rudimentary principles, in this direction causes untold agony to the millions who place themselves under treatment, and it is a shame and a blot upon our profession that in its ranks there is such a large percentage who resort to the forceps as the best, surest, and quickest cure, and as a road to a large, successful, and eminent practice in the community in which they live, and they will continue in this sure and successful treatment of all their patients, regardless of the curses of the thousands yet unborn, and who cannot understand why God gave them teeth to make life a hell upon earth.

It would be useless for any one to undertake to treat the subject in all its magnitude, as it would fill volumes; but we can, to a limited extent, examine into some of the medicinal agents we are called on to use in daily practice, such as antiseptics, anæsthetics, astringents, caustics, disinfectants, escharotics, laxatives, narcotics, refrigerants, sedatives, styptics, stimulants, and tonics.

The essayist, being a believer in the germ theory, will, in treating antiseptics, also include germicides, and so will open the battle, depending on friends of more ability to assist him, for the army of this class of medicines is so large and has such a rivalry for supremacy that I feel that I cannot keep them under control.

Carbolic acid, considered by a large percentage of our profession as a panacea for all diseases of the oral cavity, is obtained from coal tar, and chemically considered is an alcohol. Water dissolves 6 per cent., 5 parts in one of alcohol; 4 in two parts of ether. The best quality contains 2 per cent of water, and should be hard and dry, with no odor of creosote or of volatile sulphur compounds. Much of what is called creosote is nothing but impure carbolic acid. I will not here enumerate the abuse of this medicine, in its use, but will just say that it would have been better, for this reason, had it never found its way into dental practice. It is neither a very active antiseptic nor a germicide; in fact. all the virtue it possesses lies in its escharotic power. Iodoform has been almost as much over-estimated for its virtues as carbolic acid, and but for the disagreeable odor would in the hands of some have held as high a position as carbolic acid. I will not extol the many virtues which are secured to it by its forming a cuticle over and around the diseased tissue, but will touch upon its use in the treatment of nerve canals by what is now commonly known as the "stink gun" treatment. The little instrument known as the "stink gun" is constructed with a reservoir to hold a given amount of iodoform, which can be volatilized by the application of heat and the gases forced into the nerve canal by pressure upon a bulb arranged upon the instrument for that purpose. Many are as loud in their praise for this treatment as is the odor from the iodoform, in their offices, after its use, and while we are able to recognize the benefits to be derived from such treatment, this very objection to the odor will have a tendency to limit its use, yet it has opened up an avenue for the further study of the application of medicine in this way, and I must say that when we look into this one alone, we readily see the absolute necessity of our knowledge of chemistry, in the application of medicines by different modes, and also the obtaining of desired results without the objectionable features referred to above in the use of the iodoform; and in this case, I think, with a little thought, we can, for when we decompose iodoform by this process, the gaseous fumes forced into the nerve canals are condensed in the form of resublimated iodine, which is almost insoluble in water, and has the virtues of iodine. Now, if we use resublimated iodine and decompose it by the same process, we will have the same mechanical and medicinal results without such an offensive odor.

Bichloride of mercury is one of the best and most important medicines at the command of the dentist for general use in the treatment of diseases of the oral cavity, from an antiseptic and germicidal standpoint. Some of the most learned men who have made a study of this medicine prove its efficiency, and my experience has been that, after taking everything into consideration, it is one which we could not well dispense with. We should have the aqueous solution 1 to 500, put up in a 1 or 2 gallon stone jug, and we can then keep a pint bottle at hand to use from; but, for fear of accident, would color with permanganate of potash. For cleaning the hands alone it is indispensable. In extracting, cleaning the teeth, lancing the gums, disgusting breath, and in fact for any lesions caused by operations or otherwise in the mouth, about one-half ounce of this solution in a glass of water makes a most beneficial and unobjectionable mouth wash. In the treatment of abscesses and nerve canals in the anterior teeth, a 2 per cent. alcoholic solution will be found one of the best applications that can be made, and I would state that a procedure that reminds us of that of the gasous treatment by iodoform in the use of this remedy might be explained as follows: With hot air, dry out the canals, then pump in the alcoholic solution of bichloride; dry and treat again; dry with hot air again; continue this treatment as often as the severity of the case demands. The principle is that the alcohol, having a great affinity for water, will take it up and deposit the bichloride, and after several applications the crystals of bichloride of mercury are not only absorbed by the tooth structure, but the canal will be thoroughly coated with them. This treatment should not be used in the front teeth, as it has a tendency to turn them a black color.

NEURALGIA.*

BY E. F. HICKMAN, D.D.S., NASHVILLE, TENN.

THE term " neuralgia " is applied in a general way to pain, which is either of idiopathic origin or constitutes the principal and at times the only symptoms of some obscure lesion. Neuralgia is a symptom indicative of direct injury to, or altered nutrition of a sensory nerve, which in the former case is more or less persistent, but in the latter is usually paroxysmal.

Loomis describes the morbid anatomy, etiology, and symptoms in

---

* Read before the Dental Section of the Nashville Academy of Medicine.

the following language: "I say this disease may be functional or
organic, but in the majority of instances no changes can be found
after death. When neuralgia is a symptom of acute neuritis or
perineuritis the nerve trunk is hyperæmic and swollen or degenerated and atrophied.

Neuralgia is often a hereditary disease in those of a neuropathic
tendency. Any disease causing general or local, permanent or transient anæmia is a marked predisposing cause. Among exciting
causes are cold, lead, mercurial, and other states of chronic blood
poisoning and traumatism. Disease of the genito-urinary tract, especially in women, often excites reflex or sympathetic neuralgia in
remote nerve trunks. Reflex neuralgia is also induced by decayed
teeth, dyspepsia, worms, constipation, etc., and occurs very frequently in convalescence from relapsing fever.

This disease rarely occurs before puberty, but just at this time
there is a marked predisposition to it. Those between twenty and
fifty years of age suffer more frequently with it, women being more
liable than men.

Symptoms.—Before actual pain begins in a nerve there may be
numbness, slight cutaneous hyperæsthesia, or some peculiar skin sensation, which is well known by the neuralgic individual. The pain is at
first intermitting; later, it is continuous, with slight remissions. The
character of the pain varies; it may be dull, boring, stabbing, tearing, or darting, and is confined very distinctly to the course and
distribution of the affected nerve; in fact, many patients will trace
exactly the course of some nerve when pointing out the locality of
the pain. Sudden movements, as turning and coughing, often increase the pain. Increase of pain on pressure is an important point;
the exacerbation is greatest during a paroxysm, and greater in proportion to the intensity of the original pain. Some points are more
sensitive than others. These points are where the nerves make
their exit from bony canals or foramina, the point where they pass
through a muscular aponeurosis, at their bifurcation, and where ter-
minal branches become superficial. These pain points are better
marked the longer the patient has suffered from neuralgic attacks.
In connection with the pain there is generally associated with it
some vasomotor disturbance, as extreme pallor, or vivid redness,
and reflex movements and twitching of the muscles. Should the
nerves of a gland be attacked, secretion will probably be increased.
After cessation of the pain the part often feels sore and bruised,
and there is a general sensation of exhaustion and weariness.

Differential Diagnosis.—Neuralgia may be mistaken for myalgia, syphilitic periostitis, and for cerebral abscess. Myalgia is distinguished by its nonparoxysmal character, by the pain being increased by motion, and by the fact that the attachments of the muscles are the points chiefly involved. Syphilitic periostitis is to be distinguished from neuralgia by the presence or absence of other symptoms of constitutional syphilis.

Treatment.—Neuralgia has been well said to be the cry of a nerve for better blood. Should anæmia be evidenced, a generous diet, cod liver oil, the hypophosphites, or small doses of phosphorus, and the appetizers, along with quinine.

Neuralgia due to syphilis demands iodide of potash. For rheumatism, the antirheumatics; and for malaria, quinine, and especially in tic douloureux is quinine the most effectual remedy. A patient with neuralgia should be removed from exposure to cold and all kinds of irritation. Locally, blister, continuous current, chloroform, opium, belladona, and liniments, and cold or very hot water may be applied. For the immediate relief of pain, morphine is the most effectual.

I will now take up neuralgia from a dental standpoint. It has been thoroughly established by dental surgeons that dental irritation gives rise to neuralgia in many nerves, and more particularly in the trigeminal nerve. The perpetual irritation in such instances, after being reflected to the nerve centers, instead of passing over to adjoining motor centers, induces some pathological condition of the nervous structures themselves, which manifests itself in pain. Whatever the pathology of facial neuralgia may be, it seems unquestionable that the disease may be caused by dental irritation; but in many cases it cannot be distinguished from neuralgia induced by some other cause. One of the great writers has sought to lay down rules for diagnosticating a neuralgia of dental origin.

They are as follows:

1. When a tooth is in itself the seat of pain and the patient definitely specifies it as such there can be no doubt that it is the origin of the disorder.

2. When in addition to the pain there follows a swelling of the cheek or gum resulting in abscess the indication is specific; if a tooth is sensitive to percussion and seems longer than others, the indication is decisive, whether it be carious or not.

3. Even if the pain is diffused and spread over the side of the.

face and attended with distinct exacerbation, thus presenting all the
signs of typical facial neuralgia, nevertheless, if it eventually local-
izes and limits itself in the region of the dental arch, accompanied
with pain, redness, swelling, and extreme sensibility to pressure,
possibly terminating in an abscess, the disturbance is in such cases
of dental origin.

4. Another point serving to distinguish neuralgia of dental origin
from that due to other causes is the continual agitation and persist-
ence of discomfort of the patient, there being no such periods of
calm as occur in neuralgia from other causes; the pulse is accelerated and hard, and there is at times general sweating.

I think these are very good rules to guide us in making our diagnosis, but at times we will have patients suffering with neuralgia
with none of the above symptoms, and on examining the mouth
carefully we will find the teeth in what seems to be perfect order.
For instance, I will relate a case of my own. Lady about thirty years
of age, dark hair and dark eyes, suffering with neuralgia between
the eyes of about seven weeks' standing. One of our distinguished
ophthamologists diagnosed the case as strain of the eyes, and made
three pairs of glasses for her and told her to wear them and in a few
days she would get relief. On making an examination of the mouth
her teeth seemed to be in perfect order. In passing a piece of silk
floss between the teeth it would jump from the grinding surface to
the gum with a cluck, showing that she had good interdental spaces.
Her bite was perfect. None of the teeth were sore to percussion.
The gums were in a good, healthy condition. The second bicuspid
above on the right-hand side had a small decayed spot on the grinding surface which was so sensitive she could not bear for me to
touch it. I advised her to have the nerve devitalized, but she refused, and said she would wait a few days. In three or four days she
came in again, still suffering with neuralgia. I inquired into the history of the case as best I could. Her parents had never suffered with
neuralgia in any way, and the young lady herself had never been sick
a day in her life. I then told her that I thought without a doubt
that this tooth was the cause of her neuralgia. After persuading
her for quite a while she agreed to let me destroy the nerve. I excavated the cavity as best I could and applied my arsenic and told
her to call the next morning. It was about 10 o'clock in the
morning when I applied my medicine, but instead of waiting until
the next morning she came in that evening about 4 o'clock and
said she was entirely relieved of her pain. The next day I treated

and afterward filled the root canals with chloro-percha, and as far as I know she has had no recurrence of the pain.

In making a diagnosis of the mouth of a patient suffering with neuralgia we should extract all impacted wisdom teeth and all other teeth that are wedged together, and we can very easily detect the teeth that are wedged by passing a silk floss between them; and I would advise the dentist to put his attention to the teeth that have living nerves in them, for I do not believe it possible to have reflex pain from a dead tooth. I know that the nervous tissue is the only tissue that has the power of transmitting pain, and after the death of the pulp you sever the connection with the main trunk of the nerve.

In the treatment of pyorrhœa alveolaris and all alveolar diseases there is nothing that will take the place of chloride of zinc. It is, in the first place, an escharotic, a stimulant, and a powerful astringent. We have all three qualifications, and besides this it is quite a strong antiseptic, so that it gives us for certain uses around the mouth qualities that we cannot get from any other substance that I know of. It will not stain anything that you use like sulphuric acid. Sulphuric acid, if it touches your clothes or your napkins, destroys them. If chloride of zinc comes in contact, it does no harm. (Dr. Frank Abbott, in *International*.)

School Commencements. · · · ·

COMMENCEMENT EXERCISES OF THE DEPARTMENT OF DENTISTRY, VANDERBILT UNIVERSITY, SESSION OF 1892-93.

THE fourteenth annual commencement of the above institution was held in the spacious hall of the university chapel on the evening of the 23d of February. In addition to the officers of the university and the Dental Faculty, a large concourse of the students and friends of this department were gathered to witness the closing exercises and to bid farewell and Godspeed to the young gentlemen about to enter upon their professional career. Choice selections of vocal and instrumental music was furnished at intervals by the celebrated "Glee Club" of the Vanderbilt, and added much to the enjoyment of the occasion. An appropriate prayer was made by the Vice Chancellor, Dr. W. F. Tillett, who next announced, to the great regret of all present, that the venerable Dr. Garland, Chancellor of the university, would be unable, by reason of ill health, to be present, consequently it would devolve upon him as Vice Chancellor to officiate in his stead.

Dr. Tillett congratulated the Dean and Faculty upon the continued prosperity of the Dental Department, especially emphasizing, from personal knowledge, the thoroughness of instruction given and the careful training in the practical branches of the students of this school. He spoke in complimentary terms of the enthusiasm displayed by both students and Faculty, and was of the opinion that this in no small degree accounted for the extraordinary success of the Vanderbilt Dental Department. He considered the small number of the class presented for graduation as significant of the higher qualifications and greater attainments now required of those seeking the degree of D.D.S., and concluded by predicting a bright and glorious future for this department of the Vanderbilt.

Dr. Tillett introduced Mr. Neander M. Barnett, of Tennessee, as the Valedictorian of the Dental Class, who, in a few well-chosen sentences, bade a fond adieu to his beloved instructors and fellow-classmates.

Prof. W. H. Morgan, Dean of the department, presented the following gentlemen to the Vice Chancellor as worthy of graduation, and the degree of Doctor of Dental Surgery was conferred upon them : W. J. Baker, Thomas S. Brown, James D. Smith, G. A. Hughes, Alabama; James R. Beachum, Neander M. Barnett, Mississippi; William G. Downs, Illinois; John A. Johnson, Nova Scotia; John C. Minton, Texas ; Le Roy Rentz, Pennsylvania.

Dr. W. H. Morgan, as Dean, delivered the charge to the graduating class. His address was impressive, full of interesting information, and highly appreciated for its excellence by all present. He gave a succinct report of the methods of instruction and of the work accomplished during the session. One hundred and eleven matriculates were registered, coming from all parts of the United States, a large number from the North and West. The British provinces, Canada and Nova Scotia, and the Republic of Mexico combined to swell the list. Dr. Morgan referred with pride to the harmonious relations sustained by the members of his Faculty and the consequent unanimity of action in all matters affecting the welfare of the school. He spoke of the clinical advantages enjoyed by the students of this department, which he claimed were equal to the best, and excelled by none. Particular attention had been given to crown and bridge work, under the personal supervision of the ablest operators to be found in America. Several postgraduates had availed themselves of the superior advantages afforded, and were in constant attendance during the past session.

Dr. Morgan explained the inevitable changes resulting from the adoption of the three years' course of instruction. The time has come when the student of the science and art of dentistry must lay broad the foundations of his knowledge; he must not be content with a mere understanding of his specialty, but must bring into requisition the collateral sciences if he would keep abreast of the advance guard of the intellectual and scientific world. The man who stops now to look backward in this age of wonderful progress and achievements will soon be found lagging in the rear, like the stragglers of an army.

Dr. Morgan spoke eloquently of the moral obligations incurred by those offering to engage in the practice of this humanitarian profession; of the sacred nature of this relation to their patients and to their professional brethren. He exhorted them to uphold the honor and dignity of their vocation. He claimed that they were in honor bound to give the best service of which they were capable to every patient under treatment, or accepted by them, whatever his station in life. Be he white or black, rich or poor, he should be given the most skillful attention possible.

Rev. Dr. Collins Denny in a chaste and appropriate manner addressed the honor men of the class; his speech was a perfect gem of oratory, and elicited frequent and timely applause. The medals were conferred upon the following: Founder's medal for highest general average, to James R. Beachum, of Mississippi; Morrison Brothers' medal to second in contest for Founder's medal, to John A. Johnson, of Nova Scotia; Prof. H. W. Morgan's medal for best gold filling, to T. B. Ragin, of Illinois; Prof. Ambrose Morrison's medal to Senior best qualified in anatomy and physiology, to Thomas S. Brown, of Alabama; by same, medal to Junior best qualified in anatomy and physiology, to J. K. Campbell, of Mississippi.

After another exquisite song by the "Glee Club," which being vigorously encored, brought forth a response to the tune of "Auld Lang Syne," "Pull my tooth out! Pull my tooth out! Pull my tooth out! Quick!" the benediction brought the exercises of the evening to a close.

DENTAL DEPARTMENT, UNIVERSITY OF TENNESSEE.

THE Commencement exercises of the Dental Department of the University of Tennessee were held at Watkin's Hall, in this city, on the evening of February 23, 1893.

A large and attentive audience was in attendance, and the programme was interspersed with good selections of instrumental music.

The valedictory was delivered by Dr. J. H. Leland, of Michigan, and the liberal applause which greeted his effort fully attested his popularity with the student body and his marked abilities as a public speaker.

Prof. J. Berrien Lindsley, D.D., M.D., of the Faculty, delivered the charge to the graduating class in a pleasing and effective manner.

The Faculty medal, the first honor of the institution, was awarded to Dr. J. B McCormack, of Florida. Morrison Brothers' medal, second honor, was awarded to Dr. D. V. Mitchell, of Tennessee.

Gen. W. H. Jackson was presiding officer of the occasion, and, in behalf of President Dabney, who was unavoidably detained at Knoxville, conferred the degrees upon the graduates.

The class was necessarily small, owing to the adoption a few years since of a three term course in this department of the University; but in the thoroughness and efficiency of the course, and the full and careful preparation for future usefulness on the part of the graduates, the results of the year just closed were eminently satisfactory to officers, students, and public.

Selections.

EMPLOYMENT OF THE POST IN ANCHORING FILLINGS.*

BY C. J. UNDERWOOD, D.D.S., ELGIN, ILL.

I SEE the programme tells me that I am to speak on " the employ-ment of the past in anchoring fillings." If you were to judge the fu-ture by the past, you might anticipate my paper to be a worse chest-nut than it is, for they have had me on the list now three times for irregularities of the teeth, and the paper is still in embryo. But that is an error. It should be post. It may, however, not be amiss to remark that we may employ the post in anchoring fillings; in that we may profit by the mistakes of the past. At first I took it to be typographical error, and felt some resentment, but when I come to copy my efforts from the original I concluded it was a scriptographed innominata, and that the compositor was entitled to a vote of thanks for mastering so much of it.

In preparing this paper I have abjured books and journals, and anything and everything that may have been written on the sub-ject, and endeavored to adhere closely to the actual details of the operation, just as I do the work in my practice. I do this for two reasons: First, if, happily, my practice embraces aught of value, some one may be benefited by it. And, second, if my practice is faulty and unscientific, I may be benefited by your criticism.

I will consider but three cases, or three classes of cases:

1. A proximal cavity involving the cutting edge in a devitalized incisor or cuspid.

2. The same with a living and healthy pulp.

3. An interior proximal cavity in a devitalized bicuspid.

Case 1.—We find a large anterior proximal cavity in a devital-ized central incisor, involving one-fourth the cutting edge. After filling the root and cutting away frail margins we find the cone of the tooth gone and a thin plate of enamel in front, giving little promise of safe support for a large filling reaching, as it will, to the

---

* Read before the Northern Illinois Dental Society, October, 1892.

cutting edge. A post is indicated; not a screw or How post, but a triangular platinum wire post, always cemented in. And to obviate the annoyance and often disastrous consequences of the post being in the way, I bend it in such a way as to carry it well back into the cavity, down through the center of the tooth to a point near the cutting edge, where it curves outward to a point near the corner to be restored. The post is shaped before setting to an abrupt point, at the end toward the cutting edge, this being accomplished by flattening the wire at the end and then cutting off the corner at an angle of 44 to 60°.

The post is thus out of the way in the body of the filling, yet retaining its full size and strength to near the cutting edge. and here the taper is so short that the maximum amount of strength is secured, with the minimum amount of post. I then cut the usual groove at the base of the cavity, to prevent slipping of the filling; and a longitudinal groove to receive a part of the lateral strain.

The sample I have prepared is very nearly a typical case, and I trust it will serve as a key to my awkward description.

Case 2 is the same sort of a cavity in a "live" tooth. There be ing no circumference to the cavity, but only a base, resort is had to the post—or pin or lug. if you please. I take a very small bur and drill a hole nearly through the tooth toward the distal side, and at right angles with the long axis of the root, at a safe distance from the nerve and from the cutting edge. Then I enlarge with a slightly larger bur till it is as large as the thickness of the tooth would suggest or justify and cement a properly shaped pin in place. slightly bent at the point of emergence from the tooth, toward the corner to be restored, thereby affording a better grip for the gold, and also being more out of the way while building base for filling. A groove is also cut in the base of this cavity as in Case 1.

Case 3 is a large anterior proximal cavity in a bicuspid. a filling in the buccal portion of which will show. and should therefore be of gold. An all-gold filling is contraindicated both by size of cavity and extent of decay at cervix, and by the generally attenuated condition of the patient's pocketbook. I put in a compound filling, the lingual portion and body of the filling. amalgam and the buccal portion that shows subsequently, with gold. I use a post here for two reasons: to secure greater certainty for retention, and to avoid bringing the amalgam in contact with the buccal wall of the cavity, thereby discoloring it. The post is prepared as before, beveled sharply to a point from the point of emergence from the cement, and is placed near the center of the cavity, the point reaching the proximal surface of filling. In cementing it in place, the cement is carried well into the buccal portion of the cavity, the lingual portion being left free for the reception of the amalgam. It will thus be seen that the post supplies the place of the buccal wall to the amalgam filling and the amalgam affords easy retention for the gold at the subsequent sitting. The effect of an all-gold filling is thus secured at a great saving of time and trouble and in my judgment accomplishing better result.

I think this covers all the uses of the post wherein I would be likely to suggest anything new or instructive.

I wish to say before closing that I never use a screw post. The threads weaken it. It is no stronger at the point of emergence from the tooth than at the point. Besides, in screwing it to place you are liable to fracture the enamel. (*The Dental Review.*)

EXTRACT FROM AN ARTICLE ON "COLLAR CROWNS AND ROOTS."

THE most frequent cause leading to failure of collar crowns may be enumerated as follows:

1. Imperfectly prepared roots.
2. Low carat gold for band.
3. Roughly finished or porous band.
4. Band impinging on peridental membrane.
5. Failure to follow gum margin evenly.
6. Faulty articulation
7. Failure to restore anatomical contour.

If we bear in mind the anatomy of a tooth, we see that the exposed portion of dentine that projects above the gum is entirely covered by a shell of enamel, which is also carried a little below the margin of the gum and is there overlapped by cementum Now in restoring a broken-down crown we cannot do better than imitate nature as closely as possible. The enamel should be entirely removed under the gum, leaving the cement uninjured and allowing our gold band to run up under the gum and the sharp edge overlapped by the cement, as was the case with the natural enamel. In that case we would have no imitation whatever, provided our gold is twenty-three carat fine and smoothly polished.

In many cases it will be necessary to not only remove all the enamel, but also to cut into the dentine, leaving a shoulder. Then a band of considerable thickness can be used, twenty-six gauge or even thicker, thus giving additional strength and rendering a more perfect contour possible. In no case should the sharp edge of the band project toward or come in contact with the peridental membrane, but should fall inside of the line of cement as did the natural enamel. But the smooth surface of the band about one-eighth of an inch from the edge at the point where it leaves the gum margin can be brought to bear considerably against the gum without danger of irritation, and would be beneficial by keeping collections from finding their way up under the free margin of the gum.

It is an important point to preserve the anatomical contour of the crown. We will be aided materially in this by cutting and soldering the band to the shape of a truncated cone, fitting the narrowest portion to the neck of the tooth. In a late number of the *Dental Review* Dr. Black has an essay on "The Proximate Spaces and Gum Septum." The points brought out by Dr. Black are equally applicable to crowns as well as to the contour of fillings. The crown should be so shaped at the proximate spaces as to allow the gum septum to retain its natural position, and should come in close contact with the adjoining tooth near the grinding surface so as to prevent any food from being wedged down at that point and produce irritation.

In preparing roots the gum should be left in as healthy and normal condition as possible. If in reducing the root it is necessary to remove a portion of the peridental membrane, the injury will be of a traumatic nature, and will in all probability heal by first intention, and no grave results are to be anticipated. But the laceration of the gum margin by being ground or chopped up by corundum wheels and the like is an injury of considerable seriousness, the healing of which brings about an extensive cicatrix which contracts and is absorbed on the slightest provocation. Then we have the exposed band and other conditions which I have previously mentioned. (C. W. Jones, D.D.S., St. Paul, Minn., in *Dental Review*.)

--- -- --

CAUSTIC PYROZONE.

BY RODRIGUES OTTOLENGUI, M.D.S., NEW YORK, N. Y.

In my opinion, the ethereal solution of hydrogen peroxide known as caustic pyrozone will very shortly be considered one of the most valuable topical drugs in the medicine cabinet of the dentist. From my experience with it in a number of cases, I am satisfied that for several purposes it supersedes anything previously at our command. Its chief employment, of course, must be upon pus-generating surfaces or tissues. Of these, it will be more welcome in the pockets of pyorrhea alveolaris than in abscesses or other pus-yielding diseases, though useful in all.

Its most marked and valuable characteristic is its affinity for pus. Brought into contact with it, there results a bubbling and rushing forth of the pus which will astonish those who see it for the first time. We are often led to suppose that pyorrhea has not yet attacked a specified tooth, because pressure will not force an escape of pus around the neck. I am satisfied now that many such seemingly healthy individual teeth, in diseased mouths, are affected, for I have been amazed to observe the free flow of pus in such cases immediately upon the application of caustic pyrozone. I think I can best describe the manner of using this drug by giving the record of a few cases.

CASE I.—Chronic pyorrhea alveolaris. Patient a man ætat thirty-five. In good general health. Using tobacco freely to chew and to smoke. Calcareous deposits upon the inferior teeth, mainly upon the lingual surfaces. Some deposits upon the superior teeth. Gums of both jaws much inflamed, bleeding upon the slightest touch. Pus oozing around necks of all teeth save the six anterior superior. Disease in its most advanced form around the wisdom teeth, about which the processes had been almost entirely lost. Patient suffering almost constant pain, though with the exception of the wisdom teeth none were loose, the pockets all being quite shallow.

Treatment.—Calcareous deposits removed as far as possible, resulting in copious hemorrhage. Because of the severe bleeding, I decided to treat with pyrozone, at the first sitting, only the six anterior superior teeth, which were least affected, no tartar being about them. This latter fact rendered scraping with instruments avoidable, and these teeth were therefore the only ones about which no hemorrhage had been occasioned. A small pledget of cotton, rope-shaped, was moistened with the caustic pyrozone, and with a probe introduced between the cuspid and lateral, being pressed up under the gum margin as far as possible. There it was left until the appearance of foam indicated that pus had been found. This was perhaps from five to eight seconds. It was then withdrawn, and immediately there was a considerable discharge of boiling pus, finally escaping tinged with blood. The same treatment repeated about the others of the six teeth selected brought forth a foaming mass which covered the adjacent parts as with a thick lather. The mouth was then rinsed with warm water and the patient dismissed for four days. At the second visit, so great a change had occurred that an application of the agent about the same teeth produced scarcely a perceptible escape of pus, save in one pocket. The gums were remarkably improved in tone, inflammation having almost entirely disappeared. The patient reported that the night after treatment was the most comfortable passed in months. At this visit the six anterior inferior teeth were treated similarly, and the treatment repeated four days later, by which time they had so far advanced toward recovery that I proceeded to take up the posterior teeth. I would call attention to the fact that I deem it wiser not to treat too many teeth in the same mouth at the same sitting. It will be better to take those first which cause the most suffering, and make an application to them only. In this case, by the fifth visit the caustic treatment was abandoned, all pus having disappeared. The mouth is recovering rapidly under occasional dressings of an astringent nature, medicinal pyrozone being used as a mouth wash. All pain has been controlled, and the teeth can be brushed without bleeding of the gums. I do not consider that this is a permanent cure, but it is the most rapid recovery to a condition of good health that I have ever seen.

CASE 2.—Pyorrhea alveolaris complicated with alveolar abscess. Patient a woman of forty. Presented in great pain, occasioned by a well-defined abscess about an inferior cuspid. This was one of those rare cases where an abscess is present despite the fact that the pulp is alive. The pocket was not very deep, and other teeth were involved in the general disease; but I shall confine myself to this special condition. The gum at the lingual aspect was much swelled, and the discharge of pus copious. I cleansed the pocket by manipulation with the finger and by syringing with warm water until it appeared quite clean. I also packed the pocket with absorbent cotton, wiping pus from the soft tissues in this manner. I then, as in the above cases, inserted caustic pyrozone on a bit of cotton, passing it down into the pocket. In a couple of seconds the foaming was observed, and I removed the cotton. Immediately there issued forth a foaming discharge which completely hid all the neighboring teeth. At the second visit the patient reported that the pain had been greatly relieved, and I renewed the treatment, there being yet a free discharge of pus. Contrary to instructions, the patient remained away from the office for a week, and returned in pain, and with large quantities of pus escaping. I became satisfied that there must be some special cause for this condition, and after a more thorough exploration decided that there was caries of the alveolus about the tooth, especially involving the septum between it and its neighbor. With the engine-bur I operated, removing the dead bone freely. I then treated with the caustic pyrozone, and with a second application, two days later, reached a point where the tooth was as well as its neighbors, reducing it and them to a stage where they will be easily controlled. Pain, inflammation, and pus have all disappeared.

CASE 3.—Abscess without fistula. Patient presented with aching tooth. Removed old filling and found a putrescent pulp. Symptoms indicated that pus might be present about the apex of the root, but there was no fistula and no sign of one forming. Neither was there any discharge through the root. I dressed the canal with cotton slightly dampened with caustic pyrozone, and left it in for one minute. Upon withdrawal, I was surprised to see the amount of clear yellow pus which followed. In this case I think that the wonderful affinity of pyrozone for pus caused its passage through the foramen, and once having passed that point, it continued to discharge itself through this vent which offered.

From my experience in these and other similar cases I may offer a few suggestions to those who essay to use this drug. The first caution is as to quantity. A little will do all the good possible, while more will be harmful. The application is painful, producing what the patient will call a burning sensation. It will, however, be less painful applied to diseased surfaces than if placed upon healthy tissues. For this reason, and because the cauterizing of the healthy parts is undesirable, care should be taken that the cotton rope or tampon is not so saturated that when pressed into the pocket or fistula the excess will be forced out and escape upon the other parts than those that are generating pus. If this should occur, pain will follow, which may be quickly relieved by rubbing freely with tannin and glycerol. Another objection to permitting the agent to reach the healthy surface of the gum is that it will produce an ugly white stain. Whether this is a true eschar or not, I am in doubt. The eschar caused by carbolic acid, salicylic acid, and other escharotics results in the death and exfoliation of the sur-

face of the soft tissues. This does not seem to occur with pyrozone. I accidentally spilled some upon my fingers, and afterward
washed my hands, whereupon, within a few minutes, there appeared a chalky white stain, quite ugly in appearance. I feared
that there would be a slough and a sore finger; but, to my utter as-
tonishment, when I reached home it had entirely disappeared, the
cuticle being as perfect as though no caustic had reached it. This
led me to some experiments, the result of which I will state. Immediately after placing a drop of caustic pyrozone upon the finger,
a rapid evaporation is visible, with a sensation of burning. If the
cuticle is broken, the pain will be great. If left untouched, in about
twenty minutes a whitish stain will begin to appear, increasing
slightly, till in half an hour it is distinctly visible. This stain will
slowly disappear, vanishing entirely in three hours without medication of any kind. If, however, an effort be made to wash the
caustic from the fingers, the stain will appear within ten minutes
thereafter, and will be of intense whiteness and very conspicuous.
The cuticle may be scraped off, and the stain will be found not to
have penetrated beyond; but this cuticle will be seen, by the mag-
nifying glass, to be thoroughly stained throughout. Nevertheless,
if undisturbed, this stain, though deeper than that found when water had not reached the part, will all disappear within from three
to four hours. If gloves are worn or the hands placed in the pocket, the stains pass away within one hour.

A deduction from this is that any excess of pyrozone should be
removed with bibulous paper and the application of water avoided.
I think that the water simply softens the surface of the tissues, allowing a deeper penetration. (*Dental Cosmos.*)

TREATMENT OF DIFFICULT CANALS IN MOLARS.*

BY DR. W. R. BLACKSTONE, MANCHESTER, N. H.

Mr. President and Gentlemen: This subject is one on which I
have consumed a vast amount of time in experimenting, especially
with respect to the buccal roots of molars, the small, delicate bi-
rooted superior bicuspids, and the roots of inferior molars. I think
in my practice I have had more trouble in treating and preparing
the buccal roots of the superior molars. We will take, for example,

---

* Read before the Second District Dental Society of the State of New
York, November 14, 1892.

a case that was presented to me—a lady, twenty-two years old, with an aching left superior molar. On making a thorough examination, I concluded to devitalize the pulp by the use of arsenious acid. After making the application, the cavity was sealed with a pellet of cotton, saturated with a solution of gum sandarac, and the patient advised to call in two days.

She came at the appointed time, and had suffered no inconvenience in the meantime. I removed the dressing, and found the pulp completely devitalized in the palatal root, and I should judge for about one-fourth the length of the roots in the buccal canals. No trouble was encountered in removing the remains in the palatal canal, but in trying to extract the pulp of the buccal roots I made a most pronounced failure. It was utterly impossible to introduce the most delicate instrument more than one-quarter of the distance, and in doing so it would touch a very sensitive spot. At last, tired of my unsuccessful attempts, I cleansed the palatal canal, and filled it with an iodoform paste covered with gutta-percha. I made an anodyne application, sealed the cavity with cotton saturated with carbolated resin, and dismissed the patient, informing her it was impossible to extract the pulp at that sitting. I allowed the tooth to rest for a week.

To make a long story short, she became dissatisfied and changed dentists. She was advised to have the canals reamed and filled immediately, which was done. The customary time elapsed for the formation of an alveolar abscess. Dissatisfied again, she called at my office and insisted on the immediate extraction of the tooth, which was done. After examining the tooth, I found it was utterly impossible to prepare and fill the buccal roots on account of their irregularity and decided curvature.

At that time I came to the conclusion that I should adopt an entirely different procedure in the treatment and preparation of canals, especially the buccal, of the superior molars and bicuspids, the inferior incisors and molars. The method, which has proven to be entirely successful, so that I have no hesitation in recommending it in all such cases, is *not to fill them at all*. How many of such canals are absolutely filled to the apex and hermetically sealed? It is doubtful if one in one hundred is thoroughly closed. There must be an intervening space between the filling and the apex, and unless the end of the root becomes encysted, so to speak, there must be an exudation of a small amount of fluid into the canal. This becomes putrescent, and finally inflammation of the peridental membrane

results. If space necessarily exists in ninety-nine out of every one
hundred of such cases, why not have it the entire length of the
canal? for we can safely say that we will have no more trouble
with one than with the other.

The manner in which I dispose of a superior molar with a putrescent pulp which has been devitalized by arsenic, or has died as the
natural consequence of exposure, is first to adjust the rubber dam
and gain access to the pulp chamber and canals by the liberal use
of burs and chisels; but previous to opening into the canals a solution of hydrogen peroxide and bichloride of mercury is injected into
the cavity, which is kept well filled with the antiseptic solution
during the process of preparation. After the cavity and pulp
chamber are thoroughly free from carious matter, I proceed to remove the remains in the canal by the use of small broaches and
hooks made from piano wire and platinum, followed by reaming
slightly at the entrance to the canals, keeping them and the cavity
well filled by repeated injections of the bichloride solution. When
satisfied that all particles of pulp and *débris*, "as far as possible,"
are removed, the process of drying is next undertaken. This is
done by the use of the hot-air syringe and the Evans root-drier,
followed by dressing the canals with a solution of iodoform in oil
of eucalyptus, subsequently using the hot-air syringe until the
eucalyptus is evaporated, leaving the canals well lined or covered
with a coating of iodoform.

Here the operation is completed, so far as the canals are concerned. Filling the pulp chamber with an oxyphosphate cement
and the cavity with gutta-percha, the patient is dismissed, and advised to return by appointment to have the treatment completed
by filling the cavity with gold or amalgam. Theoretically and scientifically, this method may be unsound; but practically it is a perfect success. The cases are very rare that return with symptoms
of pericementitis. If inflammatory symptoms arise, the use of
cotton saturated with methyl chloride applied to the affected parts
produces very gratifying results. Should the case become obstinate, and the inflammation not subside by the use of counter-
irritants, the filling is removed, or an opening is made through
the alveolar process to the apex, which I am happy to say is seldom
necessary.

The superior bicuspids, inferior incisors, and the anterior roots of
lower molars are treated substantially in the same way, and invariably with pronounced success. The roots that have larger canals,

such as central incisors, cuspids, posterior roots of inferior molars, etc., are prepared and treated with the same care as previously stated, but in no case is the canal filled to the apex. I depend entirely on the roots being thoroughly aseptic before filling the pulp chamber and cavity. Nature completes the operation more thoroughly than the most skillful operator, by encysting the apex of the root, which is accomplished as well with the canal empty as when filled, without the injurious effects that might occur from forcing particles that may be retained through the foramen by the materials used in filling.

Of course there are exceptions to all rules and methods, and the only one in this case is the preparation and treatment of cuspids, central and lateral incisors, for crown and bridge work, where we depend on the post for strength and durability. In such cases the canal is filled as near to the apex as possible with a stiff paste, made from oil of eucalyptus and iodoform, to prevent air pressure and the possibility of cement being forced through the foramen when adjusting the crown or bridge. In using the screw post for molar and bicuspid crowns the canals are reamed to correspond with the size of the post. By the use of cement mixed very thin, it is gently inserted and held firmly until the oxyphosphate has thoroughly hardened. When the operation is completed you have a canal that is virtually empty, but one that can be trusted with the utmost confidence.

Gentlemen, if you are skeptical, and this method seems unscientific and at war with all theoretical procedures, you will find it by actual experiment, if you think it worthy of a trial, to be one of the most successful and satisfactory methods of preparing root canals. (*Dental Cosmos.*)

CROWNING FRAIL ROOTS.

A NARROW band is fitted around the neck of any root, a cap placed on the top of that, and a pivot fitted in the root and through the cap, the whole being then soldered together. One or two vent holes are then drilled through the top of the cap, and is set to place with oxyphosphate, the excess coming out through the holes. These holes are then reamed out and filled with gold, and the edge of the band under the gum is burnished to the root. The tooth is then fitted to this cup and set on the projecting pivot with oxyphos. phate. The advantage of this plan is that the root being slightly tapered with proper paring instruments, the band can be made to fit absolutely, while the excess of oxyphosphate is gotten rid through the vent holes instead of being squeezed out around the edge of the band. The crown used is similar to the Howland crown. Another method employed with these, as well as the old-fashioned pivot teeth, is to prepare the root even with the outline of the gum, and set a pivot into it with oxyphosphate. The end of the root is cut very smooth and even and the base of the crown accurately fitted. A mat is made of several thicknesses of soft gold No. 5, and a clean hole cut in the center of it, of the size of the pivot. It is then put over the pivot as a washer, and the tooth set with oxyphosphate. If in time the cement wastes, the gold remains to preserve the root. (Dr. S. G. Perry, in New York Odontological Society, reported in the *Cosmos*.)

In a paper on the "Influence of Menstruation on Incidents of Dental Origin," read before the Odontological Society of Paris, Prof. Sauvez shows that, since the augmentation of the mass of blood during pregnancy is known to produce certain influence on the jaws, similar conditions existing during menstruation, women are more liable to dental troubles at that period.

On general consideration it would seem to follow that the menstrual outflow, disturbing the nervous system and manifesting itself in congestion toward the head and the organs of the chest, would naturally produce a state of least resistance in any tooth susceptible to pulpitis or periostitis.

After observing a large number of cases, the professor sums up his experience as follows: "In her menstrual period woman is subject to affections of the teeth, the same as during pregnancy, which are principally caused by the congestion of the pulp or the periosteum. Generally, woman may be said to be disposed to diseases of the teeth when her uterus is troubled. The difficulties mentioned occur at the *commencement* of the menstrual period."

The paper concludes with the advice that special care be exercised by practitioners. and that obturation of caries in the fourth degree, for instance, shall not be attempted during the menstrual period.

Extracts . . . .

TOOTHACHE.

THE *Deutsche Medicinische Zeitung*, No. 61, 1892, recommends the following in odontalgia in consequence of caries of the teeth:

R. Creosot...
Tr. menth. pip. æther.....................................
Ol. camphorat..ää gtt. II
Ol. caryophyll...gtt. V
Cocain. mur...o. 15 (gr. IIss)
Chloroform..q. s. ad pasta
M. d. s.: Fill into the cavity.

FUSIBLE METAL.

DR. C. M. RICHMOND says that the fusible metal which he uses for bridge work is as hard as zinc, and gives the following formula for making it (the metals to be melted together in the order named):

Tin.....................................20 parts by weight.
Lead....................................19 "   "   "
Cadmium.................................13 "   "   "
Bismuth.................................48 "   "   "

This compound can be melted and poured into a plaster impression without generating steam, as it melts at 150° Fahrenheit. (*The International Dental Journal*.)

THINGS WORTH REMEMBERING.

To remove the stains of tincture of iodine, from either the hands or napkins, apply strong ammonia. The spots will immediately come out clear.

The stains of nitrate of silver, from either the hands or napkins, can be easily removed. First cover the spots with tincture of iodine, wait a few moments, then apply strong ammonia, and rub well.

In using argenti nitras in treating children's teeth one may accidentally get a few grains on one's hands and not discover it till the

hands are washed, when the black stains will be well set. By proceeding as above they will disappear at once. (Dr. Maxwell, in
International Dental Journal.)

"ABSOLUTELY WITHOUT PAIN."

A DENTIST in Vermont writes: "I would like to hear from the editor, and others in the profession, in regard to the filling and extracting of teeth absolutely without pain."

We answer: We know of no one who professes to fill and extract
teeth "absolutely without pain" by any local anæsthetic. If some
dentists and patients did not demand so much to satisfy them, adver-
tisers, and workmen who really have something good, would be
more modest in their claims. Some want everything absolutely
perfect, whereas they themselves and nothing they have are "absolutely" perfect. I really do not know who to refer you to for a
process for "filling and extracting teeth absolutely without pain,"
unless you are willing to work on "absolutely" dead men. (Editorial in *Items of Interest*.)

HOW TO OBTAIN AN EXACT IMPRESSION OF ROOT CANALS.

BY EDWARD G. CARTER, L.D.S., ENG. AND GLAS.

1. SHAPE the canal.

2. Mop it out with glycerine.

3. Fill it with pink gutta-percha—the kind supplied for taking
impressions.

4. Heat a small French nail, press it into the canal, and leave the
head projecting.

5. Take impression with modeling composition, and the gutta-
percha will come away on the pin, thus enabling one to make a
pivot with a maximum thickness of pin and a minimum quantity of
cement.

The root should be roughened just before inserting the pivot.

TOOTH AND MOUTH AFFECTIONS IN INFLUENZA.

HUGENSCHMIDT (*Correspondensblatt für Zahnärtze*, October, 1892)
found the following tooth and mouth affections during influenza:

1. A special form of contagious stomatitis associated with ulceration. The immediate cause thereof was influenza.

2. Ulcerative or simple gingivitis, accompanied by general periostitis.

3. Purulent alveolar periostitis.

4. Neuralgia of the fifth nerve.

5. Difficult eruption of the wisdom teeth.

Simple and inflammatory tonsilitis, more or less pronounced, were
also observed.

No special local treatment was employed. Antiseptic mouth
washes, thymol and solution of boric acid were used. The following was applied to the gums:

R. Salol...6.0 (℥ Iss)
    Liq. vasel.......... ....40.0 (℥ X)
M. d. s.: Apply to the gums.

Liquid food was prescribed with the above.

When pain was marked the gums were painted with

R Menthol................................. 3.0 (gr. XXXXV)
   Chloroform........................... . 5.0 (℔ LXXV)
M. d. s.: Apply to the gums.

_____

AMŒBÆ IN AN ABSCESS OF THE JAW.

FLEXNER (*Johns Hopkins Bulletin*, 1892, No. 25, p. 104) has reported the case of a man, sixty-two years old, who, nearly two years
before coming under observation, had noticed a small hard lump
within the mouth, beneath the gum, at the anterior part of the inferior maxillary bone, just to the right of the middle line. After
two months the tumor was removed by an operation. Five months
later an ulcer appeared at the site of operation, exposing the
bone. A slight discharge set in and continued thereafter. In the
course of fourteen months a hard swelling again appeared on the
floor of the mouth, gradually increasing and extending beneath the
chin to the front of the neck and below the jaw, forming a large
and prominent swelling that occupied the entire space included
within the arch of the inferior maxillary bone. The width of the
mass corresponded with that of the lower jaw. The growth extended backward to the angle of the jaw and downward to the cricoid
cartilage. It was firm, indurated, immovably fixed, and felt like
a solid growth. At the most prominent point of the tumor in
front, however, tenderness was present and fluctuation could be detected. The teeth of the lower jaw were deficient. On the upper
aspect of the transverse portion of the inferior maxilla the bone was
denuded in a situation corresponding with the two incisors and the
canine tooth of the right side. The exposed bone appeared spongy

and was evidently necrotic. It was in part covered with dirty, grayish purulent matter and shreds of greenish sloughing tissue. On making firm pressure over the tender and fluctuating area on the anterior aspect of the swelling, a small amount of dirty purulent material was observed to ooze from an opening in the mouth over the necrotic bone. The breath was offensive. There was no glandular involvement. An incision evacuated about two and a half ounces of pus and exposed a large abscess cavity, the walls of which were quite thick. The necrotic bone was thoroughly scraped, the abscess cavity packed with iodoform gauze, and a dressing applied. The man did perfectly well, and at the time of the report bade fair to recover. The pus evacuated was grayish yellow in color and extremely offensive in odor. It was not thick, but was mixed with a small quantity of blood, and contained a number of flakes and granules of lighter color and more opaque than the other constituents. Microscopic examination disclosed the presence of pus cells, red blood cells, bacteria, and amœbæ indistinguishable from the amœbæ dysenteriæ. The last were especially abundant in the flakes. Subsequently, examination of the contents of the drainage tube employed disclosed the presence of the same organisms. There was no history of diarrhœa.

A FOREIGN BODY REMOVED BY THE "POTATO TREATMENT."

Dr. J. Solis Cohen reports, in the Philadelphia *Medical News*, that a patient was brought to him several hours after having swallowed an irregularly shaped dental clasp. Exploration of the œsophagus showed that tube to be unobstructed. The patient was ordered to be fed exclusively on buttered mashed and roasted potatoes, and to examine his stools carefully for the foreign body. Within forty-eight hours it was voided, thoroughly coated with potato.

THEY DIDN'T FILL TEETH.

There is a prevalent idea that filled teeth have been found in the mouths of Egyptian mummies. It is true that Herodotus says there were "physicians for the teeth" among the Egyptians of his day, but there is no evidence that they attempted the salvation of teeth by filling. No such instance has ever been presented. Artificial substitutes have been found, which were usually human teeth, held in place by means of gold wires or bands, and extraction and cleaning of the teeth were done; but that plugging as a prophylactic measure was ever a practice there is no proof to establish. (Exchange.)

_____

TAKING AN IMPRESSION.

THE necessary condition to be obtained in the adaptation of the denture to the tissues is to have it embrace the alveolar ridge and extend backward on the palate to an extent that the entire periphery will impinge on and slightly displace lax soft tissue. This can only be definitely accomplished by securing an accurate impression of the surfaces of these lax soft tissues, which calls for an impression of more of the surface of the mouth than it is ordinarily considered necessary to obtain.

It is important that the impression material should pass upward between the alveolar ridge and the lip and cheeks to the greatest extent possible, without putting the lip and cheeks on more than a slight tension. It must be carried accurately to the extreme height of the space at the outside of the tuberosity, when such a space exists, and it should extend on the tissue posterior to the tuberosity for a short distance, and on the soft palate for a sufficient distance to allow of locating on the model the line of attachment of the soft palate to the posterior margin of the hard palate. (W. B. Ames, in _Review_.)

_____

ORGANISM OF MAN.

IN the human body there are about 200 bones. The muscles are about 500 in number. The length of the alimentary canal is about thirty-two feet. The amount of blood in an adult averages fourteen pounds, or fully one-tenth of the entire weight. The heart is six inches in length and four inches in diameter, and beats seventy times a minute, 4,200 times per hour, 100,800 per day, 36,792,000 times per year, 2,565,440,000 in three score and ten; and at each beat two and a half ounces of blood are thrown out of it, one hundred and seventy-five ounces per minute, six hundred and fifty-six pounds per hour, seven and three-fourth tons per day. All the blood in the body passes through the heart in three minutes. This little organ, by its ceaseless industry, pumps each day what is equal to lifting 122 tons one foot high, or one ton 122 feet high. The lungs of an average-sized person will contain about one gallon of air, at their usual degree of inflation.

We breathe on an average 1,200 times per hour, inhale 600 gallons of air, or 24,000 times per day. The aggregate surface of the air cells of the lungs exceeds 200,000 square inches, an area very nearly equal to the floor of a room forty feet square. The average weight of the brain of an adult male is three pounds eight ounces; of a female, two pounds four ounces. The nerves are all connected with it directly or by the spinal marrow. The nerves, together with their branches and minute ramifications, probably exceed 10,000,000 in number, forming a "bodyguard" outnumbering by far the greatest army ever marshaled.

The skin is composed of three layers, and varies from one-fourth to one-eighth of an inch in thickness. The atmospheric pressure being about fourteen pounds to the square inch, a person of medium size is subjected to a pressure of 40,000 pounds. Each square inch of skin contains 3,500 sweating tubes, or perspiratory pores, each of which may be likened to a little drain pipe one-fourth of an inch long, making an aggregate length in the entire surface of the body of 201,166 feet, or a tile ditch for draining the body almost forty miles long. Man is marvelously made. (*Scientific American.*)

CLEANLINESS.

MOST of the serious diseases to which human flesh is heir are due to filth, or are fostered and spread by filth. But it is the hardest thing in the world for the race to be cleanly. Personal cleanliness, for example, as understood by the Anglo-Saxon race, by the Dutch, and by the Japanese, is unknown to most other people. Cities go on for generations drawing their water supply from polluted sources, and lamenting at "visitations"—the impiety of it!—of fevers. Or they invite diphtheria by carelessness in regard to drainage, or reeking masses of filth in defective cesspools. Filth out of sight is so generally out of mind. By and by an awful pestilence comes—through the "inscrutable providence" of the Almighty, as people used to say, reverently but at the same time impiously. It is a "providence," but no one thinks of it any more as "inscrutable." It comes in the nature of things, as effect from cause. An immutable law is violated. Quick or slow, the inevitable retribution comes.

It is curious to reflect that this has been the usual course of the human race from the very earliest times—namely, to invite pestilence by filthiness; then, when the invited guest appears, attended by goodman Death, to charge his coming to divine wrath at something other than the true offense; and then, while seeking to appease the angered Deity with sacrifice or increased devoutness in religious service, incidentally to clean the camp, or house, or city. One of the earliest pestilences on record was that which assailed the hosts of the Achaians beleaguering the city of Troy, which Homer describes in immortal verse in the first book of the Iliad. How far back of the dawn of history it was we do not know, but how true to human life it all is after these thousands of years!

The Greeks, "child-minded men," were sure when the pestilence struck the camp, smiting the mules and dogs as well as men, that the wrath of the gods had been kindled against them. A prophet was ready to recall that their most powerful chief had only a few days before despitefully treated an aged priest of Apollo, who came grief stricken to the camp, bearing rich ransom, and piteously pleaded to be allowed to redeem his daughter. Agamemnon brusquely refused, and bade him go; and the old man had prayed for vengeance. Surely this was the cause of the pestilence! The only thing to do was to send the maid home to her father without ransom, and with her rich gifts to the outraged god. The Greeks applauded the wisdom of their prophet; public opinion compelled the reluctant chief to accede, and maid and gifts were dispatched with due ceremony to the house of the insulted priest. That done, Homer naïvely adds, as though it were an afterthought: "Atreides bade the folk purify themselves. So they purified themselves, and cast the defilements into the sea, and did sacrifice to Apollo."

And so they rid themselves of the pestilence. We have reached the stage where we begin to clean up as soon as we hear that a pestilence is on the way, seeking for congenial filth. Some time, some time we shall have in us that belief which compels practice in the truth of the old adage, "Cleanliness is next to godliness" (on which side the adage does not say), and we shall keep clean all the time. (*Public Opinion.*)

WILLIAM BLAIKIE ON WALKING.

WALKING, when properly engaged in, is of great benefit to all those who lead sedentary lives, and should be practiced every day in the year in order to keep the system in good order and full working condition. If young men and women would stop for a moment, cast their eyes around them, and behold how many people there are who are simply weakening their constitution by degrees,

from the lack of knowing how, when, and where to walk, it would
not be a great while before every man, woman, and child would re-
solve to make an effort to walk aright, and gain in a great measure
that buoyancy, freshness, and vigor which comes to all who endeavor to take up their burden anew and walk in the right way.
(*Good Health.*)

SAFEGUARDS TO PURITY.

THERE are parents who think they can bring up their children,
and especially their daughters, in a large degree ignorant of the
evil that is in the world. As the king in the fairy tale banished all
spinning wheels from his dominions, that his daughter might not
wound her fingers with a spindle, and realize the prophecy of the
spiteful fairy at her christening; even so mothers withhold useful
and necessary knowledge from their daughters, lest with it may be
mingled something leading to harm. And even as this charming
princess, notwithstanding every precaution, by accident came upon
the only spinning wheel in the realm, was wounded by the spindle,
and fell into her hundred years' slumber, so often does the young
girl unawares stumble upon experiences of whose possible existence
she never dreamed, and which are even more disastrous to her than
those that befell the Sleeping Beauty.

Boys and girls as they grow up will learn the ins and outs of this
wicked world. If their parents do not give them this knowledge,
somebody else will; and the manner in which this information is
given is in all moral respects vastly more important than the matter. The parent may instruct the child in everything it should
know, satisfy its curiosity within proper limits, and thus preoccupy
the ground that would otherwise be sown by chance cultivators
more with tares of vice than with the wheat of knowledge.

It is simply astonishing how some young children pick up slang
words, vile words, profane words, and attach to them meaning.
Objectionable words and phrases seem armed with hooked burs,
and cling tenaciously to the mind they catch hold off. It is equally
surprising how instinctively they conceal all this knowledge from
their parents.

It is not possible for the mother to cultivate too great intimacy
with her child. She should have the juvenile heart spread out before her as a mirror, reflecting every thought, every feeling, every
passion of the child. Thus she will be able judiciously to administer antidotes to vice and build up safeguards to virtue. When

there is a perfect understanding between parents and children; when the daughter feels that she can carry every thought and desire to her mother, and the son is in full sympathy and counsel with his father, there is little danger that the happiness of Christian parents will be wrecked by the profligacy of their children, little danger that the children will wander far from approved lines of conduct. (Selected.)

CONSUMPTION COMMUNICATED BY BEDBUGS.

IT has long been known that fleas are sometimes carriers of hospital gangrene and erysipelas, by conveying upon their bodies the infection from diseased wounds to healthy ones; and it has been shown that the mosquito is a means of introducing infection when withdrawing blood from the body.

A German physician recently reported the case of a youth of eighteen who contracted consumption through inoculation by means of bedbug bites, while living with a brother who died with the disease. On examination of the bugs found in the house in which the young man lived six per cent. were found to contain tubercle bacilli. Three guinea pigs injected with a filtered solution of the crushed bodies of the bugs died of tuberculosis.

This observation is an important one, as it not only brings forward another means by which diseases are communicated, but also furnishes an explanation of some of the mysterious and untimely deaths which are so frequently charged to an inscrutable Providence, and teaches in the most emphatic manner the duty of cleanliness, and the danger of contact with vermin of any sort. Vermin are not only an evidence of filth, but are of themselves dangerous enemies to life and health. (*Good Health.*)

THE APPLE AS FOOD.

MORE attention is perhaps being given at the present time to the food value of fruits and other vegetable products than at any other time since the days when Pythagoras preached vegetarianism at Krotoma. Eminent German chemists have recently been making a careful analysis of the apple, and pronounce it to be the most nutritious of all fruits. It is said to contain more phosphates than any other vegetable. Without adopting this theory, we are quite ready to recommend the apple as one of the best of fruits, and well calculated to aid in maintaining good digestion, especially in persons of sedentary habits. Not infrequently, persons whose stomachs are very delicate are able to digest ripe sweet or subacid apples without serious inconvenience, though unable to eat any other kind of fruit. The malic acid of the apple is excellent for persons who are chronic sufferers from biliousness, though, singular as it may seem, it does not add to the acidity in cases of acid dyspepsia, but rather has a tendency to diminish acidity. (*Good Health.*)

WASHING CHILDREN'S EARS.

THE foundation of chronic deafness is often laid in early childhood. Few ailments are more common among children than earache. Most mothers are unconscious of the fact that they are themselves the cause of much suffering in their children by attacks of this painful malady. In her anxiety that her children's ears shall be thoroughly clean the mother endeavors to remove every particle of earwax from the inner portion of the ear, by boring it out with a hairpin or other sharp instrument, covered with a towel, or with the corner of a towel twisted to a point. Nature knows how to care for these hidden recesses far better than does the most accomplished mother. This portion of the ear requires no attention. Nature takes care of it in the most admirable manner. The membrane lining the canal of the ear contains a great number of little glands which secrete a waxy substance having an intensely bitter taste. The purpose of this is to prevent the entrance of insects and to keep the ear clean, as the layer of wax dries in scales, which rapidly fall away, thus removing with them any particle of dust or other foreign matters which may have found entrance to the ear. Nothing more irritating than a few drops of olive oil, warmed to a temperature a little above blood heat, should ever be placed in the ear. (*Good Health.*)

TO BREAK UP A COLD.

AT this season of the year a cold is one of the most common accidents. An ordinary cold is usually cured in from two to six weeks, but not infrequently a hard cold leaves behind it relics, recovery from which may require months or even years. Sometimes a fatal disease finds its beginning in a neglected cold. One of the best means of breaking up a cold, especially if taken by getting the feet wet, is to take a hot mustard footbath, which may be made by adding a tablespoonful of ground mustard to two gallons of water as hot as can be borne, in an ordinary footbath or a wooden pail. The bath should be continued fifteen to thirty minutes, or until the skin is well reddened and tingling.

While taking the footbath, swallow one or two pints of hot water or hot tea of some sort. Catnip, wintergreen, cinnamon, or almost any herb tea will answer the purpose. It is, of course, the hot water that produces the effect, so that it is a matter of small consequence what is used as flavoring.

After the footbath, dry the feet quickly, go to bed, and have applied over the part in which the cold seems settled an ordinary towel wrung out of cold water, sufficiently dry so that it will not drip, and cover it with several thicknesses of flannel or sheet cotton, so as to keep it warm during the night.

If the seat of the cold seems to be in the lungs, the compress should be applied over the chest and also the back of the shoulders. It should be large enough to cover the whole surface of the chest— that is, the whole of the upper part of the trunk, or that portion in which the ribs lie.

If the attack is a severe one, so that a serious illness is threatened, the patient should stay in bed for one or two days, or in bad cases, for a longer time, as may be indicated. The footbath and the hot drink should be repeated each day until the patient is relieved, and the cold compress should be renewed night and morning.

In case the compress becomes cold during the night, it should be covered with oil muslin or rubber cloth, so as to prevent evaporation. If the bowels are inactive, empty them by means of a large coloclyster of water, as hot as can be borne. The diet should be sparing; it should consist of fruits and grains. Hot water should be taken plentifully. At least two or three quarts should be taken in the course of twenty-four hours. (*Good Health.*)

LEMON JUICE AS A GERMICIDE.

RECENT experiments conducted at the Pasteur Institute, in Paris, have shown that drinking water may be completely freed of cholera bacilli by the addition of fifteen grains of citric acid to a quart of water. As citric acid is an acid of lemon juice, it would appear that strong lemonade would answer the purpose equally well. (*Good Health.*)

Correspondence. · · · · · · ·

DENTAL EXHIBITS AT THE WORLD'S COLUMBIAN EXPO-
SITION.

BUREAU: CAPITOL, ALBANY, N. Y., February 4, 1893.

To the Editor of the DENTAL HEADLIGHT.

Sir: Upon January 14 last the rules and regulations to govern
the matter of awards at the World's Columbian Exposition were
adopted. These rules provide among other things that each exhibit
shall be examined by an individual judge, who must be so far as
possible a competent expert and must formulate his opinion of an
exhibit in writing and sign his name thereto, and that this report
must receive the confirmation of the Departmental Committee of
which he is a member. There are thirteen of these Departmental
Committees, one of which is assigned to each of the thirteen great
departments of the Exposition, and these thirteen committees
compose the Board of Judges. There will be foreign representation
upon each of these thirteen committees, and also one or more wom-
en judges upon all committees authorized to award prizes for exhibits which may be produced in whole or in part by female labor.

It is the desire of the World's Columbian Commission to com-
pensate judges, but this will depend upon the action of Congress, to
which application has been made for an appropriation sufficient to
defray all expenses connected with this subject of awards. Should
the appropriation be made, judges from the United States will
received $600 each, and those from foreign countries $1,000 each.
Their work should begin June 1 and last about two months.

The management of this whole matter is in the hands of this Executive Committee on Awards, which is receiving applications for
appointments as judges at their temporary offices and address as
above. After March 15 the office will be permanently established
in the Administration Building of the World's Columbian Commission, at Jackson Park, Chicago. As judges are to be so far as possible competent experts, and the duties of the position require efficiency and ability of a very high order, and it is the desire of the

Executive Committee to appoint only those absolutely qualified for the place and, whenever possible, of national and international reputation, this Executive Committee considers it its duty to call the attention of the technical journals of the country to the facts, in order that it may receive the assistance and coöperation of those who by their position and connections are best able to assist the committee. It is not possible to make this announcement in the way of an advertisement, and the information is sent you in the belief that it is of sufficient interest to your readers to obtain some form of mention in your journal, and that you will suggest the names of such experts as you know to be of the capacity required.

JOHN BOYD THACHER,
Chairman Executive Committee on Awards.

INTERNATIONAL MEDICAL CONGRESS.

To the Editor of the DENTAL HEADLIGHT.

Sir: The Eleventh International Medical Congress will be held in Rome in 1893, beginning the 24th of September and continuing until the 1st of October.

The Committee on Organization, following the precedent established in London in 1882, has provided for a section of odontology. As America has contributed preeminently to the scientific progress of dental surgery, it is hoped that the dental profession in America will be creditably represented. All reputable practitioners are entitled to membership in this section.

The time chosen is the most delightful of all the year, and to those who have never visited the "Eternal City" the meeting of the Congress will afford a rare opportunity.

The North German Lloyd Steamship Company have an established line of first-class steamers to Genoa making the passage in less than eleven days. It proposes to reduce the fare to Genoa by twenty per cent., and the return trip by ten per cent. to those attending the Congress.

The French Railway Company have also offered a reduction of fifty per cent. on its fare.

Dr. Norman W. Kingsley, 115 Madison Avenue, New York, has been appointed Member of the American National Committee for the promotion of the interests of the Odontological Section. All communications in reference to that section should be addressed to him. A. JACOBI, M.D.,
Chairman of the American National Committee.

Editorial. . . . .

THE WORLD'S COLUMBIAN DENTAL CONGRESS.

WE give space to the letter from Dr. A. O. Hunt notifying us of the change of time of the Congress, as it is a matter of importance and should be understood. The change of time and the necessity for limiting the time is to be regretted. The Executive Committee is not responsible in the least degree, the matter being solely under the control of others. The time is now definitely and unalterably fixed, and sufficient and timely notice given not to in any way interfere with the arrangements of those who wish to attend.

The following is the letter referred to:

IOWA CITY, IOWA, January 14, 1893.

To Henry W. Morgan, Editor of the DENTAL HEADLIGHT, Nashville, Tenn.

Dear Doctor: The following communication was received from President Bonney, of the World's Congress Auxiliary, which necessitates a change in the time of meeting and also a rearrangement of the order of business for the World's Columbian Dental Congress.

"The Dental Congress has been assigned generally to the week commencing Monday, August 14, 1893. The Congresses of Science and Philosophy have been assigned to the week commencing Monday, August 21, 1893. With more than a hundred Congresses to provide for, you will readily understand the extraordinary difficulty of making suitable arrangements for each, but the extra provision which has been made for the places of meeting will render practicable arrangements which under the ordinary circumstances would be simply impossible. When the Congresses were first proposed we expected to have only one large audience room, with a suitable number of small halls; but as the World's Congress work enlarged the places of meeting were also made more adequate. As the World's Congress Art Palace is now planned there will be large audience rooms capable of accommodating three thousand persons each, and more than twenty smaller halls, which will seat from three hundred to seven hundred persons each. Thus providing for no less than thirty-six large meetings and three hundred and sixty smaller meetings in a single week, by holding morning, afternoon, and evening sessions. Among the other Congresses assigned to be held in parallel with the Dental Congress are those of pharmacy, medical jurisprudence, and horticulture. For all these the accommodations will be adequate. You understand, of course, that everything in the nature of an exhibit is required by the Exposition authorities to go to Jackson Park. The Congresses deal not with things, but with men; not with matter, but with mind."

In accordance with the above statement the time of meeting will be from Monday, August 14, to Saturday, August 19, inclusive.

Please note this change in your journal.

Yours very truly, A. O. HUNT.

COMMITTEE ON EXHIBITS, WORLD'S COLUMBIAN DENTAL CONGRESS—IMPORTANT NOTICE.

THE Committee on Exhibits for the World's Columbian Dental Congress desire to obtain rare specimens of growths, abnormalities, casts, illustrations of methods, instruments and appliances, both ancient and modern, whereby the growth of the profession may be shown from its early infancy to the present time. They also desire to exhibit an ideal library, operating room, and laboratory; and to this end earnestly request all members of the profession, together with dental dealers and publishers, to loan them any specimens, instruments, appliances, books, photographs or pictures of societies and eminent men, of all countries, together with anything and everything that will be of interest to any dentist from any part of the world. They will pay all transportation charges on such exhibits to Chicago and return, and will insure the same while on exhibition if desired.

CHARLES P. PRUYN, Chairman, 70 Dearborn Street, Chicago, Ill.;

ARTHUR E. MATTESON, 370 Cottage Grove Ave., Chicago, Ill.;

E. M. S. FERNANDES, 36 Washington Street, Chicago, Ill.;

M. L. RHEIN, 104 East Fifty-eighth Street, New York;

A. W. McCANDLESS, 1,001 Masonic Temple, Chicago, Ill.;

R. C. YOUNG, Anniston, Ala.;

JAMES CHACE, Ocala, Fla.;

W. A. CAMPBELL, Gold and Fulton Streets, Brooklyn, N. Y.,
Committee.

Address all communications to Dr. A. W. McCandless, Secretary, 1,001 Masonic Temple, Chicago, Ill.

ALABAMA AND MISSISSIPPI DENTAL ASSOCIATIONS.

In this issue will be found the programme for these Associations, which meet this month. Aside from the benefit to be derived from a professional standpoint by attending these annual gatherings, the social features are such as to make every stay at home green with envy.

For geniality, cordiality, and whole-souled good fellowship and kindly feeling toward one another and visiting dentists we pit the members of these Associations against the world.

All those who were in attendance at the last meetings will be present this year, and many more. Every arrangement has been made, and an unusually interesting time is promised.

C. L. BOYD, D.D.S.

WE have the pleasure of presenting with this issue a most excellent picture of Dr. C. L. Boyd, of Eufaula, President of the Alabama Dental Association—a gentleman by birth and education, honored by his brothers, and who enjoys a large and lucrative practice. He received his degree from Vanderbilt University in 1889, and is an acknowledged expert operator.

_____

DEAD PATRIARCHS.

WE mourn the death of two more of the older members of our profession. On the 15th of February, at his home in Xenia. O., Dr. George Watt, one of the most prominent pioneer dentists, died, after years of suffering from locomotor ataxia, resulting at last in partial paralysis.

Dr. Walter W. Alport, one of the Vice Presidents of the Columbian Dental Congress, died March 21, at his home in Chicago, of meningitis.

In these deaths the profession has been robbed of two men of world-wide reputation—the one a scientist, journalist, and author; the other a scholar, a scientific operator, writer, and debater. Their love for dentistry and work in its upbuilding has endeared them to its members, and it will live long after them as monuments to their honor and an encouragement to others to higher and nobler aims and ends.

_____

GOSSIP.

DR. W. H. G. WHITE, of Smyrna, Tenn., was married on January 25 to Miss Evie Rucker, daughter of the urbane Capt. T. G. Rucker, conductor on the N. and C. railroad.

Married in Nashville, March 7, in the Christian Church, Dr. Frank W. Smith, class of 1889–90, Vanderbilt University, of Jackson. Mich., to Miss Westelle, daughter of Mr. and Mrs. Alexander Fall.

At the Presbyterian Church. Homer, La., January 19, F. W. Meadows, D.D.S., class of 1889–90, Vanderbilt University, was married to Miss Dink Brown.

J. W. Peden, D.D.S., of Dyersburg, another Vanderbilt graduate, was married to Miss Su Donia Percell, at the Rehoboth Church, Ro-ellen, Tenn.

At Goodlettsville recently, Leonard T. Hallums, D.D.S., Nashville, class of 1890–91, Vanderbilt University, was united in marriage to Miss Beulah, daughter of Mr. T. J. Phipps.

These young and happy couples are congratulated and have the best wishes of the DENTAL HEADLIGHT in taking this important step in the discharge of the duties of citizenship and the Master's wishes.

To be married in Selma, Ala., at the First Methodist Episcopal Church, South, on April 6, Dr. Rivers Ashley Rush (an Alumnus of the Dental Department of the Vanderbilt) to Miss Mary Pinkie Fowlkes, daughter of Mr. and Mrs. Samuel Augustus Fowlkes.

BOOK NOTES.

CATCHING'S COMPENDIUM OF PRACTICAL DENTISTRY FOR 1892. By B. H. Catching, D.D.S., Atlanta, Ga. Price $2.50.

This volume has been received, and any dentist who has not the time to read all the journals published (and who has), and yet desires to keep posted as to the practical articles published in them, can secure them bound in cloth, printed in clear, bold type, on beautiful and elegant paper, for the sum of two dollars and fifty cents.

The work is well done and improves each year, showing continued advance in the profession and in the judgment and experience of the riper years of the compiler.

CASSIDY'S ELEMENTS OF CHEMISTRY AND DENTAL MATERIA MEDICA. By J. S. Cassidy, M.D., D.D.S. Published by Robert Clark & Co., Cincinnati, O.

We are pleased to note and commend a contribution from the pen of Prof. J. S. Cassidy, of the Ohio College of Dental Surgery, on the subjects of Chemistry and Materia Medica. A careful perusal has confirmed our first opinion as to its intrinsic worth, and at the same time confirmed the belief that such an association of subjects would render his text-book awkward, if not quite unavailable, in courses unlike his own. Apart from its real merit, which is unquestionable, its easy and almost graceful presentation of the subjects marks it as unique. To his own students we should say it must prove a boon, carrying as it does so much of the author's personality, but the student body at large we fear can only use it as a ready, correct, and concise book of reference. We must congratulate the author, also, upon style, correctness, and arrangement.

R. D. S.

Associations. · · · · · · · ·

PERMANENT OFFICERS OF THE WORLD'S COLUMBIAN DENTAL CONGRESS.

PRESIDENT.—L. D. Shepard, Boston, Mass.

VICE PRESIDENTS.—W. W. II. Thackston, Farmville, Va.; A. L. Northrop, New York City; W. H. Morgan, Nashville, Tenn.; W. W. Allport, Chicago, Ill.; W. O. Culp, Davenport, Ia.; C. S. Stockton, Newark, N. J.; Edwin T. Darby, Philadelphia, Pa.; H. J. McKellops, St. Louis, Mo.; J. Taft, Cincinnati, O.; J. II. Hatch, San Francisco, Cal.; J. B. Patrick, Charleston, S. C.; J. C. Storey, Dallas, Tex.

SECRETARY GENERAL.—A. W. Harlan, Chicago, Ill.

ASSISTANT SECRETARIES.—George J. Freidricks, New Orleans, La.; Louis Ottofy, Chicago, Ill.

TREASURER.—John S. Marshall, Chicago, Ill.

EXECUTIVE COMMITTEE.—W. W. Walker, Chairman, 67 West Ninth Street, New York City; A. O. Hunt, Secretary, Iowa City, Ia.; L. D. Carpenter, Atlanta, Ga.; J. Y. Crawford, Nashville, Tenn.; W. J. Barton, Paris, Tex.; J. Taft, Cincinnati, O.; C. S. Stockton, Newark, N. J.; L. D. Shepard, Boston, Mass.; A. O. Hunt, Iowa City, Ia.; II. B. Noble, Washington, D. C.; George W. McElhaney, Columbus, Ga.; J. C. Storey, Dallas, Tex.; M. W. Foster, Baltimore, Md.; A. W. Harlan, Chicago, Ill.; J. S. Marshall, Chicago, Ill.; II. J. McKellops, St. Louis, Mo.

The meeting will be held in Chicago August 17–29, 1893.

OFFICERS OF THE AMERICAN DENTAL ASSOCIATION.

J. D. Patterson, Kansas City, President; J. Y. Crawford, Nashville, Tenn., First Vice President; S. C. G. Watkins, Mont Clair, N. J., Second Vice President; Fred A. Levy, Orange, N. J., Corresponding Secretary; George H. Cushing, Chicago, Recording Secretary; A. H. Fuller, St. Louis, Treasurer; W. W. Walker and S. G. Perry, New York, and D. N. McQuillen, Philadelphia, Members of the Executive Committee. Next meeting to be held in Chicago.

OFFICERS OF THE SOUTHERN DENTAL ASSOCIATION.

B. Holly Smith, Baltimore, President; R. K. Luckie, Holly Springs, Miss., First Vice President; S. B. Cook, Chattanooga, Second Vice President; L. P. Dotterer, Charleston, S. C., Third Vice President; D. R. Stubblefield, Nashville, Tenn., Corresponding Secretary; S. W. Foster, Decatur, Ala., Recording Secretary; H. E. Beach, Clarksville, Tenn., Treasurer; W. R. Clifton, Waco, Tex., and Gordon White, Nashville, Tenn., Members of the Executive Committee. Next meeting to be held in Chicago.

OFFICERS OF THE TENNESSEE DENTAL ASSOCIATION.

S. B. Cook, Chattanooga, President; W. W. Jones, Murfreesboro, First Vice President; W. J. Morrison, Nashville, Second Vice President; P. H. Houston, Lewisburg, Recording Secretary; D. R. Stubblefield, Nashville, Corresponding Secretary; H. E. Beach, Clarksville, Treasurer. Next meeting to be held in Chattanooga first Tuesday in July, 1893.

TENNESSEE STATE BOARD OF EXAMINERS.

S. B. Cook, President, Chattanooga; J. L. Mewborn, Secretary, Memphis; J. Y. Crawford, Nashville; W. T. Arrington, Memphis; A. F. Shotwell, Rogersville; R. B. Lees, Nashville. Next meeting July, 1893, at Lookout Mountain.

GEORGIA STATE ASSOCIATION.

S. M. Roach, President; N. A. Williams, Valdosta, First Vice President; S. H. McKee, Talbotton, Recording Secretary; L. D. Carpenter, Atlanta, Corresponding Secretary; A. H. Lowrance, Athens, Treasurer. Executive Committee: Drs. Wells, Sims, Rosser, McDonald, and Hinman. Permanent place of meeting in the future, Atlanta. Next meeting, Tuesday, May 9, 1893.

NORTH CAROLINA STATE DENTAL SOCIETY.

THE annual meeting of the North Carolina State Dental Society will be held in Raleigh May 23 to 26, 1893. An attractive programme has been arranged, and it is hoped and expected there will be a large attendance. The standing committees for 1893 are as follows, the first named in each case being the Chairman:

Dental Education.—Drs. J. E. Freeland, J. H. White, W. P. Moore, A. O'Daniel, W. W. Rowe, C. S. Boyette.

Dental Chemistry and Metallurgy.—Drs. I. N. Carr, E. J. Tucker, R. W. Reece, P. E. Hines, A. J. Pringle, C. W. Banner.

Dental Pathology.—Drs. J. H. Durham, A. C. Liverman. L. B. Henderson, D. L. James, F. C. Frazier, R. P. Anderson.

Dental Therapeutics.—Drs. S. P. Hilliard, H. Snell, H. V. Horton, J. M. Ayer, V. J. Burgin, J. A. Blum.

Operative Dentistry.—Drs. J. E. Matthews, J. E. Wyche, J. M. Riley, E. K. Wright, G. W. Whitsett.

Prosthetic Dentistry, Including Crown and Bridge Work.—Drs. N. M. Culbreth, T. M. Hunter, W. J. Conrad, J. H. London, C. L. Alexander.

Orthodontia.—Drs. J. F. Griffin, V. E. Turner, H. C. ˀPitts, E. E. Murray, Luther White, C. W. Brasher.

Oral Surgery.—Drs. W. H. Hoffman, A. M. Baldwin, J. N. Hester, J. A. Ballentine, T. W. Harris, W. H. Edwards.

Materials and Appliances.—Drs. J. W. Holt, C. J. Watkins, J. B. Little, J. A. Hurdle, R. L. Ramsay, J. C. Goodwin.

Dental Prophylaxis.—Drs. J. F. Ramsey, William Lynch, J. M. Parker, R. M. Morrow, G. B. Patterson, S. W. Gregory.

Anatomy and Physiology.—Drs. C. A. Rominger, J. H. Benton, J. S. Spurgeon, Frank Boyette, W. B. Murphy, H. D. Harper.

J. E. WYCHE, *Secretary*, Oxford, N. C.

MISSISSIPPI STATE DENTAL ASSOCIATION.

THE nineteenth annual meeting of the Mississippi State Dental Association will be held in the Senate Chamber of the Capitol, at Jackson, Miss., commencing on Wednesday, April 5, 1893, and continuing three days.

The following programme has been announced:

Prayer, by Rev. A. F. Watkins, Jackson, Miss.

Welcome Address, by Hon. L. F. Chiles, Mayor, Jackson, Miss.

Response, by Dr. J. B. Askew, Vicksburg, Miss.

Annual Address, by Dr. A. A. Dillihay, President, Meridian, Miss.

Calling of roll and regular order of business.

Morrison Bros., Stuart & Adams, R. I. Pearson & Co., and S. S. White, dental dealers, will be on hand, and as usual will make an elegant display. Their exhibit rooms will be promptly closed at the hour for the opening of each session, and will remain closed until the adjournment of the same.

CLINICS.—DR. ROBERT K. LUCKIE, SUPERINTENDENT.

All clinical operators will be expected to furnish their own instruments.

Dr. George B. Clements, Macon, Miss.: Combination gold and platina crown, claiming for same more perfect adaptability, and simplifying the process of soldering.

Dr. Morgan Adams, Sardis, Miss.: Gold filling, using Williams's crystalloid gold.

Dr. T. M. Allen, Birmingham, Ala.: Porcelain enamel inlay, fusing porcelain at the chair; also his method of making aluminium crowns and bands seamless.

Dr. J. D. Killian, Greenville, Miss.: Gold crown.

Dr. W. H. Marshall, Oxford, Miss.: Will demonstrate the best method for absorbing mercury from amalgam fillings.

Dr. Frank H. Smith, Water Valley, Miss.: Gold contour.

Dr. J. O. Frillick, Meridian, Miss.: Will make a bridge, using a pinless tooth of his own design.

Dr. W. T. Allen, Amory, Miss.: Treating and filling the roots of pulpless teeth.

Dr. Francis Peabody, Louisville, Ky.: Gold filling, using soft foil.

Dr. W. H. Morgan, Nashville, Tenn.: Will give several oral clinics.

Members of the profession at large are cordially invited to attend and unite with us in making the meeting one of great profit and pleasure.

For further information address

FRANK H. SMITH, *Corresponding Secretary.*

Water Valley, Miss.

Special Notice.—The Board of Dental Examiners will meet at the same place as the Association on Tuesday and Wednesday, April 4 and 5. All those desiring to procure license to practice dentistry in the state of Mississippi must apply in person before the Board on one of those days. All must have license; none are exempt. A. H. HILZIM, *President of the Board.*

ALABAMA STATE DENTAL ASSOCIATION.

Dr. C. L. Boyd, President, Eufaula, Ala.; Dr. J. H. Allen, First Vice President, Birmingham, Ala.; Dr. R. A. Rush, Second Vice President, Selma, Ala.; Dr. G. M. Rousseau, Treasurer, Montgomery, Ala.; Dr. S. W. Foster, Secretary, Decatur, Ala.

The twenty-fourth annual meeting of this Association will be held in Birmingham, Ala., April 11–14, 1893.

The following is the programme:

First Day.—Meeting called to order by President at 11 A.M. Prayer. Address of Welcome. Response on behalf of Association, by A. A. Pearson, D.D.S., Montgomery, Ala. Annual address by President. Calling roll and collection of dues.

Second Day.—Reports of committees and discussions. Afternoon: Papers on special topics, incidents of office practice, etc.

94 THE DENTAL HEADLIGHT.

Third Day.—Clinics will be given by prominent dentists in the different branches, with the latest methods and appliances. The entire day will be devoted to clinics. Those expecting to operate will bring their own instruments with them.

Fourth Day.—Fourth day will be devoted to finishing reports of standing committees. All unfinished business of Association must be attended to. Election of officers. Adjournment.

Special Notice.—The State Board of Dental Examiners will meet at the same place on Monday, April 10. All those desiring to obtain license to practice in this state must be on hand. All must have license. None are exempt.

WORLD'S COLUMBIAN DENTAL CONGRESS.
NEW ORDER OF BUSINESS.

Monday, August 14.—10 A.M.: Meeting of the General Executive Committee. 11 A.M.: Opening of the Congress. Reading of the resolutions creating the Congress by the Secretary-General. Address of welcome by John Temple Graves, of Georgia. Responses. Responses from foreign countries. Address of the President. Adjournment. 2:30 P.M.: Papers to be read in the Sections. 5 P.M.: Adjournment.

Tuesday, August 15.—9 A.M.: Clinics. 10 A.M.: Meeting of the General Executive Committee. 12 M.: Address before the whole Congress. 1 P.M.: Adjournment. 2:30 P.M.: Papers to be read in the Sections. 5 P.M.: Adjournment. 8 P.M.: Bacteriological exhibit.

Wednesday, August 16.—9 A.M.: Clinics. 10 A.M.: Meeting of the General Executive Committee. 12 M.: Address before the whole Congress. 1 P.M.: Adjournment. 2:30 P.M.: Garden party. 2:30 P.M.: Garden party. 2:30 P.M.: Garden party. 8 P.M.: Biology. Lantern exhibit. 8 P.M.: Conversazione. 8 P.M.: Conversazione.

Thursday, August 17.—9 A.M.: Clinics. 10 A.M.: Meeting of the General Executive Committee. 12 M.: General address before the whole Congress. 1 P.M.: Adjournment. 2:30 P.M.: Papers before the Sections. 8 P.M.: Public address under direction of World's Congress Auxiliary.

Friday, August 18.—9 A.M.: Clinics at hospitals and at the Art Institute. 10 A.M.: Meeting of the General Executive Committee. 12 M.: General address before the whole Congress. 1 P.M.: Adjournment. 2:30 P.M.: Papers before the Sections. 8 P.M.: Dinner to the whole Congress. (Subscriptions by members from the United States only.)

Saturday, August 19.—10 A.M.: Visit in a body to the Medical and Dental Exhibits at the World's Fair Grounds. 12 M.: Closing addresses to the Congress. Luncheon by the members in the restaurant. (Name to be supplied.)

SOUTHERN MEDICAL COLLEGE ASSOCIATION.

RULES AND REGULATIONS.

THIS Association shall be composed of delegates from Southern Medical Colleges, whose Faculties have signified a desire to become members thereof, signed these rules and regulations, and paid the membership fee of $5.

The objects of the Association are to cultivate closer and more intimate relation between medical colleges, and to elevate the standard of medical education by requiring a more thorough preliminary training and an increased length of medical study.

The Association shall be composed of one or more delegates from each medical college belonging thereto, who shall be elected annually by their respective Faculties. Each college shall be entitled to one vote in the transactions of the Association.

Every student applying for matriculation must possess the following qualifications :

He must hold a certificate as the pupil of some known reputable physician, showing his moral character and general fitness to enter upon the study of medicine.

He must possess a diploma of graduation from some literary or scientific institution of learning, or certificate from some legally constituted high school, General Superintendent of State Education, or Superintendent of some county Board of Public Education, attesting the fact that he is possessed of at least the educational attainments required of second-grade teachers of public schools. Provided, however, that if a student so applying is unable to furnish the above and foregoing evidence of literary qualifications he may be permitted to matriculate and receive medical instructions as other students, and qualify himself in the required literary departments, and stand his required examination, as above specified, prior to offering himself for a second course of lectures.

The foregoing diploma or certificate of educational qualifications, attested by the Dean of the medical college attended, together with a set of tickets showing that the holder has attended one full course of medical lectures, shall be essential to attendance upon a second course of lectures in any college belonging to this Association.

Candidates for graduation, in addition to the usual requirements of medical colleges, must have attended three courses of lectures of not less than six months each in three separate years;

Must have dissected in two courses, and attended two courses of clinical or hospital instructions,

And must have attended one course in each of the special laboratory departments, to wit:

1. Histology and Bacteriology.
2. Chemistry.
3- Operative Surgery.

These requirements shall not apply to any student who has received a course of medical lectures prior to September 1, 1893.

ILLINOIS AND IOWA STATE DENTAL SOCIETIES.

JOINT MEETING.

The twenty-ninth annual meeting of the Illinois State Dental Society will be held at Rock Island, May 9–12, inclusive.

The thirtieth annual meeting of the Iowa State Dental Society will be held at Davenport, May 9–12, inclusive.

These cities are located on opposite sides of the Mississippi River, and arrangements will be made to hold the meeting jointly, so that those in attendance at the meeting of either society will have an opportunity to listen to the papers, take part in the discussions, and witness the clinics, of both societies.

No efforts will be spared to make this union meeting one of the most interesting in the history of each society.

Members of both societies are urgently requested to attend. All dentists are cordially invited to be present.

Every one should bring models, specimens, appliances, or anything that may be of interest to the profession.

LOUIS OTTOFY.

Secretary Illinois State Dental Society, Chicago;

W. O. KULP,

Chairman Executive Committee Iowa State Dental Society, Davenport, Ia.

T͟H͟E͞ DENTAL HEADLIGHT.

| VOL. 14. | NASHVILLE, TENN., JULY, 1893. | No. 3. |

Selections. • • • • • • • • • •

OPERATIVE TECHNICS.

BY L. S. TENNEY, D.D.S., CHICAGO, ILL.

WHEN my name was placed upon the programme for a paper this evening, and I was requested to take as my subject "Operative Technics," I complied with considerable reluctance, feeling that possibly so many of you were familiar with this course that it . would prove a wholly uninteresting subject.

Upon being assured, however, that the course of study and training as conducted in this department in the Chicago College of Dental Surgery and the results attained therefrom were not generally understood, I consented to give a brief outline, believing as I do that this is one of the most essential features of the college curriculum.

This department was created by the Chicago College of Dental Surgery in the winter session of 1888 and 1889, and the class as then organized was known as "the Pioneer Class of the World in Operative Technics."

It is very true that previous to that time some preliminary technical training had been adopted by a number of the better colleges, but the work for the most part was incomplete, was in a measure optional with the student, and was lacking in many features which have made the course so successful in the college of which I speak.

Not until the session of 1888 and 1889, therefore, was the work thoroughly systematized and made a special requirement of the freshmen year.

I may say that the main objects of the course are three in number: 1. To afford a thorough study of the anatomy of the teeth as regards both their outward form, but especially the character of their central canals. 2. To familiarize the new student with the general principles in the preparation of cavities, the nature and methods of manipulating the various filling materials, and the use

7

of dental instruments. 3. To afford a manual training preparatory to his practical work in the infirmary.

In carrying out the first object of the course, the study of dental anatomy, no time nor labor is spared in making it full and complete and at the same time thoroughly practical in every particular. I would not care to take up your time with a detailed account, but will briefly describe the method of teaching adopted in this portion of our work.

At the beginning of the course we have on hand a large number of natural teeth, and with these before us we begin a careful study of their form, noting first the characters common to all, such as crown, root, gingival line, pulp chamber, root canals, and the various surfaces. Then examine the variation in form between the different classes and finally enter upon a careful study of each individual tooth as regards both its typical and unusual forms.

I have endeavored in this work to follow a system as nearly as possible, and as each tooth is taken up separately we study first the general form of the entire crown, then the general form of each of its surfaces, keeping very carefully in mind the size and location of all ridges, developmental grooves, fossæ, cusps, and sulci, noting also the points at which pits or fissures, and in consequence of these decay, are most liable to occur.

Passing to the root, the same general method of study is followed, and finally we begin our examination of the pulp chamber and canals.

You are all doubtless familiar with the dissections which are made for this purpose.

Beginning with the central incisor, we invest it in a quantity of sealing wax, and file it upon one of its lateral surfaces until the canal is fully exposed from the apical foramen to its extreme coronal portion. This gives us a labio-lingual section. Another tooth of the same denomination is then filed upon its labial or lingual surface, which gives us a mesio-distal section.

In like manner dissections are made of each tooth of the entire denture, and in addition to these cross sections of the posterior teeth are prepared showing progressively the size and form of the canal throughout its entire length.

Having thus prepared our sections, they are properly inked and a number of silhouettes made from them. With these before him it will be seen that the student may familiarize himself with the size, form, and location of each canal and tell whether it be straight or tortuous, single or bifurcated, flattened or rounded, also the distance it passes into the crown of the tooth and the thickness of its surrounding walls.

He is impressed with the necessity of accurate knowledge of this subject, the disastrous results which are apt to follow the insertion of a metallic filling too near a living pulp, and the evil consequences of leaving unfilled a single minute canal.

For this study in dental anatomy the first ten weeks of the course are required, after which work of a different character is taken up. This consists in the study of the treatment of teeth having exposed living pulps, and the management of those diseases resulting from the death and decomposition of this organ.

The student invests a number of carious teeth in a plaster model, and is then supplied with a few of the most efficient remedies used in the treatment of these diseases. Each tooth is assigned a certain diseased condition, and as the treatment of this is considered the student applies the remedies as in actual practice, and at the end of the treatment the canals are properly filled with guttapercha.

I hasten through this matter to consider for a moment the work of the latter half of the session.

Having invested a number of carious teeth, we begin the preparation of cavities.

First the character of enamel is studied, its composition and structure, the direction of its prisms, and its relation to dentine, the direction of cleavage, and the ease with which overhanging walls are fractured.

Then the general principles in the preparation of cavities are discussed, the form of cavity most desirable, the removal of diseased tissues, the formation of under cuts, and the shaping of margins. At the end of each lecture the student proceeds to the practical application of the principles laid down, and as he goes step by step through the different operations is given instruction in the use of instruments, the ones to be employed for each operation, their care and manner of handling.

To the new student who has had no practical experience it would be entirely useless, if not unnecessary, to attempt to teach the preparation of special classes of cavities. Upon his entrance into the infirmary it is enough for him to know the general and groundwork principles to be observed, and it is these that I endeavor to have fixed thoroughly in his memory.

As we proceed, then, with the preparation of cavities they are filled with the various materials used for this purpose in actual practice, and each is taken up separately, its compatibility with tooth structure is studied, its physical properties and its favorable and unfavorable qualities as a filling material.

We begin with gutta-percha and cement, follow this with amalgam, then tin, and finally tin and gold in combination, and gold alone.

Taking up amalgam, its property of conducting thermal changes is noted, its spheroidal tendency and methods of partially overcoming it, the action upon it of oral fluids and the manner of its manipulation.

A number of fillings of copper amalgam are also inserted. We prepare this in the laboratory by precipitating metallic copper from an aqueous solution of its sulphate by the use of a sheet of pure iron. Having driven off the surplus mercury, washed, dried, and molded the precipitate, a fairly good quality of amalgam is obtained. We have not laid much stress upon this material, having simply called attention to its color, antiseptic properties, solubility in the oral fluids, and its freedom from any extensive change of form.

In the handling of gold we are limited somewhat by the expense of the material. The study of the subject, however, is thorough, its physical properties being carefully considered, its qualities as a filling material, and its manipulation in both the cohesive and non-cohesive state.

Each student is required to fill a simple cavity with cohesive gold, and a compound cavity with tin and gold at the cervical wall, completing the operation with gold alone.

He is given instruction in the use of the various hand and mechanical mallets, the handling of pluggers, the methods of starting fillings, the annealing of gold, its condensation at the margins and in the main body of the filling, and the manner of finishing.

And so it seems, gentlemen, that I might go on with the subject almost indefinitely, for you are aware that this work extends over a period of six months, and in that time a vast amount of technical training may be accomplished.

I have called your attention to only a few of the more prominent portions of this work, not caring to take up your time with a description of many other features of more or less importance, such for instance as instruction in the application of the dam, the adjustment of ligatures, the disinfecting of instruments, the use of clamps, separators, and wedges, and many details of actual practice.

In closing then I wish to call your attention to the good results accomplished by this course of preliminary training. You are aware that during the past few years no subject has commanded the thought of the profession more earnestly than that of dental education.

Our colleges, our societies, and the profession as a whole, are united in the efforts to improve our methods of teaching and better our facilities for practical training, and I consider that among the rapid strides that have been made, not the least of them has been the inauguration of this work in the technical departments. It affords first a manual training, without which the brightest and most intelligent student is but a bungler at the operating chair. This is a matter which it seems to me has heretofore been sadly neglected in problems of dental education.

The imparting of theoretical knowledge and the development of practical manual training should be given due and equal consideration in every college. Of what use is it to a student to know every principle, every method, every rule regarding the preparation of a cavity or the insertion of a gold filling if he does not possess the mechanical ability to carry them into effect. It is very true that we have in every class a certain number who, having spent a few years in the office of the practitioner, have partially obtained a mastery of instruments and of the hands, but by far the larger portion is made up of those who have never had the advantage of this private instruction, and enter the college, therefore, untrained and unskilled. It is to supply the needs of such that this course is specially adapted.

Another point: It seems that it is within only the past few years that our colleges have awakened to the fact that the student of dentistry should be made familiar with the structure of those organs upon which he is to operate, before permitting him to enter into actual practice in the infirmary.

What would be our opinion of that medical college which allowed its students to perform or attempt to perform surgical operations without first having thoroughly studied the anatomy of the human body and made extensive dissections. And yet we have seen students enter dental infirmaries and have patients assigned them in whose mouths many delicate operations were needed, hardly being able to tell the difference between an incisor and a molar, and not knowing but what the root of a tooth contained a dozen canals or none at all.

We have seen them enter infirmaries with no preparatory manual training whatever, without the slightest knowledge of the structure of enamel and dentine, and not knowing one rule governing a single operative procedure. This cannot be other than an injustice to the student and an imposition upon the patient.

By means of a course of preliminary training let him first be made familiar with the instruments he is to use, the structure of the organs upon which he is to work, the rules governing the operations he is to perform, before permitting him to practice upon living sensitive tissue.

Such a course of training is afforded in the department of Operative Technics. I would not have you believe me over zealous in this matter. We do not claim to offer a dental education in this department, but we do claim to send the student from here into the infirmary with an intelligent idea of the duties which there devolve upon him, and with the manipulative ability to discharge those duties well. I believe the time is not far distant when the department of Operative Technics will occupy a prominent position among the educational facilities of every dental college.   .

INFLAMMATION.*

BY L. B. TORRENCE, D.D.S., CHESTER, ILL.

THE only apology that I have to offer for presenting this subject, if it be necessary to offer any at all, is the fact that there is no one subject that more deserves our most careful and earnest study than the subject of inflammation. It is a complication that rocks the surgeon in the cradle of sleepless nights, one that keeps friends in terrorizing anxiety, one that has caused the victims of neglect and accident, or the subject of surgical operations to suffer incomprehensible agonies. Then, indeed, is it not important that we dental surgeons, who are almost daily called upon to treat this dread disease in its milder forms, should give this subject some of our most thorough study, in order that we may be able to treat it intelligently, and avoid its most grave and deplorable terminations?

Not only should we study for purpose of treatment, but also for the purpose of avoiding its uprising when favored by its absence.

---

* Read before the Southern Illinois Dental Society October 19, 1892.

The propriety of this theme as a subject for a paper occurred to me last fall while preparing a paper on antiseptics, because in its treatment we have nothing but a veritable war between sepsis and antisepsis. Therefore in the treatment of one we have nothing more than the practical application of the other. I want to state here that I am a disciple of the theory of bacteriology, and that I accept as one of the most plausible theories of modern medicine that all morbid conditions of the body or the organs of the body are due to the presence of morbific germs.

In discussing the etiology of inflammation I shall, therefore, give as the prime cause of all imflammatory processes the presence of these morbific germs.

Inflammation may be described as a succession of changes occurring in a living tissue as a consequence of an injury and the pressure of morbific microphytes. The injury may be mechanical, thermal, or chemical; in fact, any cause that would reduce the vitality of the tissue so as to be unable to resist the action of these morbific germs. The changes which occur in a living tissue while undergoing inflammatory reaction are in part analagous to those which occur in the production of normal tissue. They consist of first the Hunterian "determination of blood," and the consequences which have been for the last quarter of a century associated with active congestion—namely, exudation of liquor sanguinis and emigration of white blood corpuscles through the attenuated walls of the capillaries.

But if this were all, if the process stopped here, nature would speedily repair the injury, healing by first intention. And were this all, the association of the microbe in the etiology of inflammation would not occur.

But the popular and accepted idea of inflammation including a termination characterized by the formation of pus. If the formation of pus is regarded as an essential part of the process, as the pathognomonic symptom of inflammation, then must we admit as the associated cause of the morbific process the presence of the pus-forming germ, it being understood that there are several species of these germs which can more or less replace each other as pus-producing agents. Up to the discovery of these microphytes it was accepted as correct pathology that pus was formed by an injury and by that alone; let it be either mechanical, chemical, or thermal, or a combination of two or more of them. But it has been proven in repeated observations by Burdon-Sanderson that tissue may be exposed to all the above conditions, and if the microphytes be excluded no pus formation results. But of course injury is a part of the process, it being necessary to reduce the vitality of the tissue before the germ can work.

The microscopical study of inflammation is therefore not materially changed. We have only to remember the catalytic influence of the microphytes while we observe under the microscope the same phenomena which occurred prior to their discovery. These germs are always present in the body and in the air, but the perfect vitality of the healthy membrane prevents their ravages.

The clinical symptoms or signs of inflammation are usually given as redness, heat, swelling, pain, and impaired function. But none of these may be termed as the pathognomonic sign or symptom, because they are found physiologically. As for instance redness exhibited in modesty's blush. swelling in the areolar tissue beneath the eye following a night's sleep.

Neither are heat and pain infallible symptoms, because the same nerve influence that causes the rush of blood to the part, through the part, and from the part would naturally cause heat and swelling. and when continued to the extent of pressure on nerve endings will cause pain. Even the infallibility of pus as a pathognomonic symptom is impeached by its presence in the normal fluids of the mouth and serous cavities. Having shown the different clinical phases of inflammation, we may clinically define inflammation to be a succession of phenomena observed in living tissues and organs analogous to those which may be produced in the same tissue and organs by the application of some irritating extraneous substance and the catalytic presence of microphytes. These changes can only be studied with the assistance of the microscope, and we will now proceed to study those changes as found to occur by our most eminent microscopists. In order to better facilitate this study we have prepared a number of charts * representing the different stages of inflammation in the different tissues, in order that we may keep before you the subject in hand, and show the occurring changes in a short, concise, and I hope entertaining manner.

First, we will take up cartilage. Cartilage is extra-vascular and always protected from the atmosphere by other tissues. Therefore it is only necessary to expose cartilage in order to set up inflammation. The cells of articular cartilage are usually scattered in the

---

* Not reproduced.

ground substance in groups of two or four, as shown in Chart 1. This disposition of cartilage cells into groups of two and four is obviously due to the fact that these groups originated from single cells which first divided into two, and these two into four, and so on.

The mode of division is that of karyokinesis. The cell action accompanying the first stages of inflammation is identical with the cell action found in normal tissue formation; the last stages are an exaggeration of that action. This is shown by a comparison of Charts 2 and 3.

In Chart 2 we find that the cell undergoes the process of karyokinesis, the cell elongates, the nucleus divides. The cell divides and forms two distinct cells. Thus far the process is perfectly physiological, there being no deviation from the formation of normal cartilage. But here the analogy between the physiological and pathological ends. Instead of dividing into separate and distinct cells as before, they divide into a large number of embryonal cells and losing the power of depositing cartilage they become enveloped in a sack as shown on Chart 3.

This proliferation continues until the sack bursts under the tension and the cells are poured out as pus cells.

Thus you see that upon the application of irritation we have at first nothing but normal cell action, and I doubt not that, if it were possible to supply nutrition in proportion to the cell action, that normal cell action would continue; but the nutrition being less than proportionately demanded by the cells to fulfill their function, the result is that they degenerate into pus cells and are suppurated off.

Now remember that in all inflammatory action the cells return to the embryonal condition, and by this fact we are able to account for most new formations. As for instance when a cell returns to an embryonal mesoderm, it does not follow that that cell will necessarily return to muscular tissue from which it may have come, but it is just as likely to become bone cartilage or any other tissue of mesodermic origin. In the formation of pus in cartilage the pus cell is the result of the degeneration of this embryonal cell.

Having shown you the changes which occur in the nonvascular tissues, we will now proceed to take up the vascular tissues.

In the nonvascular tissues we have the embryonal cells as the source of the pus cell, but in the vascular tissues we have the additional source of the leucocytes, which pass through the attenuated walls of the capillaries into the surrounding tissues, and are there degenerated into pus corpuscles or cells.

In the vascular tissues the first effect of the application of an irritant is a rush of blood to the part, through the part, and from the part, a condition known as hyperæmia or inflammatory congestion. To accommodate this we have first a dilation of the arteries which gradually extends to the veins and capillaries, accompanied by an acceleration of the flow of blood. Now Cohnheim says that this dilation with accelerated flow is purely accidental—as a result of injury. But where the dilation comes on with diminished velocity it is constant and permanent so long as the cause acts, and may be regarded as the essential vascular change of inflammations. As the stream of blood becomes retarded in the dilated vein (using the web of a frog's foot for observation) we find white corpuscles increasing in number in the plasmatic layer in the smaller veins, rolling slowly along, sticking here and there until the veins are lined by layers of spheroidal-like cells. This forms an obstruction to circulation, and the capillaries become packed full of red corpuscles with some of the white. At first the contents of the corpuscles oscillate to and fro with the systole and diastole of each pulsation of the heart. This is succeeded by stasis, in which no movement occurs, although the contents of the capillaries may remain fluid for as long as three days (says Paget). This is followed by thrombosis or coagulation, but not until the walls of the capillaries are dead. Soon after the veins become lined with leucocytes, we will observe these leucocytes sinking gradually into the walls of the vessels and finally pass into the adjacent tissues.

Various stages of their passage may be observed. At first small button-shaped elevations appear on the outer wall of the vessel as in Chart 4. These generally increase in size until they assume the form of pear-shaped bodies adherent to the vessel by their small ends. (Chart 4.) Ultimately the small pedicle gives way, and the passage is complete, the corpuscles remaining free outside the vessel. (Chart 4.) Once outside it degenerates into pus, and we have following suppuration.

Having followed the course of inflammation to the degeneration of tissues, we will now study the terminations of the same. The most frequent as well as the most favorable termination of inflammation is resolution. It consists in the cessation of the process and the restoration of the part to health. For this to occur, it is necessary, first, that the exciting cause be removed; next, that the walls be restored to their normal condition, in order that abnormal transudation may be arrested; and thirdly, that all exudation and all dead tissue be removed and the damaged tissue be regenerated.

The second termination is that of necrosis. This may be produced in three ways: 1. By severe injury continuously acting upon a part, producing inflammatory disturbances of the circulation ending in thrombosis. The tissue, being dependent upon the vessel for nutrition, of course dies as a result. 2. By an irritant conveyed to the part by the vessels, affecting the vessels primarily and the tissues secondarily. 3. By pressure of inflammatory exudation, either fluid or solid, rapidly or slowly strangulating the vessels of supply.

The third termination is that of new growth, sometimes called productive inflammation. This termination is sometimes one of the hardest to control. Included in the last we have tumors, which consist of a mass of new formation which tend to persist and grow. They originate from preëxisting cells of normal tissue and have a circulation peculiar to themselves. They are termed malignant according to the amount of embryonal tissue found in them, and benign according to the amount of normal tissue found in them. So in the third class of terminations we have included the malignant sarcoma as well as the persistent epithelioma. Such terminations as these should impress us with the depth of importance attached to the study of inflammation. And now we come to the practical part of our discourse—that is, the treatment.

In order to intelligently and scientifically treat inflammation it is absolutely necessary that we know: First, that we have correctly diagnosed it as such; secondly, the stage of the disease; and thirdly, the diathesis of the patient; for I do believe there is such a thing as inflammatory diathesis. The diagnosis has been covered in previous remarks, so as it is unnecessary to repeat. Upon the stage and diathesis depends the success of our treatment. First we will take the incipient stages in a person of favorable diathesis. In this class of cases it is only necessary to remove the cause to expedite a favorable and rapid recovery; while on the other hand in a subject of inflammatory diathesis the pathological action would continue until it reached the grave stages. In the treatment of inflammation we regard it as the most plausible omen of success that we adopt as our course of treatment one which clinical experience has substantiated, one that the scientific research of microscopy corroborates, and one which the stage of the disease would warrant us to use with the expectation of success. To do this we must know the stages of the disease.

Even in this enlightened age you can find practitioners who know, or at least use, no other treatment for inflammation except the application of cold. And on the other extreme you find another class of practitioners who advocate only the application of hot fomentations. Each are a charitable Godsend to humanity in its proper sphere, and each capable of doing a most serious harm when injudiciously and ignorantly used. The application of cold is of great benefit during the earliest stages of inflammation, at a time when exudation is only beginning and the capillary vessels are dilated and only partially obstructed.

Cold when applied under these circumstances becomes a valuable remedial agent: 1. By producing contraction of the small blood vessels. 2. By producing at least an inhibitory effect upon the microörganisms in the inflamed tissues. The contraction of blood vessels which takes place under the application of cold has a tendency to clear the capillaries of their contents and to prevent mural implantation. Microörganisms can only multiply at a certain temperature; and if this can be kept at a point low enough to prevent their increase in the tissue by the application of cold, then this agent fulfills one of the causal indications in the treatment of the inflammation. If, however, stasis has already taken place in the capillaries first affected, then the application of cold will prove harmful, as it will tend to prevent formation of an adequate collateral circulation. The sensations of the patient can be safely taken as a guide as to the length of time that the application should be continued. As we find value in the application of cold, so also in the application of heat do we find a valuable remedy. Heat in certain stages of inflammation, when the affection has passed beyond the stage where the cold exercises a favorable influence or where cold applications increase the suffering, warm antiseptic fomentations should be applied. The ordinary filthy poultices of flaxseed, slippery elm, or bread and milk are nothing but veritable hot beds of bacteria, and as such they should be discarded. Hot fomentations act as derivatives and favor the formation of collateral circulation; at the same time they relieve pain, by relaxation of the tissue, thereby relieving pressure upon the nerve endings. A good application is a number of layers of flannel cloths wrung out of some warm antiseptic solution.

(One of acetate of aluminum of one per cent. strength is a safe one under all circumstances.)

Whatever means tend to restore the equilibrium in the circulation of the part is to our mind the proper and scientific treatment of inflammation. I regard heat and cold intelligently administered as the most important agents in this dread disease. They are those furnished us by the all-wise Creator, who, in his far-seeing wisdom, has provided for our wants in the past, furnishes us with means to alleviate the pains of the present, and will provide for our comfort in the future.

Another treatment of inflammation, indulged in by some of the most able teachers and practitioners for centuries, is that known as the antiphlogistic treatment. Included under this head we have blood letting, cupping, leeching, and the internal use of cathartics. It was urged that as inflammation was attended by an increase of heat, swelling, and redness, such remedies should be employed as will reduce arterial tension. But as an unimpaired power behind (*vis a tergo*) is one of the best means to prevent stasis within the inflamed capillaries, practical experience has shown that all remedies and agents which diminish the intra-arterial tension only diminish the prospects for a favorable termination of inflammation. Let me dilate upon this. It was urged that the depletion of the system would relieve the arterial pressure and also the liability of the inflamed area to congestion and stasis. The first part of the premise is correct. When we remove a part of the fluid from a closed vessel, such as the arterial venous systems, together with their connecting links the capillaries form, there is no doubt that the tension is reduced. But in the reduction of that tension we have reduced the pressure exerted by the heart through the arteries to drive the blood through the congesting capillaries. Cohnheim showed this experimentally in the exposed mesentery of a frog. He avoided threatened stasis in inflammation by injecting into one of the veins one centimeter of a six per cent. solution of sodic chloride. But if under similar circumstances he abstracted a considerable quantity of blood, the congestion resulted in complete stasis in a short time.

Venesection is another old time remedy, and while venesection in the treatment of inflammation has been discarded, the direct abstraction of blood from the inflamed part has proved a useful therapeutic source. In order to be of benefit the scarification must be made through the inflamed part, so as to unload directly the dilated and engorged capillary vessels, and on this account thi ɛ method of treatment is only applicable when the inflammation is superficial and affects accessible parts.

Scarification is followed by great relief when the bleeding is encouraged by applying warm water.

Now a word to those who treat a person suffering with acute inflammation by sailing in on a course of antipyretics. If the rise in temperature which attends acute inflammatory affections is due to the introduction into the circulation of phlogistic substances which are produced by the action of the afore-described microörganisms in the inflamed tissues, it is not difficult to conceive that its artificial reduction by the internal use of chemical substances is not followed by any permanent benefit. The rational treatment of this fever, as in any other disease, is the removal of the cause. In our treatment of pericementitis it is often objectionable to remove the cause, which is an offending tooth, and we are compelled to adopt other measures, but we simply avoid the heroic means of extraction, by adopting elimination of the poison through absorption into the circulation and elimination through the outlets of the body intended by nature as sewers for the effete matter of the same.

While it is clinically true that antifebrin, antipyrine, quinine, and other antipyretic drugs, when administered in large doses, will usually reduce the temperature several degrees for a few hours, it is also clinically true that this reduction is always accomplished at the expense of the forces which are laboring to clear the obstructed paths, and on this account their use has resulted in vastly more harm than good to the patient. Quinine, the least objectionable of the drugs which we have mentioned, when administered in the beginning of an inflammation, by its known tonic effect on the small blood vessels when administered in large doses, has a favorable effect in preventing rapid dilation of and stasis within the capillary vessels. If used at all, it should be given in a decided dose, fifteen grains in solution—no time to wait for capsules to dissolve—and given immediately or soon after the development of the first symptoms.

Sponging the surface of the body with warm water or the use of warm baths are the most rational antipyretics, as these simple measures do not weaken the heart's action, while they have a decided effect on temperature, and at the same time add to the comfort of the patient and favor elimination of microbes through that great excretory organ, the skin. The kidneys are also known to eliminate microörganisms that reach them through the general circulation. Their function should be carefully inquired into; and if the secretion of the urine be scanty, diuretics like liquor ammonii acetatis or acetate of potash should be given. In the treatment of symptoms of sepsis in inflammation alcoholic stimulants should be freely administered to meet in time the dangers incident to heart failure.

In chronic inflammatory affections systematic massage is an exceedingly important and valuable therapeutic resource. It stimulates the surrounding vessels to increased action and exerts a potent influence in restoring the normal circulation in the affected capillary vessels and promotes absorption. Counterirritation is another treatment which must not escape our attention. This is a time-honored treatment, and no doubt will be with us for some time to come. It is a relic of antiquity, and while its near kinsman and associate, venesection, has long since been relegated to a place in the pages of past history, counterirritation reigns superbly supreme in a great many dental and medical offices throughout the world as a king in the treatment of inflammation throughout all of its stages. It has at least one good quality: it will satisfy the patient and prevent him from falling into the hands of charlatans until the time has arrived to resort to more effective and radical measures. But do not understand me to mean that counterirritation has no therapeutic value. No remedy could stand the test of long years of observation and criticism without having at least some valuable curative attribute.

The theory that we usually find advanced in regard to the action of a counterirritant is ridiculous in the extreme. The theory usually advanced is that by the application of an irritant at some point distant from the affected area we will cause a rush to the artificially irritated part, and a consequent lowering of the blood pressure of the general circulation and a consequent release of pressure in the affected part. This effect is more due to the influence of nervous connection than anything else. This is very easily proven by the immersion of one hand in cold water, and the consequent fall of temperature in the other. In order that we may successfully use counterirritation we must have a comparatively competent understanding of the anatomy of the part upon which we intend to work. There must be a continuity of the vascular connections in order that the fluxion sequent to counterirritation shall relieve the blood pressure of an affected part. In other words, for counterirritation to work successfully, it should be applied so as to rob the inflamed area of the surplus of blood flowing to the part. To do this we should make the application on the track of the supplying artery, at a point nearer the heart than the inflamed area.

But we dentists of the present age do not treat inflammation entirely by local applications. As an adjunct to our local treatment we find many systemic remedies of much value. For instance, in veratrum viride, gelsemium, arnica, and especially digitalis do we find valuable as arterial sedatives to diminish blood supply to inflamed area, and they cease to be beneficial when exudations begin. Veratrum viride acts by lessening the force with which the blood is propelled and the number of cardiac contractions. Active hemorrhage in the plethoric is sometimes stopped by a full medicinal dose of the drug. (Tinct. M. j. to M. v.) Gelsemium and arnica do the same thing. Digitalis acts by contracting the arterioles. When given in lethal doses short of a toxic dose the effect is first a rise in temperature, succeeded by dimunition no doubt due to paresis which results from over stimulation. Its therapeutic action is to slow the action of the heart and contract the arterioles.

All the above remedies are useful before exudation begins, but their function ceases when that begins. In the exudation stage we find such remedies as the alkalies, especially the potash salts and ammonia, very valuable. They lessen heat and promote the excretion of the products of inflammation. The potassium chlorate is valuable in the treatment of acute tonsillitis, aphthous ulceration of the mouth, stomatitis materna, and mercurial stomatitis. But this chlorate has become such a universal domestic remedy that its valuable qualities are much abused by injudicious use. Therefore it should be recommended with caution, and in a great number of cases its use should be discouraged.

Gentlemen, I do not feel justified in detaining you longer. This is but a brief outline of this subject. Our field of study is boundless. In wandering over untrodden ground we have found pleasurable entertainment in variegated study. The more that I study this subject the more important and interesting it becomes. In the quiet hours of reflection I review a panorama made up of unsightly and hideous deformities caused by neglect or, what is worse, caused by the imposition of an unscrupulous charlatan upon an unsuspecting innocent public. What does it mean? Read the handwriting on the wall. There carved in bold relief stands the solemn admonition: "Observe, study, apply." And no professional man is worthy of the calling he has adopted unless he does faithfully, diligently, and continuously observe, study, and apply.

FROM "ON SOME WAYS OF MANIPULATING THE FILLING MATERIAL."

BY MR. HALLIDAY.

OXYCHLORIDE OF ZINC.

THE chief uses of oxychloride of zinc are for lining cavities to be filled with amalgam or gold, to strengthen frail walls (it is antiseptic, decay never being seen under oxychloride fillings), for bleaching discolored teeth, and where the pulp is protected for obtunding sensitive dentine.

To Mix Oxychloride.—To sufficient liquid add about an equal amount of powder. Work small portions of powder into a soft cream with a stiff spatula, gradually adding more powder until it is all mixed, carefully mixing each part smooth before adding more powder, until a decided puttylike consistence is obtained. It should not be kneaded. If used for lining the cavity, place in position with the point of the spatula, and press into position with pellets of cotton-wool, taking care not to let the fibers of the wool be entangled in the mass as it is setting. It should be introduced in small portions. If the cavity is to be filled with gold, let the cement have about a day to set; if amalgam, it should be introduced before the oxychloride is quite set.

A very thin layer of oxychloride will cause little, if any, pain, which results sometimes from a bulk of this material, even in pulpless teeth. By introducing in small portions most of its shrinkage is disposed of, and to set it hard requires from fifteen to twenty minutes.

If used to obtund sensitive dentine, a layer of gutta-percha or oxysulphate should be placed over the pulp, but where the pulp is protected it is the most comfortable of all fillings.

OXYSULPHATE OF ZINC.

As a filling oxysulphate of zinc is not so useful as oxychloride, being more soluble. Oxysulphate should be mixed to little more than a milky consistence, never thicker than ordinary cream when used for pulp capping. It should then be worked with the spatula

8

until it begins to give the least perceptible evidence of thickening,
when it should be taken on the end of a spatula and placed accurately in position by being pushed off by a moderately fine probe.
It should not be worked after it ceases to flow under this instru-
mentation comparatively easy, or a very small pellet of wool dipped
in it directly it is mixed, and placed accurately in position, being
readily secured in place by touching the edges with a small smooth-
ended instrument, or it may be taken up when just mixed on a
spoon excavator, and if the convex part be placed next the pulp,
the fluid will readily flow off the instrument whenever wanted.
This takes five to fifteen minutes to set hard. It may then have
oxychloride side linings for gold amalgam or gutta-percha or all
oxychloride or phosphate fillings. It may be mixed with oil of
cloves for pulp capping, which slightly hastens its setting.

PHOSPHATE OF ZINC.

Dr. Flagg says the legitimate uses of zinc phosphate are lining cavities, strengthening frail walls, largely filling large-sized cavities
which are to be filled with gold or amalgam, and for increased ad-
hesion of fillings, where there is weak retentive power in the cavity
itself.

To Mix Zinc Phosphate.—A portion of fluid should be poured on
the slab and more than sufficient powder poured out near it. A
bulk of powder about equal to bulk of fluid should then be mixed
with the fluid gradually but quickly. This should make the mixture of a thick creamy consistency, then a little more powder
should be added and quickly and forcibly made into a mass by
thorough working with spatula. The mass should be of puttylike
consistence, though some varieties are directed to be made stiff.
The mass should then be scraped up on a spatula and taken from it
by the thumb and forefinger. The warmth of the fingers makes
the mass slightly more plastic, and kneading produces a more homogeneous mass. This should be now rolled into an oval and elongated form for filling. It should be introduced by round or flat-
ended pluggers, the pressure to condense it being on the face of
the material. If instruments be touched on an oil pad, it will prevent
the osteo powder sticking to them and so being liable to drag. Superfluous portions should be worked so as to overhang or be cut off
by the margins of the cavity, so that they will break off sharp and
leaving a clear edge to the filling at the cavity walls. This gives a
harder surface to the filling than if cut down with emery disks.

At the cervical edge the cavity should be guarded with gutta-percha, as zinc phosphate wastes at the cervical edge, though not so much as oxychloride. It should he kept dry at least five minutes after introduction into the cavity, if possible, and then varnished. If it cannot have five minutes even, varnish at once, but give varnish time to dry. If it is required to be adjusted to the bite, give half an hour to dry, and then cut down with either a perfectly dry or a very wet sharp instrument. A much better surface is given to the filling if it is burnished when thoroughly hard.

Zinc phosphate is recommended by some as a pulp capper, because it does not irritate the pulp, but a material may not be irritant and yet capable of devitalizing the pulp. Dr. Flagg and Dr. Miller agree in believing zinc oxyphosphate to have bad results on the pulp, even when the pulp is intact, and that in course of time they induce a diseased condition, and eventually death of the pulp, and although this is not absolutely proved, yet the probability is so great that it renders the precaution necessary of putting either a layer of gutta-percha on the floor of the cavity, or oxysulphate, or anything that will serve the same purpose, and only using it alone in very shallow cavities.

In mixing, osteo powder should always be added to liquid, not liquid to powder. A saturated solution of sulphate of zinc used with the powder of the oxychloride cement is recommended for the foundation of fillings and setting crowns, as it becomes very hard.

For temporary teeth some recommend very thin oxyphosphate mixed with a thick solution of gutta-percha. (*Dental Record.*)

THE TEETH DURING PREGNANCY.

THE belief is very common among dentists that the teeth of women suffer during gestation, because of the needs of the growing organism; that in some way the teeth of the mother are robbed of their lime salts, to supply the demands of the fœtus. We cannot but think this a grave error.

For a long time the late Dr. John Allen annually presented a paper before some dental body, devoted to an exposition of the wrong done by the millers in bolting flour. He reasoned after this manner: The teeth of Americans are notoriously bad; those of the savage tribes are as notoriously good. The gluten of grain lies next to the bran. The Americans bolt this out that they may get flour that will make white bread. Savage tribes do not do this.

The bad teeth of our people are due to a lack of this bone-making material in their food. Hence the millers of the country are responsible for our bad teeth.

This was ingenious, but it contained a number of serious errors. Perhaps the most important of these lies in the second predicate, for it is not true that aboriginal or savage people have good teeth. On the contrary, some barbarous tribes have worse teeth than Americans.

Again, while it is true that the gluten of our cereal grains is especially rich in the phosphates, it is not true that fine flour is without sufficient for all the needs of man. The following computation has been made: If rice flour, which contains as little of the phosphates as any other common food, were the sole nutrition of a pregnant woman, and if she consumed barely enough to maintain a healthy existence, she might obtain from that alone double the amount that would be needed for herself and the growing child. It is well known that women always excrete phosphates during gestation. Fine wheat flour contains more of the bone-making elements than does rice flour; hence it cannot be that our bad teeth are due to lack of the proper material in our food.

The corollary to the hypothesis so long held concerning the nutrition of the teeth naturally was that the lack of lime salts in the food must be artificially supplied, and hence the many preparations of calcium that were formerly urged upon the people. The truth is, that under no circumstances can the animal organize the inorganic. That function rests solely with the vegetable kingdom. All of the inorganic elements of the body must be derived from organic sources. Hence it is the wildest kind of vagary to prescribe any inorganic material for nutrient purposes. It cannot be built into the tissues, and must invariably be excreted if taken into the animal system. No inorganic matter was ever yet accepted and built up by any animal organism. Such elements may have their uses in the system, but it must always be as medicines. Their presence may induce structural changes through their medicinal action, but they themselves are never used for trophic purposes. It follows, then, that the giving of any form of the phosphates, in the expectation that it will be used in nutrition, is the result of ignorance of physiological law.

To go back, then, to the cause of the decay of the teeth during pregnancy. It cannot be due to the lack of the proper ingredients in the food, provided the mother has sufficient of that which is wholesome. If the nutrient processes are in a healthy state, they will find plenty of material out of which to build bones and teeth. Besides, if there were a scarcity of the lime salts, why should it manifest itself in the teeth alone? or, being felt, why would not the number of teeth be diminished instead of the quality? Why might it not be that there should be but four toes or fingers? or why should not some bone be deficient in length or size, if the material proved insufficient?

But that the teeth of the mother should be robbed of their lime salts to help out the fœtus seems to us the must absurd of theories. There could be but one way in which the tooth could thus be depleted. There must, in that case, be a solution of the salts and their taking up by a system of absorbents. But there are no such absorbent vessels in the tooth. It is true that under certain circumstances a tooth may be absorbed, but when this is done the whole of the tissue goes; it is taken up into the system by absorbent vessels, whence it is excreted. To secure this result a special system of cells is developed—the osteoclasts—and these do the work. There are no such, or any other, absorbent cells in the hard tissue of the tooth, and hence there cannot be any such absorption.

There is no doubt that the character of the tooth tissue changes with its nutrition or malnutrition, but not through any such process as that sometimes claimed. No, the expectant mother neglects her teeth. She has sufficient upon her mind to make her forget the toothbrush. She goes to bed with her attention fixed on other matters, and her teeth are neglected. She awakes in the morning, perhaps with nausea, and she is in no mood to brush her teeth. Besides, her appetites are apt to be capricious, and she deranges her stomach with improper food, or possibly it sympathizes with the gravid uterus. The secretions are perhaps changed, and these morbid conditions add to the trouble. Her nutrition is interfered with, and the teeth are not properly nourished. All these things produce their natural result in caries of the teeth, and the etiology is traced to her condition, and that is made the primary cause, whereas it is only secondary.

Let the mother keep up the hygienic precautions usual with her under other circumstances; let her clean her teeth often and carefully; let her food be sufficient and wholesome, and her nutritive processes be in good condition, and there is no reason why her teeth should especially decay during pregnancy, or why their nutrition should in any way be interfered with. (Editorial in *Dental Practitioner and Advertiser.*)

Extracts.

PALMAM QUI MERUIT FERAT.

WE are frequently having paragraphs sent us, clipped from small local papers, with some such legend as the following: "Mr. So-and-so has just returned from a visit to the United States of America, where he has received an honorary degree of D.D.S., conferred upon him in recognition of his labors in the dental profession for fifteen or twenty years," as the case may be. It is indeed pitiable if the recipient of such honors (?) are honest in their acceptance of these degrees, and it also shows a most lamentable ignorance of the meaning of degrees in general, and American degrees in particular. When a degree is obtained after a satisfactory test examination following a sufficiently extended curriculum—in the present state of knowledge four or five years is certainly the minimum—it is a glory and an honor to the student who obtains it, the more so when he is examined by strangers, and not by his own teachers. Under such circumstances there is a rational cause both for the conference and the acceptance and use of the letters, marking such academic distinction, but when Mr. Dick, Tom, or Harry, of whom no one has ever heard outside of his native town or village, becomes the proud possessor of a dental degree *honoris causâ* (forsooth!) something seems rotten in the "state of Denmark." A man who patiently labors at an honorable calling for half a century, and who does his duty in the state to which he is called does not receive the decoration of an honorary degree. Why should he? He has only done what some nine-tenths of his fellow-men have done, and if we were all honorary D.D.S.'s, or L.D.S.'s, or F.R.S.'s it would become quite a distinction to be plain Mr. Nobody. Besides, in an unregenerate age like the present, when the air is thick with bogus degrees, or degrees granted after inadequate examination and next no compulsory curriculum, it becomes a doubtful compliment to have an alien degree conferred upon one, especially if one's *Alma Mater* has for some unaccountable reason been so blind to one's merits as to withhold the home-made article! Of course there are honorary degrees granted to men in England, but these complimentary academic honors are bestowed in the sight of all men, and certainly only to such persons as are either graduates of other universities or to those whose names are already engraved on the " sands of time " as of men who are a head and shoulders above their peers. There is certainly no reason why the hard-toiling coster of whom Mr. Chevalier sings should not become bedoctored if all those whose career has been carried out "according to their lights " are to be decorated for their tenacity of purpose and unweariness in well doing, but the thing would be regarded as a *reductio ad absurdum*, and justly. We cannot hope for meretricious honors if we have any proper self-respect, and further we cannot expect to receive from a foreign university or college granting degrees any dental degree that is worth the paper upon which it is engrossed unless we have done some work above that achieved by our fellows, and which has obtained recognition by the profession in England. To accept mere Brummagem degrees is to lower ourselves and to bring the profession to which we belong to a level a good deal lower than the angels. *Palmam qui meruit ferat!* (*The Dental Record*, London.)

THE BRITISH DENTAL ASSOCIATION AND THE DENTAL CONGRESS.

AT the annual meeting of the British Dental Association at Birmingham, April 6–8, the following is reported by the *Dental Record* as part of the proceedings:

Mr. Cunningham asked if the President had received a communication from the executive or committee of the Dental Congress to be held in Chicago.

The President, Mr. W. H. Breward Neale, said he had received no communication whatever.

Mr. MacLeod said he also understood there was some communication on the way. Its object was to explain, or try to explain, that the circular which was sent out in the names of the dentists of America some time since did not emanate from the committee who had the management of the Congress. The present executive denied the paternity of the circular. Therefore seeing that the names of the gentlemen mentioned were those of men whom they knew as being thoroughly respectable practitioners, who had the interests of practitioners at heart, their word might be taken that in some way, not easily explained, a mistake had been made. He would therefore propose that the Association should appoint certain gentlemen, and authorize them to appear as delegates at the Congress, with the promise that the delegates so appointed should try and get further explanation of the circular and take up a firm position as regarding the dental profession in Europe. They ought not certainly to allow the Americans under any circumstances to claim either priority or superiority in the dental profession. He would put his resolution in this form: "That the Association nominate the President, with Messrs. Mummery, Baker, Coffin, and Woodruff, to represent them at the World's Congress."

Mr. Cunningham seconded the resolution.

Mr. Smith-Turner moved as an amendment that the Association should maintain the position toward the question that it did at the last General Meeting. He could not say that he was possessed of that sweet simplicity which would induce him to accept recantation of a circular which had been sent out broadcast.

The President explained that the position taken up at the last meeting was that they had received an invitation which they refused. He had now had a letter put into his hands containing six pages of matter which he could not consider then.

Mr. J. F. Colyer asked that the paragraphs in the circular referred to by Mr. Smith-Turner might be read.

Mr. Smith-Turner read the following extract from the circular: "The history of modern dentistry is covered by a period of less than two generations, and yet it has advanced from the rude operations practiced by the blacksmiths and barbers to one of the most scientific and exact of the specialties of the healing art. Scientific dentistry had its birth in the United States of America. This country has the proud distinction of having organized the first school for the teaching of dental science, and the establishment of the first periodical devoted to the interests of dentistry, whilst very many of the most useful appliances and scientific methods have originated on this side of the Atlantic." He would now beg to alter his amendment, and put it as follows: "That the British Dental Association do not send delegates."

Mr. Holford seconded the amendment.

The President said, after having heard part of the circular read, he doubted if any country would care to send delegates to a body of men holding such views.

The amendment was then put and carried by 33 votes to 3.

[NOTE.—Notwithstanding the above, all liberal and fair-minded English dentists will find a warm welcome to the Congress.—ED.]

MR. QUINBY ON AMERICAN DENTISTS.

REFERRING to Mr. Quinby's strictures on American dentists in his address at Manchester, Dr. Lord says: "It is not to be presumed that America wishes to be, or should be, held responsible for all dentists who call themselves American dentists. Some schools have no doubt graduated men who know next to nothing about dentistry, practically; and if they have thus disgraced themselves and the profession, the responsibility and the odium should be placed where they belong." After discussing and denouncing the practice of securing patents for various kinds of implements, methods, and appliances, he says: "These are, I believe, some of the reasons why the word 'American,' used as a prefix to dentistry, constitutes almost a term of reproach, for on this side of the Atlantic it has become, I am sorry to say, synoymous with the veriest chicanery and humbug; but America has not ceased, and I hope will never cease, to produce dentists who are honorable men, and who will cordially agree with the sentiments of a late letter in the *Times* by a distinguished member of this Association, who says: 'Dentistry, like medicine and surgery, is catholic, and is practiced by honest men for the public good, and therefore all its methods are made public to all members of the profession.' Mr. Quinby's address contains almost a complete record of the many objectionable features connected with the practice of dentistry all over the world by the 'camp followers' and imitators of American dentists. Bogus diplomas, patented instruments and methods, the sale of state and county patent rights, advertisements of peculiar treatment and claims of special skill in the public press, all received merited and scathing denunciation at Mr. Quinby's hands. Remembering the official action of the New York Odontological Society against the patenting of instruments or methods, and because of the exclusion of all reference to the same from our published proceedings, I feel that I but express the sentiments of at least a large majority when I commend Mr. Quinby for all he has said against such unprofessional conduct. I also take pleasure in assuring him that the true American dentist—by which I mean the skillful, conscientious, tooth-saving practitioner—joins him in condemning all such practices, and would inform him that, by both example and precept, and by stringent dental law, the effort is being made, and with fairly satisfactory results, to rid the profession in this country of these objectionable features. It is strange that so good a thinker as Mr. Quinby evidently is should have overlooked the great forces which are at work for the purification of the dental profession in the United States, among which the Dental Protective Association and the excellent dental laws of many of our states naturally come first; and the labors of Dr. H. C. Meriam in his crusade against the patent evil must not be forgotten. Unfortunately, I think, for Mr. Quinby's logic, he has found in his well-arranged list of professional sins committed by quacks and charlatans, masquerading as American dentists, an excuse for a rather wholesale denunciation of American dentistry at large. But we find that in practical life the most 'successful men have the greatest number of imitators; the best artists' pictures are copied; the most pleasing architectural results are followed in the reproduction of cheaper and less meritorious works; and all because of that never failing homage which men of all countries and professions, with native shrewdness, pay to success and skill. Were the standard of practical dentistry as high in any other country as it is in America, quacks and nonprofessionals would now be masquerading as Russian dentists or Swiss dentists, or, begging Mr. Quinby's |pardon, as English dentists." ("Journalistic Memorabilia," in *Dental Record*.)

DEATH FROM NITROUS OXIDE.

THE VERDICT.

Mrs. Margaret Mohan
  vs.   No. 675, February Term, 1893.
Dr. W. S. Yates.

On January 16 and 17, 1893, |the above case was tried in the Common Pleas Court of Allegheny County, Pa., and a verdict rendered for the plaintiff in the sum of $2,500.

The facts in the case, as appeared from the testimony given by the witnesses at the trial, were that Bernard Mohan, the husband of Mrs. Margaret Mohan, the plaintiff, had gone to the office of Dr. Yates to have some teeth extracted, having suffered for some days with a severe toothache. The first day that Mr. Mohan went to Dr. Yates's office he was placed in the chair, and put under the influence of what the Doctor called "vitalized air;" but he struggled so violently while under the influence of this "vitalized air," or nitrous oxide gas, that the Doctor was unable to hold him, and told him to come back the next day and bring several friends along to hold him. On the following day Mr. Mohan came with three friends, and was again placed in the chair and put under the influence of the gas. The three friends of Mohan held him, but being a large, powerful man—a mill worker—it was all they could do to hold him after he was put under the influence of the gas, and the Doctor was unable to get all of the tooth out.

One witness testified that Mahon's face became a purplish or bluish color, similar to a man who has become drowned or partially drowned before he is resuscitated. Mohan, after reviving, suggested that the gas be administered to him again, and the remaining piece of tooth taken out, and a few minutes after the first administration of the nitrous oxide gas (probably five or six minutes) the Doctor administered it a second time, and took out the remaining piece of the tooth; but the patient had ceased to breathe, and in a short time was dead. The witnesses testified that Dr. Yates had only a small quantity of ammonia, which he used, and had no other restoratives present, and that Dr. Yates told them he had no battery, and sent one of them for a battery, also sent his office boy to the nearest drugstore for ammonia or other restoratives. Physicians were sent for, who worked with Mohan for about an hour, applying a battery and other restoratives which they brought with them, but without effect, as Mohan was evidently dead. The testimony of Dr. Yates showed that the mixture which he administered to Mohan was nitrous oxide gas, with a few drops of chloroform added to it.

A large number of experts were called on behalf of the defendant and a few by the plaintiff. The jury rendered a verdict of $2,-500 in favor of the widow. The defendant made a motion for a new trial, which has been argued but not yet disposed of by the court.

The expert testimony sustained the fact that nitrous oxide gas only produces axphyxia and not anæthesia, and that axphyxia is but the first stage of death, the evidence of which is purpleness and blueness of the countenance as in drowning and is always a warning of death. (*Dental and Surgical Microcosm.*)

MANUAL TRAINING.

HAND culture, apart from its value *per se*, is a means toward a more effective brain culture. The dentist who has spent a short time in the practice of his profession, and gained even no more than average manual dexterity, who cannot accept the truthfulness of this statement, must admit that his experience has taught him nothing. It has become a trite saying that "experience is the best teacher," and in dental teaching, within certain limits, this is put into practice. Thus all the reasoning or didactic teaching in the world will not enable a student to successfully fill a tooth or make an artificial denture till he has seen it done; and though it may be urged that the practical part of dentistry is already taught in the clinic room and laboratory, yet, if viewed in the light of their possibilities as educational means, from the standpoint of *the manual training principle*, at the same time bearing in mind, as before stated, that any method of imparting knowledge which involves the exercise of any of the perceptive faculties, is properly included in the manual training idea. it is questionable if these departments have received the attention and elaboration which their real value as educational factors demand. It has been shown that development of the intellect is a natural physiological process presenting two distinct phases—viz., the acquisition of basis facts through the perceptive faculties, and the orderly arrangement of and reasoning on the acquired knowledge. The function of the laboratory therefore naturally precedes that of the lecture room; and for the reason that by far the greatest proportion of the labor of the dental educator is devoted to imparting to the student the knowledge of facts of dentistry, it seems clear that the laboratory and clinic room should be the arenas for his most conscientious and devoted efforts as instructor. (Extract from *Items of Interest*.)

FORMULA FOR INSECT BITES.

ONE of the very best applications for the bites of mosquitoes and fleas, also for other eruptions attended with intense itchings, is: Menthol in alcohol, one part to ten. This is very cooling and immediately effectual. It is also an excellent lotion for application to the forehead and temples in headache, often at once subduing it. (*Weekly Medical Review.*)

LEMONS VS. CHOLERA.

THE imperial health officer of Berlin has issued announcement to the effect that oranges and lemons are both fatal to the cholera bacillus. Placed in contact with the cut surface of the fruit, the bacteria survive but a few hours. They remain active for some time longer on the uninjured rind of the fruit, but even there they die within twenty-four hours. The destructive property as regards the cholera bacteria is supposed to be due to the large amount of acids contained in those fruits. In consequence of this quality, the health officer considers it unnecessary to place any restriction on the transit and sale of these fruits, even if it should be ascertained that they come from places where the cholera is prevalent at the time. Not a single instance was noted in which cholera was disseminated by either oranges or lemons. (*New Orleans Picayune.*)

WOMEN DENTISTS.

It is the intention of the *Dental Tribune* to devote considerable attention to the interests of women dentists. Probably few men are fully cognizant of the interest taken by women in our profession. From a list which is as perfect as it can be made, but which by no means is complete, we find the number of women dentists in the United States to be as follows:

California	6	Montana	3
Colorado	4	Michigan	6
Connecticut	2	Mississippi	1
District of Columbia	3	Missouri	3
Florida	1	New York	10
Georgia	1	New Jersey	3
Illinois	17	Ohio	6
Iowa	4	Pennsylvania	30
Indiana	5	Rhode Island	2
Kansas	13	South Dakota	1
Kentucky	1	Texas	6
Minnesota	3	Utah	1
Massachusetts	2	Wisconsin	5
Maryland	1		

When the new Columbus building and the new Marshall Field building, in Chicago, are completed, and when they are fully occupied, judging from the number of dentists who are going into those new buildings the greatest dental center in the world will be a small district in the heart of Chicago, not more than two blocks square, within which space are located the Masonic Temple, the leading office building in the world, the Marshall Field building, the Columbus Memorial building, Central Hall, Bay State building, Quinlan Block, 65 Randolph Street, and 96 State Street. These buildings will contain not less than 150 dentists. (*Dental Tribune.*)

Correspondence. . . . .

To the Editors of the DENTAL HEADLIGHT.

I AM thankful for an opportunity to state through your journal that I have no sympathy whatever with the move to organize an independent Texas Dental Association for those who wish to advertise with freedom, and not be in any way restricted by dental ethics.

About last February one C. S. Phillips wrote me from Waco, Tex., saying that there would be a meeting of dentists (at a given date) in Waco, Tex., to transact business of vast importance to the active members of our profession. Being conscious of coming under this head, I reread his letter. Supposing it to be a meeting preparatory to attending the International Dental Congress, to be held at Chicago this year, I at once answered the gentleman that I'd try to attend perhaps, and wishing (as did the deaf man when drinking a toast) them all the luck they had expressed for me. Later I was shocked by seeing a circular with my name at the bottom, calling this meeting. I wrote to Dr. Phillips asking him to withdraw the same, and that I was not a first-class Association man anyway, and that the original Association afforded me all the latitude desired by me, and hence I could not work in harmony with those who differed from me in sentiment. He has never yet replied to any of the several letters written him by me.

My sole aim in writing this statement is to set myself right with my many friends and brethren of Alabama and Tennessee, and also to discourage, so far as in my power lies, a similar effort in the future. At home I fear no lasting harm coming from such a source, yet with those who once knew me, and with whom I do not ever meet since the Mississippi rolls between us, I think it but right that I should make this statement. My theory has it that the " Cheap John " advertiser must move occasionally, else his baits do not even entice the "damphules." LINDLEY H. HENLEY.

Marshall, Tex. ⸺

To the Editors of the DENTAL HEADLIGHT.

WILL you kindly say to the readers of your journal in your next issue that the house of the Columbia Dental Club, of Chicago, No. 300 Michigan Avenue, is open wide to the gentlemen of the profession who [visit Chicago this summer, and a cordial invitation is extended to them to make it their headquarters while in the city. You might also say that, if it is so desired, by addressing the Manager of our Bureau of Information, R. C. Brophy, in care of the Club, they can secure such rooming accommodations as they wish.

Very truly,   FRANK H. GARDNER,

Chairman Local Committee of Entertainment W. C. D. C.

CHICAGO, April 17, 1893.

To the Editors of the DENTAL HEADLIGHT.

OWING to a change in the time of meeting of the World's Columbian Dental Congress, it seemed a necessity to make a change in the time of meeting of the American Dental Association, and at the request of the officers of both the American Dental Association and the World's Columbian Dental Congress I communicated with the officers of the former, and the vote was unanimous for changing the time of the meeting of the American Dental Association.

Accordingly we gave notice that the meeting of the American Dental Association will be held in Chicago August 12, instead of August 15.

By order of the Executive Committee.

J. N. CROUSE, *Chairman.*

To the Editors of the DENTAL HEADLIGHT.

THE twenty-third annual meeting of the South Carolina State Dental Association will be held in Columbia Tuesday, August 8, 1893, continuing four days. All members of the profession are cordially invited to be present.

C. S. PATRICK, *President;*
B. RUTLEDGE, *Recording Secretary.*

Editorial. . . . .

Two years ago Dr. L. C. Garland, the venerable Chancellor of Vanderbilt University, tendered his resignation, and has since discharged the duties of his office at the request of the Board, until his successor could be elected and installed.

At a meeting of the Board on the morning of June 20 Prof. James H. Kirkland, Ph.D., who has for several years occupied the chair of Latin Language and Literature, was promoted on the first ballot to fill the vacancy. The names of a number of distinguished educators were considered and their eligibility inquired into. These gentlemen were all eulogized warmly, but on the first ballot Dr. J. H. Kirkland was elected, receiving ten out of fourteen possible votes. One of these fourteen votes was blank. A committee notified him of his election and brought him before the Board.

Dr. Kirkland expressed his surprise at the honor conferred on him by the Board, and his sense of gratitude therefor, but requested time for reflection and consultation with his colleagues of the Faculty of the university before making a decision. It was not until the afternoon session, about 5 o'clock, that he formally signified his acceptance of the Chancellorship.

Dr. Kirkland is still a very young man, having been born at Spartanburg, S. C., in September, 1859. His father was the Rev. W. C. Kirkland, of the South Carolina Conference, and his older brother, the Rev. W. D. Kirkland, now belongs to that body, and is the editor of the *Southern Christian Advocate*. Dr. Kirkland graduated from Wofford College in 1877. In 1885 he received the degree of Ph.D., from Leipsic University, and in the fall of the next year he began his work as Professor of Latin in Vanderbilt University.

The Chancellor elect enjoys immense popularity and is one of the brainy, cultured educators who are leading the march of educational progress. He is a man of great natural ability and of elaborate cultivation. His scholarship is broad, accurate, and profound. He has given to the public within the past ten days an edition of the odes of Horace that will easily rank with the best text of the

Roman poet. Outside of his chosen field he has done a great deal of reading and thinking. As a man he is modest, affable, and attractive. Not a bigot, he is nevertheless a most earnest Christian, and has taken an active part in all sorts of religious work. He is highly esteemed by his colleagues and greatly beloved by the students, with whom his election will be especially popular. The wisdom of his selection will be shown each year, and his career must be a great one.

Dr. L. C. Garland, the venerable ex-Chancellor, we are pleased to note, will still retain his connection with the great institution, whose success is largely due to the executive ability and wisdom of his administration. He will be Professor Emeritus of Physics and Astronomy, the chair he has so long and ably filled.

HONORARY DEGREES.

Extracts found elsewhere in this issue show to what extent the practice of granting honorary degrees has been abroad. The price an American diploma is too cheap. Letters are received by some institutions from the "Golden Gate of the West" to the city of Rome, from Brazil to the British Provinces asking upon what terms diplomas can be had, and even intimating that money is no consideration if the amount of time of attendance can be shortened or altogether avoided. These men, we are persuaded, are not always bad men, nor have their opportunities for education been limited. Some one in their section has a degree obtained from some institution which granted "honorary degrees." This country is full of them, or its equivalent of men who have attended but a fractional part of a course, passed a satisfactory examination (?), and been crowned with a D.D.S.

The National Association of Dental Faculties has pronounced against the practice, and we believe it has fallen into disuse almost altogether. Let it pass forever into oblivion. The question asked by Dr. S. L. Shepard, some years ago, remains unanswered: "Is a man holding an honorary diploma in possession of a degree?" That is something conferred by a President or Chancellor of an institution by pronouncing certain words, and the diploma is the evidence of his acts. Can a woman become a Mrs. Brown or Mrs. Jones merely by the making out of the certificate of marriage if she is not present to hear the priest's words and give her promises and receive those of Mr. Jones or Mr. Brown.

9

We hold that a degree is a sacred honor, the holder of which should be required to promise to honor and keep it, and to uphold the dignity of the profession it unites him to, and to guard and protect the honor of the institution conferring it. A violation of such a pledge should render null and void such a diploma, and the Trustees should have the power by law under the charter establishing such university, college, or seminary to recall its acts and strip the recipient of its favors of his right to use the title previously bestowed.

The journals of Great Britain and even writers upon this subject at home who are making loud complaint of the laxity of American schools, in conferring these honorary degrees, must remember that the colleges are not always to be blamed, for in spite of the utmost caution on their part, they are liable, even when diplomas are conferred in the regular way, to be deceived regarding the worthiness of the individual seeking the degree.

DR H. E. BEACH.

Dr. H. E. Beach, of Clarksville, has been appointed by Governor Turney, to fill one of the two vacancies in the Board of Dental Examiners which occurred on the 25th of April, the term of office of two members expiring under the law at that time each year. Dr. S. B. Cook was reappointed by the Governor, and is President of the Board.

Dr. Beach is too well known to the dental profession of the state as a scholar, Christian, learned dentist, and all round gentleman of the old Virginia school to require any eulogy at our hands.

We desire to say that the appointment came to Dr. Beach unsolicited by him, and is a deserved compliment to one who has labored long in the interest of the profession, and it is to be congratulated. As President of the State Association, for years its Treasurer, and Treasurer of the Southern Association he has been honored by professional brothers who will rejoice that so just and good a man has been advanced to a position in this honorable Board.

"THE MORPHINE HABIT AS A FACTOR IN DENTAL CARIES."

According to our observation, extending over a considerable period and including numerous cases, we are convinced that the victims of the morphine habit are invariably the subjects of a peculiar form of dental caries. The serious impairment of the nutritive processes, the result of the toxic action of the drug, soon manifests itself by the ravages of decay upon the tooth structure. The enamel loses its integrity and the margins are rendered exceedingly brittle. The insertion of fillings is followed by a characteristic leakage, and a more than ordinary wasting of the dentine around them, and their loss. The decay begins first at the gum margin on the labial and buccal surfaces and around fillings in teeth that have already been filled. No kind of filling material withstands the corrosive action, as experiments in the change of filling material were of no avail. In one case, although repeated attempts to fill cavities had met with signal failure, yet, upon the cessation of the use of morphine, the insertion of proper gold fillings was followed by favorable results.

We invite the attention of the profession to the above subject, and will be pleased to hear from them in reference to their experience and observations upon the teeth of the morphine victim.

THE TENNESSEE DENTAL ASSOCIATION.

So well pleased was the Association last wear with the Lookout Mountain and Lookout Inn that it was unanimously agreed that the next meeting should be held there.

Tuesday, July 4, is the day of meeting. Brush up, brothers! The programme is one that will interest you and if carried out must produce much that you can carry home for future use.

The local committee of arrangements is at work and promise all a hearty welcome. Lifted to such a height above the common herd, noble and humanitarian influence will inspire the mind and soul of those who seek rest and the refreshing and invigorating atmosphere of the mountain. Let Middle and West Tennessee turn out in force. East Tennessee promises to do her part. The President, Dr. Cook, has been at work, and he knows how to work, and has secured the promise of several prominent men from adjacent states to be present and writes us that he has the assurance of those down for papers that they will surely be present.

The Board of Dental Examiners with hold its third annual session at the same time and place. Those desiring examination and license must be present Monday, July 3.

EDITORIAL GOSSIP.

Dr. James M. Murphree, Gallatin, Class 1891–92, Department of Dentistry, Vanderbilt, was recently married to Miss Maggie Glenn Siler.

Dr. R. C. Gordon, Mobile, Ala., and Miss Minnie Foster were married on the 8th of June, at McKinley, Ala.

The marriage on the 18th of April of Dr. J. S. Dalton, New Madrid, Mo., Class 1891–92, Department of Dentistry, Vanderbilt, to Miss Ella Boyd, of Jackson, Mo., is announced.

At the bride's mother's, LeRoy, Ill., Dr. W. C. Chapman, of Gibson City, Ill., was united in marriage to Miss Jennie Crumbaugh, Wednesday, June 21.

We extend our warmest congratulations to the above newly made Benedicts, wishing them and their fair brides *bon voyage.*

The many friends of the distinguished Dr. W. H. Morgan will rejoice to learn that he is now convalescing, after a severe illness of six weeks' duration. This intelligence will no doubt be particularly gratifying to his numerous students throughout the states, who look forward with eager interest to the return of college days, when they are wont to sit at his feet and drink of the stream of knowledge that flows from his lips. A. M.

BOOK NOTICE.

Morrison Bros.' Illustrated Dental Catalogue for 1893. Nashville, Tenn.

The enterprising proprietors of the Tennessee Dental Depot, Messrs. Morrison Bros., have just published a large and handsome catalogue, containing an elaborate description of everything needful to the scientific and successful practice |of dentistry. This book is an artistic production, and far excels any publication of the kind ever issued in the Southern territory. A price list of the productions of any one manufacturer does not meet the requirements of the profession, who prefer to patronize a house from which they may purchase goods produced by all the reputable manufacturers. Hence the above firm, with commendable ambition, have, at great labor and expense, compiled this magnificent volume, a copy of which will be presented free of cost to each one of their patrons whose name and address is possessed by the publishers.

IN MEMORIAM.

A PROMINENT figure has passed from the activities of the dental profession by the death of Dr. Fred A. Levy. Whoever has attended a New Jersey Dental Convention for the last twenty-five years could not fail to see the smiling face and persistent activity of the Doctor. He was quite as active and useful in the Central Dental Association as in the National Association of Dental Examiners, where he was Secretary for many years.

DR. WILLIAM F. REHFUSS.

WE are surprised to be called on to announce the death of this promising young man. He was only twenty-six years of age, and yet he had already come so prominently before the dental profession that his name and writings are familiar to us all. His magazine articles were vigorous and learned. But he was chiefly known as a writer on dental jurisprudence. His book on this subject is a standard work. What has associated him more with the readers of the *Items of Interest* is his series of articles on " Oral Diseases, Surgical and Nonsurgical," now passing through the present volume. He had just finished his portion of this series, the last of which will appear in the June number. Then the portion belonging to his associate, Dr. Brinkmann, will be taken up.

Dr. Rehfuss had hardly an hour's notice of the approach of death. It was only "a cold." And who cannot successfully manage " a cold? " But its course was rapid, developing congestion of the lungs and kidneys, and finally involving the whole system, less than a week, and he was ready for the grave. Such is life; to some so brief, and to all not long. Yet to each of us it is the most important, the most momentous, part of all eternity. (*Items of Interest.*)

Associations. . . . .

PERMANENT OFFICERS OF THE WORLD'S COLUMBIAN DENTAL CONGRESS.

PRESIDENT.—L. D. Shepard, Boston, Mass.
VICE PRESIDENTS.—W. W. H. Thackston, Farmville, Va.; A. L. Northrop, New York City; W. H. Morgan, Nashville, Tenn.; W. W. Allport, Chicago, Ill.; W. O. Culp, Davenport, Ia.; C. S. Stockton, Newark, N. J.; Edwin T. Darby, Philadelphia, Pa.; H. J. McKellops, St. Louis, Mo.; J. Taft, Cincinnati, O.; J. H. Hatch, San Francisco, Cal.; J. B. Patrick, Charleston, S. C.; J. C. Storey, Dallas, Tex.
SECRETARY GENERAL.—A. W. Harlan, Chicago, Ill.
ASSISTANT SECRETARIES.—George J. Freidricks, New Orleans, La.; Louis Ottofy, Chicago, Ill.
TREASURER.—John S. Marshall, Chicago, Ill.
EXECUTIVE COMMITTEE.—W. W. Walker, Chairman, 67 West Ninth Street, New York City; A. O. Hunt, Secretary, Iowa City, Ia.; L. D. Carpenter, Atlanta, Ga.; J. Y. Crawford, Nashville, Tenn.; W. J. Barton, Paris, Tex.; J. Taft, Cincinnati, O.; C. S. Stockton, Newark, N. J.; L. D. Shepard, Boston, Mass.; A. O. Hunt, Iowa City, Ia.; H. B. Noble, Washington, D. C.; George W. McElhaney, Columbus, Ga.; J. C. Storey, Dallas, Tex.; M. W. Foster, Baltimore, Md.; A. W. Harlan, Chicago, Ill.; J. S. Marshall, Chicago, Ill.; H. J. McKellops, St. Louis, Mo.
The meeting will be held in Chicago August 17–29, 1893.

OFFICERS OF THE AMERICAN DENTAL ASSOCIATION.

J. D. Patterson, Kansas City, President; J. Y. Crawford, Nashville, Tenn., First Vice President; S. C. G. Watkins, Mont Clair, N. J., Second Vice President; Fred A. Levy, Orange, N. J., Corresponding Secretary; George H. Cushing, Chicago, Recording Secretary; A. H. Fuller, St. Louis, Treasurer; W. W. Walker and S. G. Perry, New York, and D. N. McQuillen, Philadelphia, Members of the Executive Committee. Next meeting to be held in Chicago.

OFFICERS OF THE SOUTHERN DENTAL ASSOCIATION.

B. Holly Smith, Baltimore, President; R. K. Luckie, Holly Springs, Miss., First Vice President; S. B. Cook, Chattanooga, Second Vice President; L. P. Dotterer, Charleston, S. C., Third Vice President; D. R. Stubblefield, Nashville, Tenn., Corresponding Secretary; S. W. Foster, Decatur, Ala., Recording Secretary; H. E. Beach, Clarksville, Tenn., Treasurer; W. R. Clifton, Waco, Tex., and Gordon White, Nashville, Tenn., Members of the Executive Committee. Next meeting to be held in Chicago.

OFFICERS OF THE TENNESSEE DENTAL ASSOCIATION.

S. B. Cook, Chattanooga, President; W. W. Jones, Murfreesboro, First Vice President; W. J. Morrison, Nashville, Second Vice President; P. H. Houston, Lewisburg, Recording Secretary; D. R. Stubblefield, Nashville, Corresponding Secretary; H. E. Beach, Clarksville, Treasurer. Next meeting to be held in Chattanooga first Tuesday in July, 1893.

TENNESSEE STATE BOARD OF EXAMINERS.

S. B. Cook, President, Chattanooga; J. L. Mewborn, Secretary, Memphis; J. Y. Crawford, Nashville; W. T. Arrington, Memphis; A. F. Shotwell, Rogersville; H. E. Beach, Clarksville. Next meeting July, 1893, at Lookout Mountain.

OFFICERS OF THE MISSISSIPPI DENTAL ASSOCIATION.

J. B. Askew, Vicksburg, President; George R. Rembert, Natchez, First Vice President; J. O. Frilick, Meridian, Second Vice President; W. C. Stewart, Fayette, Third Vice President; J. D. Killian, Greenville, Corresponding Secretary; T. C. West, Natchez, Recording Secretary; C. C. Crowder, Kosciusko, Treasurer. Next meeting to be held in Natchez the first Tuesday in May, 1894.

THE GEORGIA STATE DENTAL SOCIETY.

The Georgia State Dental Society met in annual session at the Kimball House, in Atlanta, May 9 to 12. The following officers were elected for the next year: N. A. Williams, Valdosta, President; W. W. Hill, Washington, First Vice President; C. V. Rosser, Atlanta, Second Vice President; S. H. McKee, Americus, Record-

ing Secretary; O. H. McDonald, Griffin, Corresponding Secretary;
H. A. Lowrance, Athens, Treasurer.
EXECUTIVE COMMITTEE.—H. R. Jewett, W. S. Simmons, E. L.
Hanes, D. Hopps, S. B. Barfield.
EXAMINING BOARD.—J. H. Coyle, Thomasville, Chairman; D. D.
Atkinson, Brunswick, Secretary; B. H. Catching, Atlanta; A. G.
Bouton, Savannah; H. H. Johnson, Macon.

THE TENNESSEE DENTAL ASSOCIATION

WILL hold its twenty-sixth annual meeting on Lookout Mountain,
Tuesday, July 4, 1893, continuing four days, holding its sessions at
Lookout Inn. The following programme has been arranged:

TUESDAY, JULY 4, 1893.

Association called to order at 9:30 A.M.
Opened with prayer by Rev. Dr. J. P. McFerrin.
Address of welcome, Col. Garnett Andrews, Mayor of Chattanooga.
Response to address of welcome, Dr. W. T. Arrington, Memphis.
Calling roll of officers.
Reading minutes of last meeting.
Annual address of the President.

AFTERNOON SESSION.

Essays and Discussions.

"Setting Crowns and Bridges," Dr. William Crenshaw, Atlanta, Ga. Dis-
cussion opened by Dr. J. Y. Crawford, Nashville.
"Saving Teeth," Dr. B. D. Brabson, Knoxville. Discussion opened by Dr.
W. T. Arrington, Memphis.
"Cystic Tumor of Alveola Process," Dr. J. T. Crews, Humboldt, Tenn.
Discussion opened by Dr. H. W. Morgan, Nashville.
"Noncohesive *vs.* Cohesive Gold," Dr. I. G. Noel, Nashville. Discussion
opened by Dr. J. U. Lee, Chattanooga.

WEDNESDAY, JULY 5.

MORNING SESSION.

Forenoon Devoted to Clinics.

"Filling Teeth by Mewborn's Reïnforcement Method," J. L. Mewborn,
Memphis.
" Bridge Work," Dr. Walker G. Browne, Atlanta, Ga.
" Filling Teeth with Bonwill Mallet," Dr. Thornton, Chattanooga.
"Filling Teeth with Noncohesive Gold," Dr. C. V. Rosser, Atlanta, Ga.
" Chronic Fistula, Amputation of Root, and Sponge Grafting," Dr. Gordon
White, Nashville.
" Logan Crown," Dr. R. R. Freeman, Nashville.

Essays and Discussions.

"The Dental Law and Its Relation to the Student," Dr. W. W. Jones, Murfreesboro, Tenn. Discussion opened by Dr. P. D. Houston, Lewisburg.

"The Progress of Dentistry," Dr. W. F. Fowler, Greenville. Discussion opened by Dr. W. J. Morrison, Nashville.

"Professional Courtesy," Dr. F. A. Shotwell. Discussion opened by Dr. R. N. Kesterson, Knoxville.

THURSDAY, JULY 6.

MORNING SESSION.

Essays and Discussions.

"Dental Hygiene," Dr. D. R. Stubblefield, Nashville. Discussion opened by Dr. William Crenshaw, Atlanta, Ga.

"Reflex Neurosis," Dr. Dupree, Chattanooga. Discussion opened by Dr. W. H. Morgan, Nashville.

"The Health of the Dentist," Dr. R. B. Lees, Nashville. Discussion opened by Dr. B. S. Byrnes, Memphis.

AFTERNOON SESSION.

Essays and Discussions.

"Dead Teeth, How Best Diagnosed," Dr. H. E. Beach, Clarksville. Discussion opened by Dr. W. F. Fowler, Greenville.

"Crown and Bridge Work," Dr. J. Y. Crawford, Nashville. Discussion opened by Dr. A. F. Claywell, Lebanon.

"Dental Caries a Contagious Disease," Dr. W. J. Morrison, Nashville. Discussion opened by Dr. W. B. Spencer, Jackson.

"Regulating Teeth," Dr. H. W. Morgan, Nashville. Discussion opened by Dr. S. B. Cook, Chattanooga.

FRIDAY, JULY 7.

This session will be devoted to miscellaneous papers and discussion of same, election of officers and place of next meeting, and miscellaneous business.

All dentists throughout the state, as well as those in our sister states, are earnestly invited to attend this meeting, and help us make it the most interesting and profitable meeting ever held by this Association. From present indications this will be the most largely attended meeting held in many years.

Come and be one of us and have a short recreation from your office duties. "INVITATION COMMITTEE."

____

MEETING OF STATE BOARD OF DENTAL EXAMINERS,

THE Tennessee State Board of Dental Examiners will meet Monday, July 3, 1893, at Lookout Inn. Applicants for certificates will please be on hand promptly. In this connection the Board wishes to call attention of the dental profession to the fact that the constitutionality of the present law, regulating the practice of dentistry in Tennessee, has recently been tested before Judge Galloway, of Memphis, who declared it constitutional in all points tested.

It is earnestly requested by the Board that all registered dentists throughout the state take special interest in the work of the Examining Board, and lend all assistance in their power, by promptly reporting all violators of the law. Send names of violators to Secretary J. L. Mewborn, Memphis, Tenn.

Four hundred and forty-three dentists have been registered in the state.   S. B. COOK, D.D.S., *Pres.;*
J. L. MEWBORN, D.D.S., *Sec.;*
J. Y. CRAWFORD, M.D., D.D.S.;
F. A. SHOTWELL, D.D.S.;
II. E. BEACH, D.D.S.

THE WORLD'S COLUMBIAN DENTAL CONGRESS.

To the Dentists of the United States of America, Canada, Mexico, Central America, and South America, Greeting:

THE movement to hold a Dental Congress in Chicago, Ill., August 14–19, 1893, inclusive, received its official *status* from the joint action of the Southern Dental Association at its meeting in July, 1890, held at Atlanta, Ga., and the meeting of the American Dental Association held at Excelsior Springs, Mo., in August, 1890. The undersigned General Executive Committee was appointed by the two Associations to adopt rules and regulations, fix the time for convening the Congress, secure the place for holding the sessions, and make such other preliminary arrangements as it deemed necessary.

The work of appointing committees to promote the success of the Congress is finished, the permanent officers have been chosen, the honorary officers have been appointed in all foreign countries, and the time and place of meeting fixed.

A general invitation has been issued, asking the coöperation of the resident dentists of the civilized world to meet with the dentists of the United States of America at the time and place fixed, for the presentation of papers, both scientific and practical, covering the entire range of theory and technology. It is believed that the newest investigations, discoveries, and methods in physiology, histology, bacteriology, pathology, oral surgery, chemistry, materia medica, therapeutics, orthodontia, operative dentistry, prosthesis, and deontology will be presented to this Congress in a manner not heretofore attempted in any international gathering of a similar character.

It is with pleasure, therefore, that we appeal to the dentists of America to assist in this great undertaking, which promises so much for the future of dentistry and dental surgery, in placing its practical and humanitarian objects before the public at large. This Congress will be an educator of such vast proportions to the practitioners of dentistry, that few can realize the direct benefits which will accrue, not only to those participating, but to those who deny themselves the opportunity to make history for the generations yet to follow.

The transactions, when printed, will be a permanent record of scientific development that may well serve as a starting point in future professional advancement, education, legislation, and prophylaxis.

Nothing will be omitted to provide for the comfort and entertainment of those who lend their presence for the furtherance of the objects of this Congress, and a programme of such literary merit will be presented as shall reflect in the clearest manner the past history and present development of dental science, including also the practical demonstration of every phase of operations known. These demonstrations will be made by those best fitted by native ingenuity, education, and technichal skill in bacteriology, histology, pathology, oral surgery, and other more directly practical subjects, such as orthodontia, prosthesis, electricity, and mechanical operations on the teeth, jaws, and associate parts.

The facilities for meetings and clinical demonstrations are ample to accommodate all who are entitled to admission to the Congress. The Memorial Art Palace is situated near the center of transportion, it is isolated from traffic, and is well lighted and ventilated.

The general headquarters will be located at 300 Michigan Avenue, within ten minutes' walk of the assembly rooms. All communications to the Secretary of the General Executive Committee to be sent to this address after July 15.

The profession in America must now assume the responsibility of making this Congress a success, on the lines laid out by the General Executive Committee. This can only be accomplished by the immediate response of those who contemplate being present in person, or by contribution, financial or otherwise.

The committee urgently requests an immediate decision from those purposing to attend, in order to facilitate the work of the various departments, and reduce to a reasonable certainty the attendance from America.

Contributions of money should be made directly, and at once, to the Chairman of each State Finance Committee, for transmission to the Treasurer, who will issue his receipt for the same. Accompanying this circular are Codified Rules and Regulations of the Congress, and instructions for the guidance of all.

Read this circular carefully, and preserve it for future reference. Adherence of the Congress will address letters of inquiry to the Secretary of the General Executive Committee, in order to receive an official reply. Cordially and fraternally yours,

W. W. WALKER,

Chairman of the General Executive Committee, 67 West Ninth Street, New York City, N. Y.;

A. O. HUNT,

Secretary of the General Executive Committee, Iowa City, Ia.;

L. D. SHEPARD,

President of the Congress, 330 Dartmouth Street, Boston, Mass.;

A. W. HARLAN,

Secretary General of the Congress, 1000 Masonic Temple, Chicago, Ill.;

JOHN S. MARSHALL,

Treasurer, Venetian Building, Chicago, Ill.;

W. J. BARTON,

Paris, Tex.;

L. D. CARPENTER,

Atlanta, Ga.;

J. Y. CRAWFORD,

Nashville, Tenn.;

M. W. FOSTER,

9 Franklin Street, Baltimore, Md.;

H. J. McKELLOPS,

2630 Washington Avenue, St. Louis, Mo.;

G. W. McELHANEY,

Columbus, Ga.;

H. B. NOBLE,

New York Avenue, Washington, D. C.;

JOHN C. STOREY,

Dallas, Tex.;

C. S. STOCKTON,

Newark, N. J.;

J. TAFT,

122 West Seventh Street, Cincinnati, O.;

Members of the General Executive Committee.

FINANCES.

Desiring that every reputable member of the dental profession shall be identified with the Congress,

1. *Resolved*, That a payment of ten dollars ($10) shall entitle one to the Transactions and to membership, if eligible.

2. That a payment of twenty dollars ($20) shall entitle one to the Transactions and to membership as above, and to the Commemorative medal.

3. That a payment of thirty dollars ($30) or upward shall have all the advantages of the twenty-dollar ($20) subscription, and also recognition as a contributor to the financial success of the Congress.

That student presenting a certificate from the Dean or Secretary of a reputable Dental College shall be entitled to Student Membership, and also to a copy of the Transactions on the payment of five dollars ($5).

RULES AND REGULATIONS.

All public announcements for the General Executive Committee shall bear the signatures of both the Chairman and the Secretary,

The admission fee to the World's Columbian Dental Congress shall be fixed at ten dollars, to be collected only from residents of the United States.

All papers to be read before the Congress shall be in the hands of the Committee on Printing Transactions not later than July 1, and shall not exceed forty-five minutes in the time of presentation. Said committee shall have full power to accept or reject any paper, to revise, or suggest a revision by the authors, and to publish or not in the Transactions the whole or parts of papers read, or abridgements thereof.

The official languages of the Congress shall be English, French, Spanish, and German, and the papers shall be printed, in the Transactions, in the languages in which they are read.

After a paper has been accepted, the committee shall prepare a brief synopsis, to be published in the official languages of the Congress.

The Chairman of each committee shall send reports of its progress to the Chairman and Secretary of the General Executive Committee at such frequent intervals as will keep them informed of all the work accomplished.

All circulars issued by any committee must be sent to each member of the General Executive Committee, and they shall be of uniform size—viz., that of the minute forms issued by the Secretary.

The Dental Congress offers a medal for the best popular paper on Dental Hygiene, for public distribution; to be referred to Committee No. 23, to be called the Committee on Prize Essays.

All matters of business presented at the general sessions of the Congress shall be referred to the General Executive Committee, and must receive the indorsement of the committee before they can be entertained by the President of the Congress.

The management of the World's Congress Auxiliary of the Columbian Exposition have offered suitable accommodations in the Memorial Art Palace, on the lake front, in Chicago, for the sessions of the World's Columbian Dental Congress, August 14, 1893.

INVITATION.

The duties of the Committee on Invitation shall be to invite such scientific persons residing in the United States and foreign countries who are not members of the profession, but who by their recognized attainments in special departments of science would add interest to the meeting. They shall also have the authority to invite such dentists of high standing and reputation in foreign countries as may be agreed upon by a majority of the committee, and a card from the Chairman of said committee to the Chairman of the Committee on Registration shall be deemed evidence of the reputability of the holder thereof to entitle him to membership in the Congress, and they shall also furnish the Committee on Membership with a list of the names and residences of those invited.

MEMBERSHIP.

The duties of the Committee on Membership shall be to pass upon all applications for membership which may be referred to it by the Committee on Registration or the Treasurer.

The membership shall consist of legally qualified and reputable dentists (as defined in the Code of Ethics of the American and Southern Dental Associations) residing in the United States, and such other scientific persons as may be invited by the Committee on Invitation; each and every member to be entitled to one copy of the Transactions.

All dentists residing in foreign countries who desire to acquire membership in the Congress will file their application with the Honorary President or Vice Presidents of their respective countries, who are empowered to pass upon their eligibility.

When applications are satisfactory to the Honorary President or Vice Presidents, or a majority of them, in said country, the names so agreed upon shall be transmitted by July 15, 1893, to the Chairman of the Committee on Registration, who will proceed to issue a membership card without further reference.

THE WORLD'S COLUMBIAN DENTAL CONGRESS.

SPECIAL NOTICE.

To the Officers of Dental Societies in the United States and Foreign Countries.

Gentlemen: The Committees on Membership and Registration of the World's Columbian Dental Congress will be saved much trouble and the applicants for membership much vexation if the members of dental societies in good standing are furnished with credentials or certificates of membership, so that they may be presented at the desk where intending members apply for their membership cards.

Advanced membership cards will be furnished on application to the Secretary of the General Executive Committee, or the Secretary-General of the Congress, when the membership fee ($10) accompanies the application. A. O. HUNT,

Secretary of the General Executive Committee, Iowa City, Ia.;

A. W. HARLAN,

Secretary General of the Congress, 1000 Masonic Temple, Chicago, Ill.

OFFICERS OF THE SECTIONS.

SCIENCE.—DEPARTMENT "A."

Section 1. Anatomy and Histology.—Chairman, R. R. Andrews, Cambridge, Mass.; Vice Chairman, E. P. Beadles, Danville, Va.; Secretary, F. T. Breene, Iowa City, Ia.

Section 2. Etiology, Pathology, and Bacteriology.—Chairman, G. V. Black, Jacksonville, Ill.; Vice Chairman, George S. Allan, New York, N. Y.; Secretary, E. S. Chisholm, Tuscaloosa, Ala.

Section 3. Chemistry and Metallurgy.—Chairman, D. R. Stubblefield, Nashville, Tenn.; Vice Chairman, J. S. Cassidy, Covington, Ky.; Secretary, E. V. McLeod, New Bedford, Mass.

Section 4. Therapeutics and Materia Medica.—Chairman, F. J. S. Gorgas, Baltimore, Md.; Vice Chairman, N. S. Hoff, Ann Arbor, Mich.; Secretary, George E. Hunt, Indianapolis, Md.

APPLIED SCIENCE.—DEPARTMENT "B."

Section 5. Dental and Oral Surgery.—Chairman, T. W. Brophy, Chicago, Ill.; Vice Chairman, M. H. Cryer, Philadelphia, Pa.; Secretary, J. F. Griffiths, Salisbury, N. C.

Section 6. Operative Dentistry.—Chairman, William Jarvie, Brooklyn, N. Y.; Vice Chairman, Daniel N. McQuillen, Philadelphia, Pa.; Secretary, Henry W. Morgan, Nashville, Tenn.

Section 7. Prosthesis, Orthodontia.— Chairman, C. L. Goddard,

San Francisco, Cal.; Vice Chairman. T. S. Hacker, Indianapolis, Ind.; Secretary, E. II. Angle, Minneapolis, Minn.

Section 8. *Education, 'Legislation, Literature.*—Chairman, J. J. R. Patrick, Belleville, Ill.; Vice Chairman. II. L. McKelloy s, San Francisco, Cal.; Secretary, W. H. Whitslar, Cleveland, Ohio.

WORLD'S COLUMBIAN DENTAL CONGRESS.

NEW ORDER OF BUSINESS.

Monday, August 14.—10 A.M.: Meeting of the General Executive Committee. 11 A.M.: Opening of the Congress. Reading of the resolutions creating the Congress by the Secretary-General. Address of welcome by John Temple Graves, of Georgia. Responses. Responses from foreign countries. Address of the President. Adjournment. 2:30 P.M.: Papers to be read in the Sections. 5 P.M.: Adjournment.

Tuesday, August 15.—9 A.M.: Clinics. 10 A.M.: Meeting of the General Executive Committee. 12 M.: Address before the whole Congress. 1 P.M.: Adjournment. 2:30 P.M.: Papers to be read in the Sections. 5 P.M.: Adjournment. 8 P.M.: Bacteriological exhibit.

Wednesday, August 16.—9 A.M.: Clinics. 10 A.M.: Meeting of the General Executive Committee. 12 M.: Address before the whole Congress. 1 P.M.': Adjournment. 2:30 P.M.: Garden party. 2:30 P.M.: Garden party. 2:30 P.M.: Garden party. 8 P.M.: Biology. Lantern exhibit. 8 P.M.: Conversazione. 8 P.M.: Conversazione.

Thursday, August 17.—9 A.M.: Clinics. 10 A.M.: Meeting of the General Executive Committee. 12 M.: General address before the whole Congress. 1 P.M.: Adjournment. 2:30 P.M.: Papers before the Sections. 8 P.M.: Public address under direction of World's Congress Auxiliary.

Friday, August 18.—9 A.M.: Clinics at hospitals and at the Art Institute. 10 A.M.: Meeting of the General Executive Committee. 12 M.: General address before the whole Congress. 1 P.M.: Adjournment. 2:30 P.M.: Papers before the Sections. 8 P.M.: Dinner to the whole Congress. (Subscriptions by members from the United States only.)

Saturday, August 19.—10 A.M.: Visit in a body to the Medical and Dental Exhibits at the World's Fair Grounds. 12 M.: Closing addresses to the Congress. Luncheon by the members in the restaurant. (Name to be supplied.)

T·H·E

DENTAL HEADLIGHT.

VOL. 14.   NASHVILLE, TENN., OCTOBER, 1893.   NO. 4.

Original Communications. · · · ·

THE DUALITY OF PROFESSIONAL LIFE.

BY J. A. ROBINSON, D.D.S., JACKSON, MICH.

IT is generally agreed that experience is the basis of all that we know; and progress is simply climbing the ladder of experience, round after round, until we reach the top. "All life is dual," says Mr. Emerson.

The duality of professions are theory and practice, and the duality between members of the professions are fellowship and justice. The truly professional man is seldom wealthy. All his discoveries and improvements are only steps by which his profession can be raised to greater excellence and usefulness—his highest life is only a preparation for a life that is higher. The patience, fidelity, and courage of yesterday makes the road smooth he travels to-day, and the conscientious work of to-day has a bearing and influence on the work of to-morrow. Good dentistry lengthens and sweetens life. It touches the border land of our spiritual natures, it refines our sensibilities, and elevates taste, like looking at beautiful pictures or cultivating beautiful flowers; for it beautifies and makes lovely the human face divine. In this process of beautifying, by the art of dentistry, the person is in the way of truth, for he is practicing truth, and is imbued with truth, and hates fallacy, until finally his whole character is elevated above duplicity and wrong. In the duality of life one part is employed in the things necessary to sustain life, and the other is intellectual or what he owes to society and the

10

world, and that is the reason why professions are ranked higher than some other callings in society.

The dentist has more and better opportunities for education and culture, and dwells more in the ideal than the actual. All civilization, as a whole or in parts, must be judged by its *uses*. Let us look at some of the uses performed by good dentistry. Does dentistry lengthen life? The reports of life insurance companies show that within the past forty years the average length of human life has increased from thirty-two years to more than forty years, and this notwithstanding the great loss of life caused by our late civil war. Of course diminished hardships, a better understanding of the laws of health, and shorter hours of labor have all helped; but modern dental science has exerted a greater influence than all other causes in reducing mortality, which has exemplified itself more in the United States, where dentistry is more thoroughly practiced, than in any other country in the world. Dentistry is an art and a profession. It is based on a thorough information and understanding of the laws of life and health. It is also mechanism of the finer sort, whereby oral deformities and facial hideousness can be obliterated or concealed, that will improve society and add untold blessings and comfort to individuals, society, and the world.

It has been charged by some against our profession that the prices for dental services are extortionate; but there is no work done by mortals in any department of life that is so serviceable and permanent as good dentistry, or will give so much comfort and satisfaction as good dental operations. It is a labor of love, and love is the fulfilling of the law. Society improves first by blind obedience to good laws that are directed by the monitor placed in the human heart. The ideal and the actual are the dual conditions of human life. We are so organized that the good, the true, and the beautiful commend themselves to our understandings through our affections or the love of right without effort or exertion, and every point gained is a stepping-stone to greater excellence in individual life and in society. We follow wholesome laws by the ideal developed within us, the altruism that teaches us to abandon egotism or the love of self for the good of society as a whole, and that is the law that raises us above barbarism, which has made our civilization, our profession, and our Columbian Exposition possible. The culminating point in our profession is the introduction of *painless dentistry*, so the fear and dread of dental operations are taken away from fill-

ing teeth when persons would have the work done but for the dread
of the pain during the operation.

In using new remedies in medicine or dentistry great care should
be exercised not to injure the teeth and do more hurt than good,
as in all surgical operations in using the knife we should know when
to stop. There are a number of remedies in which acids have been
combined to prevent pain that insiduously destroy dentine or bone
structure that the patients know nothing about until it is too late
and the teeth are lost. It is stated on good authority that quite a
number of such cases have occurred where the fillings had to be removed in less than a year or the teeth would have been sacrificed,
and the cases were almost irreparable. It is like the morphine
habit: the last end is worse than the first. Dentists should use
remedies intelligently and know what they are about. It requires
the highest skill and art to preserve the teeth and hermetically
seal the cavities and prevent their decay when the dentine and enamel are frail, and it is a crime to use anything that will soften and
weaken the dentine and enamel and hasten their destruction.
Among physicians society meetings are held to talk about remedies
and relate their experience in practice, and they refuse to use se-
cret remedies until the formulas are published, and in this way they
are bound together in friendly intercourse. If a man cares for
nothing in his profession but himself and the money he can make
out of it, and is bound up in egotism, and uses society simply for the
worldly gain that comes from it, he is of very little good in the
world.

Since application of heat has become known in the use of
remedies for obtunding sensitive dentine all remedies are more effective than before. Heat is life; it is a stimulant. It drives the
obtunder into the tubuli and coagulates the serum and cuts off all
communication from the nerve fibers and deadens the sensibility
while preparing the cavity for filling.

Another advance in dental science is *hypnotism*, which I have
practiced a long time in removing the dental pulp from nerve canals. When an individual is able to concentrate his mind upon a
single object and exclude all others he is master of the art called
hypnotism. Carlyle said matter exists only spiritually and to represent some idea or body it forth. It is here that the spiritual becomes master and can overcome and destroy the physical.

I learn from the *Dental Tribune* that our English brethren have refused to join us and take part in our Columbian Exposition. This is to be regretted. We think they have made a mistake—as much of a mistake as the person does who refuses to attend to his teeth and have them put in good condition by a competent dentist when they are out of repair. It corresponds to tooth-extraction rather than applying to the dentist to have them made serviceable and saved. Of course we are somewhat at fault in the multiplication of dental colleges in almost every town and granting diplomas to many persons who are unfit to do honor to the profession. It is exceedingly doubtful whether it is a healthy state of society when a single state is supporting a dozen dental colleges. and with teachers oftentimes who have not much to recommend them but a little money and their diplomas. Not every one who is professor is fit to occupy the professor's chair. The money-making desire and love of ease is running riot in America, and the excessive love of money gives very little moral tone. It is the egotism of the age in which we live, where everything bends to the love of money and show. It is the aristocracy of professions. The mission of the dentist is to make perfect the physical man. With all our beauty of form and expression, with all the highest ideals of what we ought to be, with all the education and culture of our present high civilization, the most beautiful and expressive portion of our sublime organism is often a charnel house of corruption from neglected growth and development of this priceless ornament of our physical life. All hereditary wrong contains within itself the germs of self-destruction. The ideal is a God-given faculty that comes down from above to give us hints of what we ought to become, and so to purify and make clean the organs of mastication and speech till they shall harmonize the voice that gives utterance to the sublime thoughts of our present exalted civilization, and make us worthy children of the father whose image we bear.

PRESIDENT'S ADDRESS.

BY S. B. COOK, D.D., CHATTANOOGA, TENN.

Gentlemen of the Tennessee State Dental Association: This occasion brings to my mind a feeling of both joy and sadness. I rejoice that an all-wise Creator has guarded and watched over us for the past twelve months and to-day brings so many of us face to face to feast once more upon the pleasures of these annual occasions. On the other hand a chord of sadness is touched when we recall the fact that we have added another year to our past and taken one from our future lives. While we should all rejoice in the full realization of the blessings we have enjoyed in a thousand ways during the past twelve months, and especially should we be grateful for the present privilege of assembling together as an Association, we should not be unmindful of the fact that there are sad hearts among us. Death has visited our ranks and taken from us those whom we have delighted to honor.

Dr. James Ross, first Vice President of the organization, and afterward President of this Association, has been called from "labor to refreshment." One other from our number, although not a member of this Association, but one of our clinicians for this meeting, has passed away. I refer to Dr. R. E. Thornton, whom many of you knew.

In presenting this, my annual address, I have chosen no particular subject—have taken this course for two reasons. First, in view of the fact that we have before us an elaborate programme, embracing many subjects pertaining to dentistry, it becomes rather a difficult task to select a suitable subject upon which to write for your entertainment. Secondly, it is not my desire to occupy your valuable time with any hackneyed crankism of my own, but rather to recommend a few points to the Association which I deem proper to be mentioned at this, the very threshold of our deliberations.

First, it has been suggested to me several times during my term of office, and with good and valued reasons for the same, that the time and place of holding the annual meetings of the Association be legislated upon. Therefore I recommend that in your deliberations you appoint a suitable committee to consider the same, May being the time, and the capitol of our state the place, for holding such meetings.

Secondly, I would, from my personal experience, recommend that the appointment of the Executive Committee for the ensuing year be left in the hands of the incoming President of the Association. By this suggestion I do not wish to find fault with our present Executive Committee; but if left in the hands of the President, he could, at the proper time, call to his assistance aid that would be very much more efficient in the arrangement of the programme and the work necessary thereto than a committee scattered throughout the state. I would further recommend that the Association at once appoint a Committee on Publications, whose duty it shall be to take oversight of all proceedings intended to be inserted in our daily papers. It is my observation that more dissatisfaction and discord has grown out of the publication of the proceedings of our meetings than any other one thing. Our reporters are always very polite and willing to report our meetings as fully as possible, but in the nature of our profession it is very difficult for them, without the aid of some one who is experienced in the phrases and technicalities of our calling to formulate a report that will in every way accord or sound in perfect harmony with the highly attuned ethical ear of the highly educated, nonadvertising D.D.S. One will object to the head lines; another will say that the whole thing is an advertising scheme, that he wrote it himself, got his best friend to do it, or paid the reporter; and thus it is we are all more or less disgusted with the entire business. I insist that if you appoint a committee for the work, simply to indicate to the reporter the matter suitable to be published, we will be very much gratified at the results, and the reading public greatly interested and edified thereby.

I would next call attention to our dental law. On this occasion we properly celebrate its second anniversary. Just now allow me, in behalf of the entire Board of Examiners, to extend its grateful thanks for the faithful manner in which many of you have rallied to its support in trying to enforce it. It is gratifying to know that a large majority of our profession cheerfully complied with the requirements without giving the Board the least trouble, and have done all in their power to induce others to do likewise. Some were at first a little obstinate, insisting that the Board wished to work a hardship upon them, but soon yielded to persuasion and complied with its requirements, as all high-toned, law-abiding professional gentlemen will do. Then there were those upon whom the full force of the law had to be applied. Still there remains another class who stand in open violation of the law and defy its enforcement. It is to this class the Board earnestly solicits your assistance and interest.

Among the number who have not registered their names there are several who hold diplomas from reputable colleges, and boast of great professional superiority, claiming this as the best reason why they should be more highly favored than you who have shown your allegiance by obeying the law of your own state. While all must admit that great good has been accomplished, still the law is not just what it should be, as we all know; as some few changes or amendments are necessary, and the time is near at hand when such amendments will be asked for, and the Board earnestly requests that you come nobly to the rescue, as we feel sure you will in an unbroken rank. In this connection I must call your attention simply to a point of business of like interest to all. To secure the passage of this law none were so zealous as Dr. J. Y. Crawford, of Nashville, spending as he did his time, his energy, and money to the amount of $440.

Dr. Crawford has received from private sources $90, and from the Secretary of this Association $150, leaving a balance of $200 which I insist that we, as members of the dental profession, are honor bound to pay. I feel that you will take steps at this meeting to raise the amount and pay it off before adjournment. It is but just to Dr. Crawford that I should state in this connection that he especially requested me not to mention this matter to the Association; but feeling as I do, that we are honor bound to reimburse him, I insist that you take action at once on the subject.

I would urge the importance of more attention being paid to clinical work at these annual meetings. Many come wholly to see operations of a difficult character performed by hands more skilled than their own, and go away disappointed because of lack of interest in this particular line.

Everything possible has been done to secure an interesting programme; and if carried out, this will necessarily be a meeting of profit and interest to all. The Association will, at the proper time, fix the hours of meeting and adjournment; this being done, in view of the extra amount of matter on the programme, we will be compelled to utilize all the time. Therefore I shall expect each member to be in his place promptly, as I shall open the meetings on schedule time.

In conclusion allow me to thank you all for the honor you have bestowed upon me by placing me in the highest position within your gift.

DENTAL CARIES AS A CONTAGIOUS DISEASE.*

BY W. J. MORRISON, D.D.S., NASHVILLE, TENN.

I FEEL that before you get through the discussion of the paper which I am about to read, if it prove worthy of discussion, I will realize that, with all my convictions upon the subject, there will be much to learn from those gentlemen who would not know a white elephant from a microbe if they should meet one while walking over the mountain, and before they get through with me

> I will not feel that I live in a land
> Where coffee grew on white oak trees,
> The rivers flowed with brandy,
> The streets all covered with white loaf sugar,
> And the girls as sweet as candy.

But when such papers are worthy of discussion they usually bring out something new and original, and that is what we want; therefore I am willing that this paper be a martyr to the cause to obtain this end.

There are no diseases the ravages of which are so far-reaching and wide-spread, the effects of which are so deleterious upon the organism and so antagonistic to human happiness, as diseases of the oral cavity; and yet those guardians of the public health ignore this form of disease, and the general public continue in ignorance of the power of a subtle enemy that holds high carnival under the very thrones of their boasted intelligence and reason, and, in hundreds of thousands of cases, cause that very reason to be dethroned.

Could the public, by some great power, at once be made to realize the awful ravages and dreadful effects of that one disease in the oral cavity, Dental Caries; could they be made to realize the long line of misery that they endure, and the predisposition to the same defects that they will stamp their posterity with, I think that the general public, in their indignation, would want to take the law into their own hands, and brand as frauds the M.D.'s who will make such a to do over the prospect of a disease that is four thousand miles away entering our country, and expend millions upon millions of the people's money for quarantine and quarantine officers, and when the disease does enter, after all precautions have been taken, they will spend millions in local quarantine and armies of officers. We will have long articles in the papers from eminent practitioners

---

* Read at meeting of Tennessee Dental Association.

upon the treatment of those attacked with the dreaded malady. States, cities, and towns will pass any laws that may be recommended by their respective Boards of Health; officers will go from house to house to see that these laws are enforced, and, if necessary, the army will be called out to assist in their enforcement.

In the case of diphtheria, scarlet fever, and smallpox, we see schools suspended, the homes of the patients posted, quarantined, and every precaution taken to prevent the spread of these diseases; and yet the ravages of all these, the suffering endured, and the reduction of the average length of life, would not equal those due to the diseases and the disturbances of the oral cavity.

I will not touch upon the troubles in this respect that, directly and indirectly, prove fatal to the thousands of children undergoing the processes of first dentition, but will only touch upon that disease with which seventy-five per cent. of the people are afflicted: dental caries. When we look into the oral cavity of some of these people it reminds one of the opening of the door of an old, damp, and moldy vault containing decaying skeletons, or gazing upon a leper in the last stages of disease; and these very persons would die of horror if they were compelled, as a punishment, to hold in their mouths for ten days an old, decaying bone half as large as one of these decaying teeth, and take into their systems the putrid matter given off by it; and yet they will go year after year with from one to thirty-two rotting bones in their mouth, and by the use of these decaying bones grind up their food, mixing with every particle of it foul and putrid matter to be a poison to their systems, thereby breaking down their health insidiously, slowly, but surely.

At one time the inflammatory theory was in vogue as explaining the nature of this disease; but after a while some of these thick-headed scientists found that, by cleaning out the decay and filling the tooth with some lasting substance, it was very effective, and the more skill displayed in inserting the filling, the better were the results. Then they commenced to conclude that if the decay of the tooth was the result of vital forces resident within its substance these remedies would tend to increase the mischief they were designed to cure.

In 1835 Robertson, of Birmingham, England, advanced the theory that caries resulted from chemical disintegration of the tooth substance, and was bold enough to deny the agency of inflammation. The destruction, he contended, was accomplished by the action of an acid which was generated by decomposition of alimentary particles or fluids of the mouth suffered to lodge about the teeth.

In 1838 Regnard, of Paris, defined caries as destruction of the teeth by decomposition, and his arguments were, in effect, identical with those of Robertson.

The subject of fermentation has been one of the most difficult with which the intelligence of man has had to grapple, and was evidently not understood by those who conceived the fermentation hypothesis for the origin of caries. Now, after the work of so many years has been added to the effort to explain the nature of fermentation, and when the labors of such men as Lister, Koch, Pasteur, and Miller have made us acquainted with the agency of microörganisms in the processes of fermentation and putrefaction, this seems to be regarded as another of the new theories which have sprung to the front demanding a hearing; but, upon careful consideration, you will see that this is but a further explanation of Robertson's and Regnard's theory.

I might quote to you page after page from Tomes, but he only brings forward arguments that had a powerful effect in drawing the thought of the profession away from the inflammation hypothesis, as an explanation of the active cause of caries of the teeth.

There are many authors such as Magitot, Weld, Slater, Coleman, and Taft, who have written well on this subject, and from whom I might quote; but as only slight shades of difference in the views presented would be obtained, we will at once proceed to the consideration of the agency of microörganisms in caries, and we will look over the investigations of such men as Koch, Miller, Miles, and Pasteur.

In 1861 Miles and Underwood, of London, by the use of Koch's improved staining methods, fully verified the findings of Leber and Rottenstein as to the presence of microörganisms in the tubules of carious dentine, and they announced the conclusion that the decalcification of the teeth in decay is accomplished by an acid secreted by the organisms. The brilliant experiments of Pasteur conclusively prove that none of the fermentations or putrefactions could progress without the presence of organic germs, and that each one of these is dependent on the presence of a *special form of organism peculiar to it and to none other.*

Dr. Miller began his observations with a series of experiments on saliva, with a view to ascertaining whether, organic germs being excluded, it contains anything capable of setting in motion such a process of change as would produce an acid at isolated spots where it or food might be detained about the teeth. In this search, the ordinary phenomena of the conversion of starch by ptyalin was observed, but with this conversion the process ended. Sugar was formed, but no acid. This agrees perfectly with all that was known of the fermentative powers of this fluid. From many sources we have learned that the further decomposition of food particles lodging about the teeth must be in accord with decompositions in general: *it must be accomplished by the life processes of an organic ferment.* With this in view, culture mediums were infected by fragments of softened dentine taken from deeper portions of the carious mass, with precaution to prevent the ingress of any germs from other sources. These were kept in an incubating apparatus at the temperature of the blood. The usual controls were placed with them. Fermentation took place promptly in the infected tubes. This occurred with sufficient uniformity to demonstrate conclusively that the ferment was derived from the carious dentine. The fermentation was constantly accompanied by acidification of the culture medium, as shown by the use of litmus paper. Other mediums were then infected by transferring to them a minute portion of one of those that had undergone the fermentative process. These new cultures promptly underwent the same process. This was continued for a sufficient number of generations to show conclusively that there was present an *organic ferment* capable of the *continuous propagation of its kind.* This much being determined, the question of the capability of the acid generated to decompose the elements of the tooth without other concentration than that attained in the culture was tried. For this purpose, fragments of fresh and sound dentine were introduced into the cultures. These were promptly softened by the solution of the lime salts. Thus was found a ferment within the carious dentine, showing itself capable of *continuous propagation in a certain line;* hence a living ferment. In connection with these cultures there constantly appeared a microörganism in appearance the same as that constantly present within the carious dentine.

I know that those men who have never been anything to scientific progress but a clog, and who claim to be intensely practical, will say that all this is too finespun. Suppose we come nearer home for proof to bear out all that has been said. We know that our mothers were always extremely careful about scalding, sunning, and rerinsing the milk pails to prevent the sweet milk from becoming sour rapidly. They knew that milk, under the same surroundings, remained unchanged for a longer time in crockery and tin vessels than in wooden ones; and soon the discovery was made that those who churned and handled the clabber must not be allowed to come near the sweet milk, or the milk room, until after a complete change of garments. We know now that this was done to exclude, as far as possible, the microörganisms productive of lactic fermentation. We have all heard the farmer complain that his cider would not turn to vinegar, but would rot, and that he had been mighty careful to use nice, clean barrels; while his neighbor, who used the same old barrels every year, never had any trouble about getting vinegar. In the former case we had putrefactive fermentation, while the old barrels contained the microörganisms for the production of acetic fermentation. I might continue upon this line of home observations, as in the making of wine, the canning of fruit, and the continuous propagation of the ferment in the making of leaven breads. I think I have given you overwhelming proof that dental caries can be, and is, produced by a microörganism, and is therefore contagious, and, had I the time, I could show that it is more so than many of the diseases that are guarded against. If the public could be made to understand this, we would not have our schools, colleges, and asylums of all kinds crowded with human beings with mouths containing enough of these microörganisms to infect all who breathe the contaminated air, and which is as disastrous to the teeth of those who are predisposed to the disease as the bacillus of tuberculosis is to the lungs of those who are in a condition to receive it, and this is also an explanation of the fact that, owing to the crowded manner in which we live in the higher state of civilization, these diseases are so rapidly on the increase.

Take one of our modern public schools, within whose study hall four hundred children, from all the walks and conditions of life, are breathing the same overheated air, rendered fetid, and laden with the disease-producing microörganisms escaping from, at a low average, two thousand decaying teeth which would, in weight, equal about thirteen pounds of decaying bone. Is not the question answered why teeth are destroyed so early in life, and that school teachers are more prone to this disease than others?

Now suppose we should be conquered by some barbarous nation, and they would collect all the children in the different localities into a room called a school, and keep the temperature of this room from eighty to one hundred degrees, and place within a cylinder from five to fifty pounds of decaying bones, and with a powerful bellows force a current of warm, damp air through this mass into the room where the children were congregated. What would be the feeling of the community? Why every man and every mother would be willing to die in the attempt to save the children from such a fate, and the civilized world would come to their rescue, when they would do so for no other cause.

Gentlemen, this state of affairs exists to-day all over this country, only under a more subtle form, and the necessity for the government recognizing that fact and the condition of affairs is as imperative as in other contagious diseases of a like character; and if we do our duty, these facts will be laid before the people, and men from our ranks who are suitable for the position will be placed upon our Boards of Health, who will see that laws are enacted tending to check the disastrous march of this disease. It is for you to accomplish these ends, and maintain that standard hurled forth by the Hercules of the profession, who, upon being introduced at a banquet in New York as the "Gentleman from the South," said that it was with pride that he pleaded guilty to the charge; that America led the world in dentistry, and the South led America.

SETTING CROWNS AND BRIDGES.*

BY WILLIAM CRENSHAW, D.D.S., ATLANTA, GA.

In attempting to write you a paper on the subject above given it will not be my purpose to offer the setting of new crowns or new bridges, and perhaps I shall offer nothing new in the matter or manner of setting either the one or the other, but shall criticize some of the methods of setting crowns and bridges, and offer that which seems to me to be best and that which has proven most satisfactory in my own experience.

That method for setting porcelain face crowns which seems to me to be most objectionable and unsightly is that which employs the gold ferule or band placed around the neck of the root disclosing the band to view. On the score of unsightliness alone I would

---

*Read at Tennessee Dental Association July 4, 1893.

condemn it.   But an objection of greater importance, and one which
condemns it once and for all with me, is the creation of a charnel
house under the band between it and the root, a haven for bacteria,
microbes, cheese, swill, and sweetness.   I cannot, Mr. President,
stand any method, or any principle, or construction of a thing the
smell of which I cannot stand.   As practicians of dentistry we have
all of us met our Waterloos in the shape of breaths—those strong
enough to go through you at ten feet, and that are nearly stiff and
thick enough to be sawn off, and which are the result usually of
thirty-five to forty years' collection of salivary calculus on the teeth.
These odors, however, command a premium over those that emanate from under the average band used as a cap over the butt of a
root.   To take the position, however, that this operation of band
fitting cannot be performed so as to prevent the dreadful state of
affairs hinted at I do not.   But with the band well fitted, if ever it
is, such advocates of this method are apt to depend too much on the
security and the strength as they regard it of this band, usually of
pure gold, and I believe never of baser plate than 22K, which, when
taking into account the thickness used, is strained out of proportion
to its strength, and is drawn or stretched out of shape, and which,
from the peculiar construction of the band with the tooth mounted,
makes a closed cap, and one cannot tell exactly when the crown
has tilted or left its original position.   The length of bite of this
band (by which I mean that length of band lapping on the root),
in proportion to the leverage of the crown from that part (the
band) to the cutting edge, is too short, and is out of proportion to
the leverage it must meet.

It is therefore evident that this method, if it be employed as a
means of strength, is a snare and a deception.   Those advocates
who set about burnishing the band up to the root to secure a close
adaptation immediately after setting the crown are in every such
case adding insult to injury, and really make matters worse.   In
the first place, this burnishing operation cracks up the cement under
the band, and thus in this the setting is imperfect from the beginning; and in the second place, it stretches the band and so makes
it larger, and a looser fit is the result.

My observation is that the fitting of a band to the depth it
should go to secure any strength endangers, in a large majority of
cases the normal articulation and adaptation of the root in its
socket.   The peridental membrane, the organ of primest importance,

is encroached upon, a degree of inflammation superinduced which seals the fate of the particular root sooner or later.

The lack of perfect fitting of the porcelain to the gold cap—that blackened line of juncture between the porcelain and the gold—is distressingly objectionable, and is a point that I have never seen decently overcome.

In contradistinction to this method of setting a crown I would offer the following as an operation more easily performed, and one which, when done well (and one which I believe the large majority of operators do perform well), is locally known with us at my home as the "Morrison crown;" not "Joe Morrison," Mr. President, but permit me to say that it could not be called after a better man, but for D. W. N. Morrison, of St. Louis, who first made and set this as a clinic operation for us at Atlanta seventeen years ago.

The root is taken off a line or so beneath the gum, the inner portion of dentine removed so as to make room for the pin. The opening in the butt of the root is to be funneled out abruptly—a short taper—so that when the crown is ready for setting it may be moved in various directions from the position of perfect fitting; by which I mean that there may properly be a goodly open space around the pin where it enters the root. In this space cement is more valuable than the tooth substance. With cement thus well distributed around the pin the strain is removed from the neck of the pin, the point where the pin passes into the root, and that portion of the root coming into contact with the tooth crown. The strain upon the pin, with the root funneled out and filled around with cement, is more evenly and properly distributed throughout the length of the pin, and particularly along that portion of it which enters the root and which must take an unwarranted strain if the root is squared off and a small opening only large enough to admit the pin.

The manufacture of such a crown is easy of execution. With root shaped as desired a pin of about the form of the Logan pin is run through a thin bit of platinum plate large enough to be shaped to the end of the root. This plate is formed, burnished to the end of the root, and may be slightly countersunk into the opening in the root. This finished, the pin and platinum are tacked together by soldering with a bit of pure gold. This finished, the tooth, a plain plate tooth, is ground to neatly fit against the outer portion of the root, and fastened to the pin now in position in the root by a small portion of wax. This is removed and invested in marble dust and plaster, and when sufficiently set, removing wax, gold plate or gold solder from 18K to 22K may be used, which may be blown down and finished ready for setting.

Much the same result as to fitting *may* be obtained in the neat adaptation of the Logan crown, with the addition in its favor of giving a better and more lifelike shade, and a crown which, if it is broken, as all porcelain ones are liable to be, admits of much easier removal of the pin than in the case of removal of the pin of the Morrison crown.

In the setting of bridges, particularly of all gold work, I have found within the past twelve months a method, new to me, which so greatly facilitates and simplifies the work that I give it here, even at the risk of making this paper too long, and also at the risk of going over something that may be adopted and used by others.

Accurate filling and adjustment of the bridge, no less than the single crown. determines its usefulness and its success as a thing of practical value. The most exquisitely wrought work poorly fitted upon the teeth that must carry the bridge cannot be expected to yield to the possessor that satisfaction which would be experienced were it poorly made and well fitted.

After having reduced the pier teeth to proper form, and after having constructed the cap crowns and the dummies to be suspended between them, the matter of occluding these with the opposing teeth comes next. Models obtained of the pier teeth and the space or spaces articulated are valuable only as approximating the occlusion, and helpful mainly in determining the size, width, depth, etc., of the crowns to be employed. Otherwise this portion of the work might be dispensed with.

But with these crowns in position on the articulated model they may be transferred, cemented together with wax, to the mouth and set in position. We should never be surprised at the twisting into position which we must observe when the teeth are closed. Seeing that each tooth is in its proper position, a cup adapted to this work may be filled with a half-and-half mixture of marble dust and plaster and water strongly primed with a saturated solution of sulphate of potash and inverted over the crowns, which in from two to three minutes may be removed, the teeth having exact relation to each other as they must in the mouth, and which, when removed from the mouth are invested ready for soldering.

In placing the crowns in position on the model the dummies, or suspended ones, should be bolstered up—supported upon a lump of wax resting on the model, gum in the mouth—to prevent the teeth from sinking and thus deranging the occlusion. When the invest-ing in this manner done is removed from the mouth the bolster of wax may be removed, and the under surfaces of the teeth are ex-posed ready for soldering.

In removing porcelain face bridges the same method may be em-ployed. The shaping of the roots to hold this class of work I hold should be the same as that for any single crown.

The important question of cements and how to work them might easily fill a paper of greater length than this. My preference so far as my experience goes is for Justi's. Still I believe that there are many others that are reliable if worked judiciously. It is of im-portance in warm weather to have the slab on which we mix cooled before mixing, as the warmth absorbed by our glass slabs causes the cement to set too rapidly, especially when the powder is added to the fluid faster than it is thoroughly moistened or digested. The consistency of the cement at the time for setting should be some-what stiffer than a pure cream, if I remember anything about pure cream—I live in a boarding house.

In the setting porcelain crowns using pins in the root I regard the barbing of the pins very desirable, but not to be cut deep.

Absolute dryness in the root opening, the cutting of recess pock-ets on the inner wall of the root, and the carrying of the cement well up into the canal are points that must be well executed.

The hot air syringe, the nozzle of which is introduced well up the root and a number of blasts in this manner blown up the canal, is of great value. The serum exudations and blood must be pre-vented from entering even the orifice of the root.

A small-shanked nerve canal plugger with a shred or so of cotton wrapped upon it and moistened with the phosphoric acid, though only moistened and not wet, is an admirable means with which to carry up the cement.

THE DENTAL LAW AND ITS RELATION TO THE STUDENT.*

BY W. W. JONES, D.D.S., MURFREESBORO, TENN.

SINCE the enactment of our dental law regulating the practice of dentistry I have been impressed with the fact that much has been

---

* Read at meeting of the Tennessee Dental Association.

done to elevate the standard of dentistry in this state, and too much praise cannot be extended to the originators and to those who so nobly assisted in the passage of such a law.

But as it now exists, the students who are not blessed with a liberal amount of finances must necessarily have a very hard struggle, for there is no provision made for them except that they pass a favorable examination before the State Board, or remain with their preceptors during vacation.

I am informed by the Secretary of the Examining Board that not one of the students in attendance at our dental schools has been examined for practice, and realizing as I do the trouble a young student experiences in securing a preceptor, it can safely be said that two-thirds of our students have none. Now what is to be done with those so unfortunate? If they practice their chosen profession, it is in violation of the law, and they lay themselves liable to prosecution, while at the same time they cannot afford to be idle. Must they return to the farm or behind the counter, thereby getting out of the train of thought the practice of their profession would suggest, and return to college the following fall as rusty as a plow which has laid out for a season.

To me this appears as a hardship to the struggling young student of our profession, and while I am not opposed to our law, I am a friend of the student, and if possible would like to see some means devised whereby a certificate could be issued to the student, granting him the privilege of practicing during his vacation, and becoming void seven months after date of issue. Then when he returns to the college of his choice, instead of being rusty, he would have made progress in his profession on the old principle that "practice makes perfect," and in my mind he will be bettter prepared to receive instructions.

In conclusion, I trust no one present will mistake the position I have taken, but I earnestly feel that favors of some kind should be granted the student, and if what I have said fails to meet your approval, but is the means of some one wiser than myself suggesting a better plan, I shall feel that my effort in the preparation of this paper has not been in vain.

Selections. • • • • • • • •

A FEW THOUGHTS ON THE COMBINATION OF FILLING MATERIALS.*

BY DR. BENJAMIN LORD, NEW YORK, N. Y.

Mr. President and Members: A great deal has been said and written, from time to time, on the subject of combining in some way different filling materials.

The idea or the practice is not exactly new, but more has been said about it of late than formerly; and, as all are aware, the subject has been discussed not a little before this Society in the last few years.

There has been less said about the combination of foils than about the combining of amalgam and gold, and the use of oxyphosphate with gold, and also with amalgam. For some reason which I do not claim to understand, we find that when amalgam is put by the side of gold, in cases where decay has extended beyond the gold filling, it does better; that it changes less in the way of shrinkage, and that the edges or margins are maintained much more perfectly than when amalgam only is used.

I have not tried the packing of gold and amalgam together, and so cannot speak from experience about it; but I should apprehend a good deal of embarrassment and risk in the attempt to pack gold by the side of or into a soft material. I have felt that in cavities in which, for any reason, I wished to use gold and amalgam in combination, a much better practice would be to make the gold filling first, not being particular to pack the gold solid at that point or part of the cavity where I propose to put amalgam; and then, after condensing and finishing the surface and margins, to remove the gold from that part of the cavity intended for the amalgam, and to pack the latter by the side of the gold. The effect of this would be the same as the packing of amalgam by the side of gold to arrest further decay, as before alluded to. I believe that

---

* Read before the New York Odontological Society, May 16, 1893.

such a method of combining the two materials will be found highly satisfactory.

The combining of alloy filings with oxyphosphate I find in my practice to be very useful. The filling lasts longer and the color is better.

My belief is, I may say, that the use of oxyphosphate in the starting or retaining of gold fillings, as recommended by some, could not be a certain or reliable practice, for very good and logical reasons.

A combination in which I find great interest is in the use of soft or noncohesive gold with tin foil. This is no novelty in practice; but I think that, for the most part, too great a proportion of tin has been used, and hence has arisen the objection that the tin dissolved in some mouths. I am satisfied that I myself, until recently, employed more tin than was well. I now use from one-tenth to one-twelfth as much tin as gold, and no disintegration or dissolving away of the tin ever occurs. I fold the two metals together in the usual way of folding gold to form strips, the tin being placed inside the gold. The addition of the tin makes the gold tougher, so that it works more like tin foil. The packing can be done with more ease and certainty; the filling, with the same effort, will be harder; and the edges or margins are stronger and more perfect.

The two metals should be thoroughly incorporated by manipulation. Then, after a time, there will be more or less of an amalgamation. By using about a sixteenth of tin, the color of the gold is so neutralized that the filling is far less conspicuous than when it is all gold; and I very often use such a proportion of tin in cavities on the labial surfaces of the front teeth.

If too much tin is employed in such cases, there will be some discoloration of the surface of the fillings; but in the proportion that I have named no discoloration occurs, and the surface of the filling will be an improvement on gold in color.

There is another combination in which I find great interest and advantage. It is the using of noncohesive and cohesive gold, by folding the two together in the proportion of one-third cohesive to two-thirds noncohesive. I first fold the noncohesive once, then lay it on the strip of cohesive gold, and fold the two together. The folding thus secures the cohesive always on the outside.

Gold prepared in this way should be used as soft gold, and works almost exactly like soft or noncohesive gold in those qualities in which soft gold is superior to cohesive; but it is tougher, packs more readily, makes a more solid filling, and gives stronger and better margins.

SERIOUS EFFECT OF NITROUS OXIDE ON THE KIDNEY.

A SYNOPSIS OF FIFTY CASES, BY J. SIM WALLACE, B.SC., M.D., C.M.

DESIRING to find the cause of the after effects which are occasionally complained of from the administration of nitrous oxide gas, I resolved to make a clinical report of a number of cases, and naturally among other things included an examination of the urine.

In each of fifty cases on which special notes were written, I examined the urine before administering the nitrous oxide, and made a few notes as to the general health, pulse, etc. I then proceeded with the administration of the gas in the ordinary way, the patient in each case having a tooth or teeth extracted. When the patient recovered from the operation and the anæsthesia I made a full report of the case, and examined the urine again. A further examination of the urine was made when the patient returned on the following day, and at other intervals.

A synopsis of the results derived from these examinations is as follows:

In about half the cases there was no appreciable change in the urine taken before the anæsthesia and that taken after.

The postanæsthetic urine was often of a deeper color than that which was taken before the anæsthesia, and sometimes deposited urates on standing when the preanæsthetic urine did not. Probably this was simply due to concentration, the urine being excreted in the one instance shortly after the imbibition of fluid, in the other after a longer interval.

In fifteen of the cases the urine taken after the anæsthesia showed a turbidity decidedly greater than that taken before the anæsthesia. I examined the turbid urine in some of these cases microscopically, and found numerous cells of various kinds, chiefly small round cells and large nucleated granular, principally squamous cells. Of course in the sample which contained blood (to be more fully mentioned below) the blood corpuscles were also observed, though not in great number.

In seven cases phosphates were indicated in quantity (by being precipitated on boiling, but dissolved on the addition of a drop of acetic acid) in the urine taken before the administration of the gas, whereas phosphates were not so indicated before the anæsthesia. In one case phosphates were thus indicated both before and after the administration.

In nine cases the urine which contained no albumen before the anæsthetic was given contained albumen after the anæsthesia. In two cases the albumen showed itself even after a week's time. In one case it was diminished in amount after twenty-four hours. In the remaining cases after a day it had disappeared from the urine. Three cases were only examined once after the anæsthesia.

In one of the cases in which albumen continued to show itself for a week the patient was addicted to eating highly nitrogenous food, and mentioned that he seldom ate bread.

As the following case is interesting, I shall give a few notes of his case.

W. K., aet. 32. Glass maker in good health. Had taken food about one and a half hours before the anæsthetic was administered. Took gas, sat during the operation, and recovered consciousness quietly. After regaining consciousness felt in a particularly happy frame of mind. Stertor was just brought on, he became slightly livid, and the conjunctival reflex was almost abolished.

The pupil, before the anæsthesia, was decidedly contracted; on beginning to lose consciousness it dilated very considerably; toward the end of the anæsthesia it again contracted, but not to the degree at which it was at the commencement of the anæsthesia.

The urine immediately before and fifteen minutes after the operation was free from albumen. So far this case may be considered normal. The following note, however, taken seven days after the operation, shows a remarkable development of the case.

W. K. had headache the day following the anæsthesia, which increased in severity for the next twenty-four hours, when it was very intense. (Headache following nitrous oxide anæsthesia is not uncommon, but I have failed to learn from the literature on the subject what the cause of the headache may be. In this case it was possibly associated with some kidney trouble.) He also complained of pain in the middle of the lumbar region, it being of a peculiar darting character. It came on twenty-four hours after the anæsthesia, and increased in severity till the following day, when it was most severe, and on one occasion made him fall. He had never had an attack of this kind before. He mentioned that his urine on the third and fourth day after the anæsthesia had been almost as clear as water, but that it had since returned to its usual color. (It may be remarked that on the two days mentioned it was decidedly colder than the days preceding and succeeding them.)

The urine, on being tested, was deep in color, acid, and showed a trace of albumen.

In the following case, although the urine was normal before the anæsthesia, besides a trace of albumen, blood was also detected. It was present, however, in very small quantity, and on the day following the albumen and the blood were both absent.

This case is rather interesting, so I shall give a few extracts from the report on the case.

J. B., a delicate-looking lad of 16 years. Two teeth were extracted. During the extraction of the second there was muscular rigidity and opisthotonos. The pulse on beginning the inhalations soon became only indistinctly perceptible, but on continuing it became more distinct. On discontinuing the anæsthetic it became very weak again, and remained in an unsatisfactory state for some time. Just after the anæsthesia he felt sick, and was so, more or less, for the next twenty minutes. He vomited a little blood which had apparently been swallowed. The treatment he got was the recumbent posture and ammonia to inhale.

The point, however, of special importance in connection with the case, over and above the fact of his having albumen and blood in the urine, is that he complained of a severe pain in the left lumbar region. The pain was described as sharp, lasted about one or two minutes, and did not return.

The fact that both the cases in which pain in the lumbar region of this suspicious character was complained of were associated with albumen and changes in the urine points rather directly to an unsatisfactory action on the kidney. That nitrous oxide has a distinct and probably serious action on this organ is, I think, obvious. It is remarkable that most books on the subject of nitrous oxide anæsthesia do not refer to the kidney; and after effects, when discussed at all, are not spoken of in connection with its action on that organ.

The cause of the action on the kidney is not very obvious. At first I thought that it might be due to a sudden transitory dilatation of the renal arteries, but that would not account for the continuance of the albumen in the urine. Besides, it is stated that the blood vessels in the splanchnic area contract, while the blood vessels of the surface of the body dilate, the sphygmographic pulse tracings showing a lowering of blood pressure.

The administration of nitrous oxide sometimes causes perspiration apparently unaccountable by the slight transitory dilatation of the vessels of the surface. It seems as if the nitrous oxide may have a direct action on the sudatory glands independent of its indirect action through the circulation.

Possibly it has also a corresponding action on the kidney, both organs being probably engaged in the elimination of the nitrous oxide. This is theory, however. The gravity of the facts I have recorded appeals to anæsthetists without my dilating upon them. I trust further investigation may give these results a more definite significance. (*Dental Record.*)

SOME OF THE RIGHTS AND DUTIES OF DENTISTS AT COMMON LAW.

BY BABSON S. LADD, BOSTON.

As preliminary to the view that I am to attempt to present some of the rights and duties of dentists at common law, I shall venture to remind you that in all employments demanding special skill and knowledge, be such employments classed as professional or mechanical, the general principles of law defining the civil responsibility and duties of persons thus employed are the same. Taking, then, for illustration the list put by Dr. Elwell in his work on medical jurisprudence, it may be said that physicians, lawyers, engineers, machinists, ship builders, and brokers, as well as all other classes of men whose employment is of the character I have indicated, are bound by the same general rules of law.

The law requires a dentist in his conduct toward his patient to exercise that reasonable degree of skill, coupled with learning and experience, which is ordinarily possessed by others of the dental profession.

States regulating the practice of dentistry simply operate to exclude from practice those persons (quacks and charlatans the courts have called them) who are incapable, from lack of the learning and experience ordinarily possessed by dental practitioners, of treating their patients with a reasonable degree of skill. All of this is for the protection of the public. This is common law.

Any one who assumes to be qualified for the exercise of any profession, art, or vocation is responsible for any damage that may result to those who employ him for the want of the necessary and proper knowledge, skill, and science which such profession demands.

He impliedly contracts with those who employ him that he has such skill and knowledge as will enable him properly to perform the duties of his calling. If he should be deficient in these respects, he violates his part of the contract, and must account in damages for any malpractice by which those who employ him sustain injury.

The failure of a course of treatment is not by any means conclusive of that want of professional skill by the practitioner. If his acts have been in accordance with the best known authority and skill, even though they should have been wrong, he will, in the eyes of the law, have done all that he could have done.

The standard of professional skill is never stationary. It must be kept up to and even with the constant advance of professional knowledge. Therefore the law demands *qualification* in the profession practiced; not extraordinary skill such as belongs to only a few men of rare genius and endowments, but that degree which ordinarily characterizes the profession.

In keeping up with the march of the profession it is well to advance cautiously in experimental practice, for if the experiment in any particular case is rash and contrary to knowledge and usage of the profession, you will be liable for malpractice, as it matters not in such cases what your general skill may be.

Some standard by which to determine the propriety of treatment must be adopted, otherwise experiment will take the place of skill, and the restless experimenter the place of the educated, experienced practitioner. When, therefore, the treatment or operation is experimental, the practitioner will act at his peril, unless he first obtains the consent of the patient or some one in authority acting for the patient. In the wrongful act of the student or servant, the preceptor or employer would be wholly responsible for the evil results which might follow if performed while in his employment, the law holding that the omission is the omission of the principal. In regard to the claims of dentists for professional services, the law recognizes the difference in the value of time. In the absence of any agreement as to the amount to be paid for services rendered, the law recognizes an implied promise to pay so much as the dentist deserves to have, which amount will be governed by the scale of prices in general use among dentists of good standing.

On the other hand, he cannot exact compensation for services necessitated by his own lack of skill and care.

The patient, however, is not without some responsibility, and must faithfully comply with professional order or advice. Nothing can be clearer than the duty of the patient to coöperate.

In giving evidence in court the expert cannot be compelled to testify, and the only way to procure his testimony is to pay for it, the obligation to pay resting upon the party who calls the expert.

FOREIGN DIPLOMAS IN GREAT BRITAIN.

It is a common observation that nowhere in the world have foreigners more liberty, both civil and religious, than in old England. Occasionally some malicious boor who construes this word into the privilege of spitting, smoking, and generally misbehaving wherever he may desire, sends forth to the world his impressions of the old sod as the most downtrodden land on the face of the earth; but no people have been more just and generous in their judgments. more fair or frank in their estimate of the British character, than American writers and travelers.

Some—not all—of our contemporaries over the border, however, seem unable dispassionately to discuss the action of the General Medical Council in revoking the privilege heretofore awarded certain foreign dental colleges, while refusing any similar favor to foreign medical schools. The worthy editor of the *Review* makes it a personally national matter, and heads one of his recent editorials: "No Americans need apply."

It was always a mystery to us why the Council was so invidious as to discriminate in favor of two of the American schools against others which were equally reputable, especially when the Council was well aware when the choice was made that not even the two favored came up to the required standard. It looked like favoritism, yet we do not see, in spite of the inconsistencies exhibited, that the action of the Council was specially against American colleges. It happened to be brought about by reason of the preference given in the past to two American schools; but it is possible to reason the matter coolly, and accept it in the spirit of justice to the British student and practitioner, who certainly have the first claim to consideration, otherwise it imposes a penalty upon the Briton, and holds forth a premium to foreign education.

It may surprise our American cousins to know that we Canadian dentists have no recognition whatever in the old home, and this in spite of the fact that in two of the provinces our matriculation is fully up to the standard required; and that for over twenty years we have had precisely the same system of indentured apprenticeship of four years of twelve months each year, with compulsory attendance upon anatomy (theoretical and practical), physiology, and chemistry, and primary and final examinations before boards of examiners elected by the profession. Brother Jonathan was getting favors refused to John Bull's own kids.

One instance may here be mentioned as affecting Canadians, Australians, and other loyal subjects of the empire. One of our Canadian students, born within the sound of Bow-bells, and desiring to return to England to practice, passed the very highest matriculation recognized in England, and graduated after three full courses at Harvard. He selected Harvard because he believed his diploma would be recognized. He then proceeded to London, presented the necessary papers to the Council, but was told that the law gave him no "privilege," *as he was a British subject.* In fact, as a further instance of the unfairness of the act to British subjects, he was positively refused a privilege granted a moment after to a friend from New York. He could get over the difficulty by forswearing his allegiance to his sovereign and becoming a citizen of a foreign nation!

Elsewhere we publish an editorial from the *Journal of the British Dental Association,* which presents a view of the question not fully disclosed on this side of the water. While we think it was a foolish action on the part of the Council to show any preference, we do not see that it is acting unjustly to our cousins. If the action is considered a rebuke, it is wiser to sit down calmly and see if it is not deserved. (*Dominion Dental Journal.*)

Correspondence. · · · · · · ·

LOOKOUT MOUNTAIN INN, July, 1893.

To the Editors of the DENTAL HEADLIGHT.

The meeting of the Tennessee Dental Association which convened here on July 4 was one of the largest in attendance of any meeting in the history of the Association; and to say that this is one of the most magnificent and delightful places at this season of the year for such conventions is but to express in the mildest language the sentiment of those present; and it is due to the place of meeting and the untiring and successful efforts of President Cook and the efficient Executive Committee that this meeting will be to the members one of the most pleasant and profitable, yet mingled with a sadness when there was no one in this world to answer to the names of the loved and honored Drs. Ross, of Nashville, and R. E. Thornton, of Chattanooga, Tenn.

President Cook's address was the subject of much discussion, as it touched upon many subjects of importance to the profession and Association. Following out his line of policy, the future President will appoint the Executive Committee, who will, with the President, decide the time of meeting of the Association; and a standing Publication Committee will be composed of the retiring President, First Vice President, and Secretary. The clinical feature will be improved to the fullest extent.

Dr. William Crenshaw, of Atlanta, who is here with his son and daughter, favored the assembly with a paper on "Setting Crowns and Bridges." In his replies during the discussion that followed the doctor proved that wit and humor can be used to good advantage in teaching his fellow-men; but I fear that, like nearly all men of this character, he is just a little sensitive when others undertake to follow his example.

Dr. Brabson, of Knoxville, read a paper on "Saving the Teeth."

Dr. J. L. Mewborn, of Memphis, gave a demonstration of the use of his reenforcing mallet, and Dr. C. V. Rosser taught us his method of using noncohesive gold in filling teeth.

Dr. R. R. Freeman demonstrated the rational surgical procedure of filling the nerve canal immediately after the removal of the nerve.

Dr. W. W. Jones, of Murfreesboro, read a paper on the "Dental Law and Its Relation to the Student." This paper was not very long and very modest, which was characteristic of the author, but it converted the Association into an oratorical assemblage. Every man in the house had something to say, and the discussion grew from being warm to a red heat.

Dr. J. Y. Crawford championed the law as far as it was for the good of the people, and in so doing he swept all the mist from the sky and from around the horizon. If the spiritualists had been in session at their hotel, Patrick Henry and Henry Clay would have applauded.

Dr. D. R. Stubblefield, of Nashville, sent in a paper on "Oral Dental Hygiene."

I would like to make the suggestion to those who send papers to be read before the convention that they have them typewritten, or printed, in order that the members may get the benefit of the author's labors.

After the reading of this paper Dr. J. Y. Crawford offered a resolution that should have been published in every paper throughout the country, and the objects carried out by the dentists in every county. It is as follows:

Be it resolved, That it is the sense of the Tennessee State Dental Association that the management of the public schools in this country should require an annual examination of the mouths of the pupils before they are permitted to enter school; and that all caries and other diseased conditions of the mouth amenable to treatment should be attended to before the pupil is permitted to be enrolled in school.

Dr. A. F. A. Shotwell, of Rogersville, read a paper on "Professional Courtesy."

Dr. W. J. Morrison was to open the discussion, but he acknowledged that the paper reached so far into the domain of psychology, and was so profound in all its bearings, that he was unable to worthily acquit himself of the honor that had been conferred upon him. This paper was an effort to open the gates to a professional millennium.

"Health of the Dentist" was the subject treated by Dr. R. B. Lees, of Nashville; while Dr. W. H. H. Thaxton, of Farmville, Va., sent in a paper entitled "The Health of the Dentist, and How Best

Preserved." Both papers were reported by the committee for joint discussion. William Crenshaw, J. L. Mewborn, A. F. A. Shotwell, Chisholm, Billmyre, and W. W. Jones offered some excellent suggestions in this direction.

It is greatly to be regretted that the Association does not employ a stenographer, that those who cannot attend the meeting could get the benefit of the discussions upon such important matters.

Dr. B. N. Dupree, of Chattanooga, read a paper on "Reflex Neurosis," which was discussed by Dr. J. Y. Crawford and Dr. Chisholm, of Alabama. Upon a motion of Dr. J. Y. Crawford, Dr. Morrison was requested to give to this Association, at its next annual meeting, the history of a remarkable case of reflex neurosis that came under his observation.

"Treatment and Filling of Dead Teeth" was considered by Dr. R. D. Crutcher, of Lewisburg, Tenn. He also took up sponge grafting, and treated the subject well, giving many interesting cases of success and failure that had come under his observation; and while the young doctor apologized for his maiden effort in the rôle of an essayist, I must say that if we had more such papers treating of subjects in a plain matter-of-fact way, and less of the big "I," they would make more instructive reading.

We regret that Dr. Gordon White was compelled to return to his office before the reading of Dr. Crutcher's paper, as he is so much interested in this subject, and has had much experience in this direction.

The subject of a paper by Dr. W. J. Morrison was "Dental Caries as a Contagious Disease." This was somewhat of a scientific extravaganza, and Dr. J. U. Lee, of Chattanooga, who was to open the discussion, said that it was too much for him.

It is to be regretted that Dr. H. W. Morgan's paper on "Regulating the Teeth" arrived so late, as the discussion which followed was necessarily very limited.

Nashville was chosen as the next place of meeting of the Association.

The following-named gentlemen were elected as officers of the Association: W. J. Morrison, Nashville, President; C. H. Smith, Chattanooga, First Vice President; B. D. Brabson, Knoxville, Second Vice President; W. W. Jones, Murfreesboro, Recording Secretary; H. E. Beach, Clarksville, Treasurer; S. B. Cook, Chattanooga, Corresponding Secretary.

Editorial. . . . .

As we go to press we learn of the death of Dr. G. W. McElhaney, of Columbus, Ga. He was one of the most prominent dentists in the South, having served as Vice President of the American Dental Association, and was also one of the Vice Presidents of the Columbian Dental Congress.

Dr. J. Y. Crawford, of Nashville, in a half hour's notice delivered the address of welcome to the Congress instead of the Hon. John Temple Graves, of Georgia, who was prevented from being present on account of sickness. Dr. Crawford's remarks were in his usual vein, and were well received, as was evidenced by the "Niagara of applause" which greeted him at the introduction and close.

The Delta Sigma Delta Fraternity added eleven to its members from those attending the Congress: L. D. Shepard, President of the Congress, Boston; J. E. Grevers, Amsterdam, Holland; George Cunningham, Cambridge, England; Thomas T. Moore, Columbia, S. C.; L. E. Custer, Dayton, O.; W. J. Younger, San Francisco, Cal.; T. E. Weeks, Minneapolis, Minn.; Alfred Burne, Sidney, Australia; G. C. Daboll, Paris, France; Henry W. Morgan, Nashville, Tenn.; and one other whose name has escaped us. The members of the fraternity are thus scattered all the world over. It is a very exclusive college organization which originated in the Dental Department of the University of Michigan a few years ago.

JOURNALISTIC ENTERPRISE.

The S. S. White Dental Manufacturing Company, under the management of Dr. Edward C. Kirk, its editor, published a daily issue of the *Dental Cosmos* in its usual style, containing the papers or abstracts of them read, and a stenographic report of the discussions at the Congress.

The *Dental Review*, of Chicago, published by H. D. Justi & Son, was also issued daily, containing the programme, editorial, and short extracts of the discussions.

Members of the Congress were thus informed as to what was going on in the various sections. A complete file of the daily *Cosmos* is almost equal to a copy of the transactions.

THE WORLD'S DENTAL CONGRESS.

CHICAGO, ILL., August 20, 1893.

To the DENTAL HEADLIGHT.

"Chicago" has been on the lips of every one since the first of May; has been the center of interest of the Western world for many months previous. A most wonderful untertaking had been projected, and all the world had been informed through the medium of the various arts of the printer and artist of the plans and proposed buildings for the World's Fair. Millions of money had been expended on buildings, artificial lakes, adorning the grounds and bringing from every corner of the earth its hidden treasures: works of science and art, manufactured articles, men and women, and visitors had viewed the result with one verdict: "It is good."

One feature of this Fair arranged by the World's Fair Auxiliary was to be a series of congresses. That of most interest to the dentist was the Columbian Dental Congress.

The immense work done by the Executive Committee was not without its full reward in what must be admitted was a great meeting—the greatest gathering of dentists in the history of the profession.

A full and complete measure of the work cannot be had until the printing of the proceedings is completed. We can promise a volume the equal of which has not been seen.

That every one should be satisfied was not to be expected. Human efforts are only human! Everything that could be anticipated was thought of, and a great majority of those present went away from Chicago rejoicing and praising those who had the management of the Congress.

The Art Palace, unfortunately, is located too near a railroad to make it a favorable place for such meetings. Speakers were obliged at times to cease until the noise of a passing train would subside.

The enrollment of members showed nearly one thousand names of American dentists, and one hundred and twenty-five foreigners, and as a representative body it compared favorably with other congresses that had preceded it.

The social element of the profession was out in force, and the reception tendered by the Chicago dental societies to the visitors at Kimsey's on Saturday evening, August 12, brought together about five hundred ladies and gentlemen from all parts of the world.

The banquet at the Chicago Beach Hotel Friday night, August 18, given by the American to the foreign members, was an occasion that will be remembered by those present. The rooms were beautifully decorated. Dr. James A. Swasey, of Chicago, presided, and two hundred and forty-two guests sat down, including nearly every foreign representative in attendance at the Congress. In the responses, which were of a highly congratulatory character, twenty-seven countries and nations were represented, as follows: Australia, Brazil, Paraguay, Uruguay, United States of Colombia, Chili, Cuba, Mexico, Philippine Islands, China, Japan, Hawaii, Denmark, Italy, Russia, Greece, Turkey in Europe, Spain, Austria-Hungary, Switzerland, Germany, France, Great Britain, and the three provinces of Canada (British Columbia, Ontario, and Quebec).

On Saturday afternoon, August 19, President Shepard gave a lunch to the one hundred and nineteen representatives of foreign lands in attendance at the Wellington Café, within the World's Fair grounds. A delightful time was had, speeches of felicitation and greeting were exchanged, and those present thoroughly enjoyed this particular feature of the social side of the Congress that called them from labor to refreshment.

The Woman's Dental Association of the United States held its second annual meeting in the Art Palace on Friday, August 18. Between thirty-five and forty women dentists were present. Dr. Herschfield, the first woman graduate in dentistry, was present and addressed the meeting.

The closing hours were most impressive. The remarks of the venerable Dr. Taft and President Shepard were listened to with great interest amid the awe of silence, and when the words were pronounced which adjourned the Congress *sine die* there was an audible sigh, and the expressions of those present indicated that they realized that the Congress " had been." H. W. M.

12

NOTES FROM THE CONGRESS.

There was a noted absence of Vice Presidents from the South.

The Chicago papers did not know the Dental Congress was in session.

The Alumni of the University of Pennsylvania had a reunion and banquet.

Dr. Hattie E. Lawrence, of Chicago, deserves much credit for having secured the large attendance of women.

Dr. Coroden Palmer, fifty-four years a practicing dentist, was one of the favored few who addressed the Congress.

Graduates of the Philadelphia Dental College and Pennsylvania Dental College also met and many kind words were interchanged.

From the way some of the members talked on all subjects and in all sections they must have been laboring under the impression that they were " Congress men."

The Executive Committee offered a gold medal for a prize essay on " Dental Hygiene." The award was made to Dr. George Cunningham, of Cambridge, England.

The Dental Club Rooms at 300 Michigan Avenue, presided over by Dr. Frank H. Gardner, President, were always open and filled, and contributed much to the pleasure and convenience of members.

One of the happiest men of the Congress was Dr. John S. Marshall, its Treasurer. His face was always aglow with good fellowship, as if it reflected the luster of the eagles gathered into his keeping in membership fees.

The work of the Section on Operative Dentistry was earnest, enthusiastic, and the section well attended; in fact, was one of the largest of the meeting, but neither the papers or discussions were quite up to the occasion.

Dr. Cunningham, in speaking of Dr. W. C. Barrett, of Buffalo, said that he used to be something of an " Independent Practitioner," but has become a " Dental Advertiser." Wonder if he (Doc.) saw a copy of a Buffalo paper that circulated somewhat last spring ?

Dr. Charles E. Blake, Sr., of San Francisco, says that the first obturator ever made was the work of Dr. Van Camp while practicing in Nashville, Tenn., and that models of the case and a duplicate are in the museum of the Baltimore College of Dental Surgery.

The assistant Secretaries, Drs. J. Bauer, of New Orleans, Louis Ottofy and E. M. S. Fernandes, of Chicago, rendered services that placed the entire membership under obligation in translating into English speeches and papers from the French, German, and Spanish.

Dr. George Cunningham, of Cambridge, England, was the most versatile man at the Congress. He made speeches, read papers, won the medals, gave a series of clinics in low-fusing continuous gum, gave lantern exhibits, and did many other things to entertain those present.

The address of Dr. L. D. Shepard, President of the Congress, was a well-prepared document, and was well read. While it was probably a little long, the matter and occasion justified an exhaustive review of the subjects treated. The applause it won indicated clearly that the Congress approved the position taken.

Among the distinguished stately gray-haired veterans who attracted general attention wherever they went were Drs. Isaac Weatherbee, of Boston; A. L. Northrop, of New York; J. Trumen, of Philadelphia; J. Taft, of Cincinnati; H. J. McKellops, of St. Louis; W. C. Barrett, of Buffalo; W. O. Kulp, Davenport, Ia.

The attendance was large. The great World's Fair failed to attract enough to reduce the perceptibly larger audiences that daily assembled in Washington Hall at twelve o'clock. The group that was photographed on the last day was too large to make a good photo, yet contains the leading spirits of the Congress, and will be treasured as a memento of the greatest gathering of dentists up to date.

The following women dentists attended the Congress: Henriette Hirstfeld, Tiburtius Berlin; Adolfine Peterson, Hamburg, Germany; Anna F. Reynolds, Boston, Mass.; E. Phillips, Martine Magnus, Christiana, Norway; Anna T. Focht, Hannah M. Miller, Emily W. Wyeth, Elizabeth A. Davis, Philadelphia, Pa.; Sarah M. Townsend, Denver, Colo.; Helen D. Searle, May Weston, Kansas City, Mo.; M. L. Woodward, Boston, Mass.; Ida Gray, Cincinnati, O.; Louise Peterson, Vida A. Latham, Hattie E. Lawrence, Chicago, Ill.; Louie H. Cuinet, Brooklyn, N. Y.; Eliza Yerkes, Philadelphia, Pa.

Tennessee was well represented in the published list of members. The following names were noted: J. Y. Crawford, W. H. Morgan, R. R. Freeman, Henry W. Morgan, W. J. Morrison, Gordon White, L. G. Noel, D. T. J. Thomas, Nashville; H. E. Beard, Clarksville; S. B. Cook, U. D. Billmeyer, J. D. Billings, O. H. Smith, Chattanooga; W. H. Richards, S. P. Sharp, Knoxville; J. P. McDonald, Shelbyville; P. D. Houston, R. D. Crutcher, Lewisburg; W. W. Jones, Murfreesboro; W. K. Slater, H. H. Pierce, Union City; H. K. Davidson, Dyer; William Townes, Memphis. All of these we had the pleasure of seeing and greeting except Drs. W. H. Morgan, W. W. Jones, and S. B. Cook, who were unable to attend, though registered and members of the Congress.

BOOK NOTICES.

LETTERS FROM A MOTHER TO MOTHERS ON THE FORMATION, GROWTH, AND CARE OF CHILDREN'S TEETH. By Mrs. "M. J. W.," the wife of a dentist.

We have secured from the publishers, the Wilmington Dental Manufacturing Company, of Philadelphia, a copy of this admirable little treatise on the above important subject. Certainly the wife of a dentist, if an intelligent, observing woman, is peculiarly qualified to undertake the authorship of a work devoted to the education of mothers concerning the care of the teeth of the rising generation. "Mrs. M. J. W." has proven herself eminently fitted for the task, as the great and continued popularity of this little book sufficiently attests, necessitating the publication of this the fourth "Columbian" edition. This number has been thoroughly revised, and should find its way into every household.

Associations. · · · · · · · ·

OFFICERS OF THE AMERICAN DENTAL ASSOCIATION.

J. D. Patterson, Kansas City, President; J. Y. Crawford, Nashville, Tenn., First Vice President; S. C. G. Watkins, Mont Clair, N. J., Second Vice President; Fred A. Levy, Orange, N. J., Corresponding Secretary; George H. Cushing, Chicago, Recording Secretary; A. H. Fuller, St. Louis, Treasurer; W. W. Walker and S. G. Perry, New York; and D. N. McQuillen, Philadelphia, Members of the Executive Committee. Next meeting to be held in Old Point Comfort, Va., the first Tuesday in August, 1894.

OFFICERS OF THE SOUTHERN DENTAL ASSOCIATION.

B. Holly Smith, Baltimore, President; R. K. Luckie, Holly Springs, Miss., First Vice President; S. B. Cook, Chattanooga, Second Vice President; L. P. Dotterer, Charleston, S. C., Third Vice President; D. R. Stubblefield, Nashville, Tenn., Corresponding Secretary; S. W. Foster, Decatur, Ala., Recording Secretary; H. E. Beach, Clarksville, Tenn., Treasurer; W. R. Clifton, Waco, Tex., and Gordon White, Nashville, Tenn., Members of the Executive Committee. Next place of meeting to be selected by the Executive Committee.

TENNESSEE STATE BOARD OF EXAMINERS.

S. B. Cook, President, Chattanooga; J. L. Mewborn, Secretary, Memphis; J. Y. Crawford, Nashville; W. T. Arrington, Memphis; A. F. Shotwell, Rogersville; II. E. Beach, Clarksville. Next meeting July, 1894.

NORTH CAROLINA STATE DENTAL SOCIETY.

C. A. Rominger, Reidsville, President; H. D. Harper, Kinston, First Vice President; E. K. Wright, Wilson, Second Vice President; J. W. Hunter, Salem, Treasurer; Isaac N. Carr, Tarboro, Essayist;

J. E. Wyche, Oxford, Secretary. The next meeting will be held at Durham, N. C., the first Tuesday in May, 1894.

OFFICERS OF THE MISSISSIPPI DENTAL ASSOCIATION.

J. B. Askew, Vicksburg, President; George R. Rembert, Natchez, First Vice President; J. O. Frilick, Meridian, Second Vice President; W. C. Stewart, Fayette, Third Vice President; J. D. Killian, Greenville, Corresponding Secretary; T. C. West, Natchez, Recording Secretary; C. C. Crowder, Kosciusko, Treasurer. Next meeting to be held in Natchez the first Tuesday in May, 1894.

ALABAMA DENTAL ASSOCIATION.

The twenty-fourth annual meeting of the Alabama Dental Association was held in the parlors of the Caldwell Hotel, Birmingham, April 11–14. The meeting was well attended, and many subjects of interest were ably discussed, and valuable clinics were presented.

The officers elected for the ensuing year, are: T. M. Allen, President, Birmingham; R. A. Rush, First Vice President, Selma; W. E. Proctor, Second Vice President, Sheffield; G. M. Rousseau, Treasurer, Montgomery; S. W. Foster, Secretary, Decatur.

The State Board of Dental Examiners are: E. S. Chisholm, Chairman, Tuscaloosa; G. M. Rousseau, Montgomery; C. P. Robinson, Mobile; G. P. Whitby, Secretary, Selma.

The place selected for holding the next meeting was Troy, on the second Tuesday in April, 1894.   S. W. FOSTER, *Secretary.*

AMERICAN DENTAL ASSOCIATION.

THE American Dental Association held its thirty-third annual session in Kindergarten Hall, Chicago, on Saturday, August 12, 1893.

The meeting was called to order at 10:30 A.M., President J. D. Patterson in the chair.

The Executive Committee reported the following resolutions:

Whereas the date of our meeting, which was fixed for August 15, under the expectation that it would immediately precede the opening of the World's Columbian Dental Congress, has been changed because of the change in the date of holding the Congress; therefore,

Resolved, That the unanimous action of the Executive Committee in calling the meeting in advance of the day selected is hereby approved and declared to be legal and binding.

Whereas it has been generally understood by the members that in order that more interest and work should be concentrated in the Congress, the meeting of the Association this year should be as nearly as possible of a merely formal character; therefore,

1. *Resolved,* That the dues for the current year be remitted and the Treasurer be instructed to give receipts in such form that a single payment shall cover the dues for the current and the coming year.

2. That the meeting this year be adjourned without any election of officers, as under the Constitution the effect of such nonelection will be to make all officers elected last year hold over.

3. That all records and transactions of this year be considered as merged in the proceedings for 1894 and so published, in order that in spirit and in name the officers elected last year shall not be considered to have held office and exercised their functions for two sessions.

4. That the Treasurer be instructed to pay all properly authenticated bills.

5. That Old Point Comfort be selected as the place of meeting for next year.

The resolutions were considered separately and adopted except that selecting the place of meeting.

The following places were put in nomination: Old Point Comfort, San Francisco, Niagara Falls, Saratoga Springs, and Lookout Mountain, and as the result of the balloting Old Point Comfort was chosen. The Association then adjourned to meet at Old Point Comfort the first Tuesday in August, 1894.

SOUTHERN DENTAL ASSOCIATION.

The Southern Dental Association met Friday, August 11, in Kindergarten Hall, Chicago, Dr. B. Holly Smith, President of the Association, in the chair; S. W. Foster, Secretary.

The meeting was called to order at 11 A.M.

On motion, the reading of the minutes of the last meeting was dispensed with.

The report of the Executive Committee being called for, the Chairman said that under the unusual circumstances it was thought best by the Executive Committee that the Association should adjourn till the next annual meeting, allowing the present officers and committees to hold over till then. He reported that the books of the Secretary and Treasurer had been examined and found correct, and suggested that no dues should be required of the members for this year.

A motion to remit dues for the year 1893 prevailed unanimously. The Executive Committee reported that charges had been brought against Dr. E. B. Marshall, of Rome, Ga., of unprofessional conduct in having advertised, and the Secretary read the following advertisements:

I.

DR. MARSHALL'S ANCHOR PLATE

for artificial teeth is the most satisfactory denture known, spoken of in the highest terms by the most eminent dentists, worn with comfort without covering the roof of the mouth, no injury to adjoining teeth, remains immovable in masticating food, can be easily removed and replaced at will. This denture is covered by letters patent. Made only in North Georgia by Dr. E. B. Marshall. Office, 302 Broad St., Rome, Ga.

II.

Rome has been distinguished among the dental profession by the invention of Dr. E. B. Marshall. The Anchor plates have been patented by the doctor, and patients are coming from a distance to get the benefit of them. It is the best denture known, and held firmly in place without clasps, and can be removed and replaced at will.

On motion, Drs. Gordon White, N. A. Williams, and M. W. Foster were appointed a committee to correspond with Dr. Marshall, and instructed to report at the next meeting.

Bills for stationery, etc., were ordered paid.

The Secretary and Treasurer was ordered to correct the list of members and prepare a new book. Carried.

On motion, the officers were allowed to hold over till the next annual meeting.

Moved and carried, that the selection of the place of the next annual meeting be left to the Executive Committee, and that the committee be instructed to report within six months.

The Association adjourned subject to the call of the Executive Committee.

NATIONAL ASSOCIATION OF DENTAL FACULTIES.

THE tenth annual meeting of the National Association of Dental Faculties was held in Kindergarten Hall, Van Buren Street, Chicago, Thursday and Friday, August 10 and 11, 1893.

The Association was presided over by Prof. J. D. Patterson, of the Kansas Dental College; Prof. J. E. Cravers, of the Indiana Dental College, Secretary.

Representatives from twenty-two colleges presented credentials.

The Ad Interim Committee had been called to pass on two questions during the year. The first, a student who had attended a full term at one college failed to present himself for examination and consequently received no certificate, applied for admission to advanced grade in another college. Upon the right to examine such student without certificate the committee ruled that the Dean of the second school could use his judgment. This decision was overruled by the Association.*

The second question was one submitted by the Dean of the Ohio College of Dental Surgery, ",Whether a student who regularly completed a course at a recognized college whose six months' term ended in June may enter the class of another recognized college the following October as a regular student?" The committee had held that such second entry would not be in conformity to the rules of the Association. The ruling was sustained unanimously.

A resolution was adopted restricting to one delegate from each college the privileges of speaking, voting, or acting on committees.

The application of the Western Dental College, of Kansas City, for membership was laid over for another year.   ·

The University of Maryland, Dental Department, is considering the adoption of separate lectures for the three classes.

The Department of Dentistry, Vanderbilt University, has added dental technics and abandoned the preliminary course in September, and instead would give a postgraduate and practical course at the end of the session.

The College of Dentistry, Department of Medicine, University of Minnesota, has adopted as a preliminary course a quiz for conditioned students. The degree of D.M.D. will hereafter be conferred instead of D.D.S.   .

The University of California, Dental Department, has added Latin to the requirements at entrance examination. Freshmen take the elements of pharmacy. Each student performs a series of experiments in metallurgy, and for Seniors a practical course in orthodontia has been in operation for several years. Winter sessions hereafter begin in September.

Applications for membership were made by the following dental

---

* Permission of the school must be first obtained.—ED.

schools, which lie over a year under the rules: University of Buffalo, Buffalo; Western Reserve University, Cleveland.

The following resolutions were adopted:

Resolved, That colleges of this Association may admit to the Junior Class graduates of recognized schools of pharmacy, subject to the examinations of the Freshman year.

Resolved, That any college of this Association failing to have a representative present for two consecutive years, without satisfactory explanation, shall be dropped from the roll of membership of the Association.

The Detroit College of Medicine, Dental Department, was elected to membership.

Action on the application of the Homeopathic Hospital College, Dental Department, rejected it, and the matter was referred back to the Executive Committee for further investigation. The Committee reported adversely. Adopted, rejecting the application.

The application of the Howard University, Dental Department, Washington, D. C., was left in the hands of the Executive Committee another year at the request of the committee.

The application of the United States Dental College, of Chicago, was reported on adversely by the Executive Committee, and the report was adopted.

The resignation of the Royal College of Dental Surgeons, Ontario, was presented by the Executive Committee, and ordered that it lie on the table until the next annual meeting, and that the college be asked to send a representative.

A resolution was introduced to repeal the rule admitting undergraduates in medicine to the Junior grade. Laid over.

Dr. Hunt moved that the rule upon the standing of graduates in medicine be amended to read as follows:

A diploma from a reputable medical college entitles the holder to enter the second or Junior grade in colleges of this Association, and he may be excused from attendance upon the lectures and examinations upon general anatomy, chemistry, physiology, materia medica, and therapeutics.

Laid over under the rule.

The Executive Committee reported the following resolution, which was adopted:

Resolved, That a committee be appointed to formulate a series of subjects and questions for preliminary examinations and a minimum standard to be reached before admitting students to colleges.

The election of officers resulted as follows: H. A. Smith, Cincinnati, President; C. L. Goddard, San Francisco, Vice President; J. E. Cravens, Indianapolis, Secretary; Henry W. Morgan, Nashville, Tenn., Treasurer; A. O. Hunt, Iowa City, Ia.; J. Taft, Cincinnati; Frank Abbott, New York, Executive Committee; James Truman, Philadelphia; Thomas Fillebrown, Boston; W. H. Eames, St. Louis, Ad Interim Committee.

The newly elected officers were installed, the retiring and incoming Presidents each returning thanks briefly and gracefully.

The following committees were appointed:

Committee on Schools.—J. A. Follett, Chairman; F. J. S. Gorgas, Louis Ottofy, C. N. Peirce, Truman W. Brophy.

Committee on Text-books.—S. H. Guilford, Chairman; J. D. Patterson, Thomas Fillebrown, A. O. Hunt, J. Hall Lewis.

Special Committee to Prepare Subjects and Questions for Preliminary Examinations.—Francis Peabody, W. Xavier Sudduth, Henry W. Morgan.

Adjourned to meet at the call of the Executive Committee.

The following Southern colleges were represented at the Association: Dental Department, National University, by J. R. Walton; Baltimore College of Dental Surgery, M. Whilldin Foster; Louisville College of Dentistry, F. Peabody; University of Maryland, Dental Department, F. J. S. Gorgas; Vanderbilt University, Department of Dentistry, Henry W. Morgan; Dental Department, Southern Medical College, L. D. Carpenter; School of Dentistry, Meharry Medical College, G. W. Hubbard.

THE NATIONAL ASSOCIATION OF DENTAL EXAMINERS.

THE twelfth annual meeting of the Association was held at the Columbian Dental Club, 300 Michigan Avenue, Chicago, Friday, August 11, commencing at 10 o'clock, Dr. W. E. Magill, President, presiding, and Dr. Edgar Palmer, Secretary *pro tem.*

The following is the result of the roll call: California, J. D. Hodgen; Indiana, M. H. Chappell, S. T. Kirk; Kentucky, C. S. Edwards; Louisiana, Joseph Bower; Maine, D. W. Fellows; New Jersey, G. Carleton Brown, F. C. Barlow; Ohio, L. E. Custer, James Silcott; Pennsylvania, C. V. Kratzer, Louis Jack, W. E. Magill; Tennessee, H. E. Beach, J. Y. Crawford; Wisconsin, Edgar Palmer;

Massachusetts, J. Searle Hurlbut; District of Columbia, Williams Donnally, H. B. Noble; Illinois, C. Stoddard Smith; Kansas, A. W. Callaham; Mississippi, W. E. Walker.

The following resolution, laid over at the last annual meeting, was taken up:

Resolved, That it is the sense of the National Association of Dental Examiners that, when a member of the dental profession presents a certificate of registration from a state Board of Dental Examiners, duly created by law, the same should entitle the holder of such certificate to registration without an additional examination in any state of the Union having a law to regulate the practice of dentistry.

Dr. C. Stoddard Smith offered the following amendment:

Provided such certificate was obtained on examination.

Discussed by Drs. Donnally, Jack, Noble, Kirk, Smith, Crawford, and others.

The amendment was accepted, and the resolution was then laid over till the next annual meeting.

A report from the Committee on Dental Colleges was adopted establishing as a preliminary condition to the reception of applications to be placed upon the list of recognized colleges, membership in the National Association of Dental Faculties.

The Committee on Colleges presented its final report, which stated that of the recognized schools for the session of 1892–93 the number of students was: Freshmen, 1,429; Juniors, 927; Seniors, 433; graduates, 320; post graduates, 44; one school not having reported. Of the unrecognized schools the number of students was: Freshmen, 111; Juniors, 54; Seniors, 22; graduates, 20.

The committee also reported, through its Chairman, Dr. Jack, the following list of colleges recognized by the National Association of Dental Examiners as reputable, as reported by the Committee on Colleges for 1893 and 1894:

1. Baltimore College of Dental Surgery, Baltimore, Md.
2. Boston Dental College, Boston, Mass.
3. Chicago College of Dental Surgery, Chicago, Ill.
4. College of Dentistry, Department of Medicine, University of Minnesota, Minneapolis, Minn.
5. Dental Department, Columbian University, Washington, D. C.
6. Dental Department, National University, Washington, D. C.

7. Northwestern University Dental School, formerly Dental Department of Northwestern University (University Dental College), Chicago, Ill.

8. Dental Department of Southern Medical College, Atlanta, Ga.

9. Dental Department of University of Tennessee, Nashville, Tenn.

10. Harvard University, Dental Department, Cambridge, Mass.

11. Indiana Dental College, Indianapolis, Ind.

12. Kansas City Dental College, Kansas City, Mo.

13. Louisville College of Dentistry, Louisville, Ky.

14. Missouri Dental College, St. Louis, Mo.

15. New York College of Dentistry, New York City.

16. Northwestern College of Dental Surgery, Chicago, Ill.

17. Ohio College of Dental Surgery, Cincinnati, O.

18. Pennsylvania College of Dental Surgery, Philadelphia, Pa.

19. Philadelphia Dental College, Philadelphia, Pa.

20. School of Dentistry of Meharry Medical Department of Central Tennessee College, Nashville, Tenn.

21. University of California, Dental Department, San Francisco, Cal.

22. University of Iowa, Dental Department, Iowa City, Ia.

23. University of Maryland, Dental Department, Baltimore, Md.

24. University of Michigan, Dental Department, Ann Arbor, Mich.

25. University of Pennsylvania, Dental Department, Philadelphia, Pa.

26. Vanderbilt University, Dental Department, Nashville, Tenn.

27. Western Dental College, Kansas City, Mo.

28. Minnesota Hospital College, Dental Department, Minneapolis, Minn. (Merged into No. 4.)

29. St. Paul Medical College, Dental Department, St. Paul, Minn. (Merged into No. 4.)

30. American College of Dental Surgery, Chicago, Ill.

On motion, the report of the committee was adopted.

The election of officers for the ensuing year resulted as follows: C. Searle Hurlbut, President; M. H. Chappell, Vice President; J. D. Hodgen, Secretary and Treasurer, 917 Sutter St., San Francisco, Cal.

Adjourned to the time and place of the next meeting of the American Dental Association.

Marriages and Obituaries. . . . .

MARRIAGES.

O. P. Hope, D.D.S., to Miss Mollie Knauer, at Pocahontas, Mo., June 14, 1893.

J. C. Montgomery, D.D.S.. to Miss Daisy Watkins, at Elizabethtown, Ky., September 7, 1893.

S. W. Johnson, D.D.S., to Miss Lillian Freeman, at Llano, Tex.

F. A. BADGER, D.D.S.

THE announcement of the death of this well-known dentist at his residence in Waverly Place was a sad surprise to this community, and especially to his professional brethren, as few were even aware of his serious illness.

He was born in Maury County, Tenn., April 22, 1833, and was the eldest son of Dr. F. A. Badger, one of the most noted dentists of the United States of his time. He received his education at Jackson Academy, in Columbia, and at the early age of eighteen began the study of dentistry under his distinguished father, who spared no pains in his training. Dr. Badger commenced practice in Hopkinsville, Ky., and afterward practiced in Florida and Georgia. He entered the Confederate Army at the breaking out of the war under Gen. N. B. Forrest, served until the close, and again entered the profession of dentistry, locating at Cadiz, Ky., and Charleston, Mo. Subsequently he moved to Nashville, in 1873, where he was continuously in active practice until he was taken sick in July last. His wife and four children, two sons and two daughters, survive him. Dr. Badger was decidedly of a literary turn of mind, spending much time in reading standard works. He stood high in his profession, and his death will cause general and deep regret among his friends and professional brothers.

Dr. Badger has for a score of years been a leading member of the profession of dentistry in Nashville, and he has always been universally respected. He was a member of the Christian Church and of the Royal Arcanum, and through the latter his widow will receive $3,000.

At a meeting of the dentists of the city, held in the rooms of the Academy of Medicine, Drs. J. P. Gray, Henry W. Morgan, and Gordon White were appointed a committee, and reported the following tribute of respect:

Whereas death has again entered our ranks and taken from us one of our honored members in the person of F. A. Badger, D.D.S.; and whereas Dr. Badger was an earnest, laborious colaborer, and won the respect and esteem of his patrons and professional brothers; therefore be it

1. *Resolved,* That we humbly bow to the will of the All-wise, "who knoweth and doeth all things well," and extend to his sorrowing family and friends our sincere sympathy.

2. That we attend the funeral and send a copy of these resolutions to his family and the papers and DENTAL HEADLIGHT for publication.

J. Y. CRAWFORD, Chairman;
E. F. HICKMAN, Secretary.

The following gentlemen acted as pallbearers: Drs. Henry W. Morgan, J. Y. Crawford, Gordon White, R. R. Freeman, W. J. Morrison, D. R. Stubblefield, R. M. Bogle, J. P. Gray.

DR. WILLIAM C. WARDLAW.

DIED suddenly, at the residence of Mr. Lewis Thompson, in Atlanta, Ga., Sunday, September 3, 1893, of heart failure, William Clark Wardlaw, M.D., D.D.S., of Augusta, Ga., on the fifty-seventh anniversary of his birth.

Dr. Wardlaw was born in Abbeville, S. C., Sunday, September 3, 1833; took his A.B. degree from the South Carolina College in 1857; graduated in medicine from the Charleston Medical College in 1861; served in the Confederate Army until the close of the war, a good soldier, loyal to the truth as he saw it; graduated in dentistry from the Pennsylvania Dental College in 1866, and the next year attended the New York Dental College, and was again awarded the degree of D.D.S.; had been President of the Georgia State Association and Southern Dental Association; a member of the Georgia Board of Dental Examiners many years; was an elder in the Presbyterian Church at Augusta, and Superintendent of the Sunday school. He had been elected to fill the chair of Anatomy and Physiology in the Atlanta Dental College, and had been chosen Dean by his colleagues.

Dr. Wardlaw was twice married: in 1862 to Miss Mary Josephine Thompson, of Kershaw County, S. C., who died June 10, 1886; and on November 14, 1889, to Harriet Chumley Adams. Six children by the first marriage, and his wife and two children by the second, survive him.

His life was one of rare purity, almost spotless. He was noted for his generosity, charity, and disposition to help others less fortunate than himself.

The following from the *Atlanta Constitution* of September 4 will be read with sorrowful hearts by those who knew him:

Dr. W. C. Wardlaw, one of the most distinguished members of the dental profession in this state, died suddenly at the residence of Mr. Lewis Thompson, on Crumley Street, yesterday afternoon. The circumstances of the sad occurrence are such as to render it a most peculiar dispensation.

Dr. Wardlaw came to the city from Augusta, Ga., last Saturday evening. He was apparently in the best of health, and his spirits were unusually gay and animated. He chatted pleasantly with his old acquaintances, and impressed the new friends to whom he was introduced with the charm of a most agreeable personality.

He had come to the city for the purpose of assuming the chair of Anatomy and Physiology in the Atlanta Dental College, in the capacity of both a dean and professor. The college was fortunate in securing the services of such an able practitioner, and the success of the department, under his able administration, was well assured.

No one thought for a moment, as the doctor moved along the streets Saturday afternoon, that the angel of death was hovering so near him, and that ere the sun of another day should touch the horizon his spirit would be beyond its setting.

And yet such was the case. When the announcement was made on the streets last night that Dr. Wardlaw was dead it was like the muttering of thunder in the starlit heaven. Many refused to believe the report, and belief was staggered until the statement was confirmed.

His death occurred about 5 o'clock in the afternoon, and was due, it was thought, to heart failure.

Dr. Wardlaw, in addition to his professional admirers, had many warm personal friends in this city and throughout the South. The news of his unexpected death, in the zenith of his fame and usefulness, will be in the nature of a shock to all who read it, and many will be the tears of sorrow that will flow in response to this afflicting dispensation.

An added circumstance, which renders the death of Dr. Wardlaw unusually distressing, is that he died on the anniversary of his birth. It seems, however, like a beautiful providence that such a useful and well-rounded career should be terminated on the day that gave it to the world.

Dr. Wardlaw attended the Columbian Dental Congress in Chicago in August, serving as Secretary of one of the sections; and, while he was quite feeble from recent illness, no one thought so soon to hear of his death, and it will prove a great shock to his many friends who saw him there.

The dentists of Atlanta and the Faculty of the Atlanta Dental College met and passed resolutions paying to his memory the highest tribute of respect. Probably no man in the South was more universally beloved and honored for his high professional standing and thorough Christian character.